A World in Crisis?

A World in Crisis?

Geographical Perspectives

Second Edition

Edited by

R. J. JOHNSTON and P. J. TAYLOR

Basil Blackwell

©Basil Blackwell Ltd 1986; 1989

First published 1989

Basil Blackwell Ltd
108 Cowley Road, Oxford OX4 1JF, UK

Basil Blackwell Inc.
3 Cambridge Center,
Cambridge, Massachusetts 02142, USA

First edition 1986
Reprinted 1987, 1988

British Library Cataloguing in Publication Data

Johnston, R. J.
   A CIP catalogue for this book is available from the
   British Library.

Library of Congress Cataloging-in-Publication Data

A World in crisis? : geographical perspectives/edited by R. J.
Johnston and P. J. Taylor. – 2nd ed.
      p.   cm.
   Includes index.
   ISBN 0-631-16271-2
   1. Geography, Economic.   2. Space in economics.   3. Natural
resources.   4. Economic history – 1971–   I. Johnston, R. J. (Ronald
John)   II. Taylor, Peter J. (Peter James), 1944–
HC59.W658   1989   89-924
337'.09'048 – dc19   CIP

Phototypeset by Dobbie Typesetting Limited, Plymouth, Devon
Printed in Great Britain by T. J. Press Ltd, Padstow, Cornwall

Contents

The idea of 'crisis' is at the heart of all serious discussions of the modern world

J. O'Connor, *The Meaning of Crisis*. Oxford: Basil Blackwell, 1987, p. 49.

The crisis through which international relations and the world economy are now passing presents great dangers, and they appear to be growing more serious.

Brandt Commission, *North–South: A Programme for Survival*. Cambridge, MA: MIT Press, 1980, p. 30.

There is a crisis in the world today, now felt even by those of us who enjoy the power and privileges at the top of the world. There is a crisis of *violence* and threat of violence . . . There is a crisis of *misery*, and threat of poverty . . . There is a crisis of *repression*, and threat of repression, of all human rights . . . There is a crisis in the environment . . . What is the root of these crises? Some people would have us believe that the root crisis is primarily a resource crisis . . . There are some others who would have us believe that the root crisis is primarily a price crisis . . . others would have us believe that what we face primarily is a population crisis . . . At the root of the crises is not resource scarcity or price increases or population pressure, *but the world structure*. The naive approach to the crises of our world would be to conceive of them as a list of problems, pairing each item with a set of technical solutions . . . I do not want to belittle these technical solutions, as there are valid elements in all of them; but they become mystifying, even dangerous, when they are not accompanied by an analysis of the world structure and an even higher dedication to a basic change of structure.

J. Galtung, *The True Worlds*. New York: Free Press, 1980, pp. 1–3.

vii

Preface

The word 'crisis' is frequently used today to characterize some difficulty – local, national or international, particular or general – relating to human societies and their interrelationships with the environment. Indeed, the word is used too frequently, often as a synonym for 'problem', so that its proper definition – 'time of acute danger or difficulty' – is forgotten among so much indiscriminate usage. Perhaps this is understandable: many of the problems called 'crises' are very severe, suggesting to contemporary observers that they might be major turning-points in the course of world history – and the observers will not know whether they were or not until much later.

As with so many aspects of human society and the ways in which it operates, if something is identified as a crisis then crisis it is – at least to the identifiers. Thus we feel justified in using the word in the title to this book. Many people do perceive a *world* in *crisis* – with the dual emphasis indicating that the crisis is believed to be global, not national or local.

Assuming that there is a perceived crisis, what is the justification for offering a series of geographical perspectives on it? That it is recognized as a global crisis is almost sufficient justification in itself, for the portrayal of global processes and patterns is central to the geographer's activity. Further, as the chapters in this volume show, many aspects of the crisis are intrinsically geographical, relating as they do to both society–environment relationships and society–society interrelationships over space. Geographical perspectives provide insights into some of the main parameters of the crisis.

This book has been written for a wide audience. It is especially suited to university courses on global problems – courses which are becoming increasingly popular throughout the social sciences and not only in geography. Through it, we hope to sensitize students to the nature and extent of the problems and difficulties that go to make up the crisis, and to show them the value of geographical perspectives on those issues.

In producing this book, we are grateful to the authors for readily agreeing to contribute and for producing such stimulating essays. We are also very grateful to John Davey for his encouragement and assistance.

R.J.J.
P.J.T.

Preface to the Second Edition

The response to the first edition of this book was extremely gratifying, reflecting the quality of the contributions and the growing interest of geographers in global issues. Hence we were delighted to be asked to produce a new edition.

The call for a new edition, less than four years after the appearance of the first, reflects the rapidity of change in the world, as efforts are made to lift the capitalist world-economy out of its recession and new geopolitical initiatives are taken. In the early discussions about the revision, it was suggested that we remove the question mark from the title. Ultimately, we decided not to do this. Many problems in the world remain unresolved – many, indeed, seem irresolvable. But whether they add up to a crisis, to a major turning-point in the organization of human society, both globally and locally, remains uncertain – both to us and to most of our contributors.

Needless to say, the basic structure of the book and its contents remain the same. All authors were asked to consider revising their chapters in the light of changes since the early 1980s; some clearly saw the need for more revision than others, and in the case of chapter 4 Simon Carter joined Phil Bradley to broaden the geographical coverage on that topic. In addition, a further contribution was solicited, to complement John O'Loughlin's chapter on 'World Power Competition and Local Conflicts in the Third World'. The relationships between the super-powers, in the context of their economic difficulties, became a major issue on the world political agenda in the mid-1980s, and this is covered in the chapter by John Agnew and Stuart Corbridge.

Once again, we are extremely grateful to our colleagues for their production of such excellent material and their ready collaboration in the production of this new edition, and to John Davey for his encouragement, support and advice.

R.J.J.
P.J.T.

1

Introduction: A World in Crisis?

R. J. JOHNSTON and PETER J. TAYLOR

At some time during the late 1960s and early 1970s the world we live in seemed to change. The clearest, most abrupt sign of that change was the trebling of oil prices by the Organization of Petroleum Exporting Countries (OPEC) in late 1973, an action which sent political and economic shock-waves right through the western world. With hindsight, we can see that this action was more a consequence than a cause of the new situation. It was the culmination of many tendencies which destroyed post-war optimism and replaced it with a 'new realism'. At the most general level, this book is itself a consequence of that destruction – a reaction of geographers to the need for a 'new realism'.

For most of the 1950s and 1960s economic, social and political policies were based on an assumption of continuous economic growth; any outstanding problems – underdevelopment, for example – were seen as temporary aberrations, since it was only a matter of time before they would be solved by a combination of rational planning and political goodwill. The market would triumph, freedom, democracy and prosperity would be achieved for all, and the forces of evil, i.e. communism, would be defeated. The only cloud on the horizon for the western world was the USSR and the cold war, but at least that never hotted up. The future was rosy, and crisis was very far from the minds of all but a few diehard Marxists who refused to bring their analyses 'up to date'.

Then things began to go wrong. This was shown in myriad ways. Most fundamentally, the economic growth on which the whole optimistic edifice was built began to show signs of faltering; perhaps the post-war good times were not infinitely sustainable after all. Student unrest, city riots and wars in Southeast Asia and the Middle East in the late 1960s all illustrated the increasing fragility of the world. Suddenly crisis seemed to be the word on everybody's lips, and the adjectives applied to it (ecological, environmental, demographic, urban, rural, debt, food, energy etc.) were so wide-ranging that they contributed to a general sense of despondency. The 1970s became the 'decade of world

1

Table 1.1   'World Conferences' in the 1970s

1	United Nations Conference on the Human Environment (UNCHE), Stockholm, Sweden, 1972
2	Third United Nations Conference on the Law of the Sea (UNCLOS III), New York, USA/ Caracas, Venezuela/ Geneva, Switzerland, 1973–1982
3	World Population Conference, Bucharest, Romania, 1974
4	World Food Conference, Rome, Italy, 1974
5	World Conference of the International Women's Year, Mexico City, Mexico, 1975
6	United Nations Conference on Human Settlements (HABITAT), Vancouver, Canada, 1976
7	United Nations Water Conference, Mar del Plata, Argentina, 1977
8	United Nations Conference on Desertification, Nairobi, Kenya, 1977
9	World Disarmament Conference (Special Session of the General Assembly of the UN), New York, USA, 1978
10	World Conference to Combat Racialism and Racial Discrimination, Geneva, Switzerland, 1978
11	United Nations Conference on an International Code of Conduct on Transfer of Technology, Geneva, Switzerland, 1978
12	World Conference on Agrarian Reform and Rural Development, Rome, Italy, 1979
13	United Nations Conference on Science and Technology for Development, Vienna, Austria, 1979

conferences' as the world's political leaders attempted to devise rational responses to one crisis after another (table 1.1). Towards the end of the decade several reports began to appear, the most famous being the Brandt Report on material inequalities (Brandt Commission, 1980), confirming our worst fears of crisis. It was but a short step to put all these events together and identify a general or global crisis.

It is, of course, extremely disconcerting to shift rapidly from a sure world full of optimism to an uncertain one full of pessimism and thoughts of crisis. From this perspective the most important events of the crucial late 1960s were the student uprisings of 1968. It was the first 'post-war' cohort of young adults that fatally pricked the bubble of optimism that had dominated society since 1945. Although concentrated in France, we can see that the challenge to existing mores was genuinely global in scope (figure 1.1). The key feature of this challenge was that it rejected existing world models *from both right and left.* Although the immediate targets of protest tended to be either right-wing or centre governments, the students did not accept orthodox left-wing prescriptions of either the reformist or the revolutionary variety. It is for this reason that '1968' has come to be put alongside 1789, 1848 and 1917 as being of world-historical

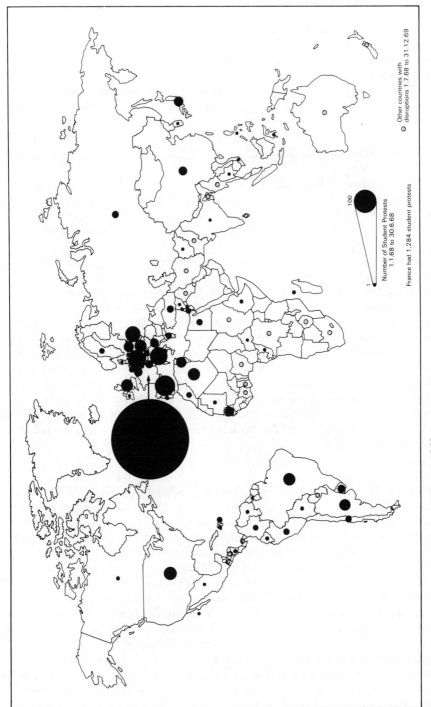

Figure 1.1 Student disturbances in 1968 and 1969
Sources: Katsiaficas, 1987; Kidron and Segal, 1981

Number of Student Protests
1.1.68 to 30.6.68

France had 1,284 student protests

○ Other countries with
disruptions 1.7.68 to 31.12.69

importance (Katsiaficas, 1987). Before 1968 the challenge to the existing social order came from an 'old left' representing exploited classes and nations. After 1968 they were joined in their task by new social movements representing previously neglected sections of exploited people such as minorities, women and young people (Arrighi et al., 1988). Although the immediate political effects of the world-wide uprisings were minimal and the medium-term fall-out has been the political success of the 'new right' in the 1980s, the full implications of 1968 have yet to unfold. The changing power relations between status groups that were placed on the political agenda by 1968 are integral to our world in crisis.

The upheaval of 1968 had repercussions beyond politics. In the world of academia it provided a fresh impetus to 'new left' intellectuals. If the right has won the media politics, the left has been much more successful in the medium term in the universities, the hearth of the initial uprisings. In geography new radical approaches date from this time (e.g. Peet, 1977). More generally, a world in crisis is a particularly exciting prospect for all critical students since it throws up new intellectual puzzles and challenges. Scientists and scholars are forced to revise their thinking. Inevitably, this leads to academic debate and infighting as society's crisis is reflected in the disciplines that study society. Hence the social sciences have suffered internal turmoil in the last two decades as they tackle the new questions: 'Is there a crisis?' 'What are the likely consequences?' 'Can we get out of it?' 'How?' In human geography the optimistic certainty of the 1960s has been replaced by diversity. After more than a decade of this, we need to assess our situation. We cannot expect to eliminate the current diversity – nor would we want to – but we should be able to map out some of the main parameters of the 'world in crisis' both empirically and in the revision of theory. The book is a modest contribution by geographers to that task.

WHAT KIND OF CRISIS? WHAT KIND OF RESPONSE?

Use of this word crisis does not imply a consensus definition of that term. In general usage 'crisis' implies something more severe than a 'problem' (or even a cluster of problems), in which the whole is greater than the sum of the parts. For every problem there is a solution, although we may not be able to find it or put it into operation. A crisis implies something qualitatively different (Frank, 1982). In its original Greek form, 'crisis' means 'decision' and so identifies a critical juncture in a process when a critical decision must be made. Its original application was in medicine. A breathing problem becomes a crisis when breathing stops; it is transformed from a problem of the lungs to a crisis of the whole organism, and the point of decision is reached when the existence of the organism itself is in the balance. But decision implies that catastrophe is not inevitable. The Chinese expression for 'crisis' uses two characters, depicting 'danger' and 'opportunity'. Galtung (1980, p. 4) argues that we must all learn from this Chinese conceptualization: crisis is fundamentally dialectical in nature as it opens up the possibilities of alternative futures.

It is the technical definition of *crisis as a decision point* that is used by writers, Marxist and non-Marxist alike, when they refer to a 'world in crisis'. A world that has not adapted to a new pattern of energy production faces an energy problem: a world that is running out of energy faces an energy crisis. In this context, Marxists use the term crisis to describe the point when the contradictions of the capitalist mode of production finally prevail and the system that it supports collapses, to be replaced by socialism. Wallerstein (1982), for example, is very careful to distinguish the long-term crisis of capitalism from the current phase of economic stagnation, as he describes it. Our problems are part of the long-drawn-out crisis of capitalism, but they do not constitute *the* crisis. For instance, the 'crash' on the New York, London and Tokyo Stock Exchanges of 16 October 1987 seemingly presaged a repeat of the 1929 collapse that heralded the recession of the 1930s. As yet, it has not come, though uncertainty continues. Hence it has been argued that the capitalist world-economy is stronger than ever, as governments in many countries – from the United Kingdom to New Zealand – seek to unfetter market forces and reduce the 'culture of dependency' that they associate with the inefficiencies of the welfare state. According to their interpretations, economic health has been restored (on the United Kingdom in this context, see Allen and Massey, 1988). However, short-term 'successes' must not be used to mask longer-term systematic trends. Wallerstein fully expects most of the current problems to be resolved throughout the world-economy in a new era of growth during the 1990s, but he believes that capitalism itself will remain terminally ill.

Within the present book, the authors have interpreted the term 'crisis' in a variety of ways, and some, such as Piers Blaikie (chapter 5), discuss its usage in their particular context. In general, their emphasis is on the current situation rather than on the long-term perspective implied by Wallerstein. This is entirely valid, for as social scientists we can only analyse the recent past as a guide to the future. Hence it is the vernacular use of the term 'crisis', reflecting the shock stimulated by the severity of recent changes, which predominates in these chapters. From Wallerstein's perspective, they emphasize just one phase of the genuine crisis of the world-system.

In identifying a crisis we are, implicitly at least, assuming the existence of a system, just as those who write about an 'urban crisis' assume that there is an 'urban system' which is breaking down in some sense. Today, it is common to refer to a wider global crisis: there is no urban crisis *per se* because there is no independent urban system; rather, there is a global crisis which is manifested in cities. Hence there are urban problems, but not urban crises. Such a perspective is part of a very strong tendency in modern social science to analyse social change on a much larger scale than heretofore. The move from optimism to the 'new realism' has involved a distinct shift in the geographical scale of emphasis from national to global, which characterizes social science as a whole and increasingly the discipline of geography in particular.

In the optimistic social science of the first three post-war decades most analysts, Marxist and non-Marxist alike, assumed that national society and the nation-state should be the basic unit of change. Probably the best known of such analyses was Rostow's (1971) 'stages of growth' model, in which every country could be allocated to a rung on a ladder leading to 'high mass consumption'. This and similar models have been criticized as part of a general critique of a modernization school that advances the comparative study of societies (i.e. nation-states) in terms of their 'modernity' (Taylor, 1989). That school of thought accepts the existence of 100–200 separate societies in the world, and assumes that change in those societies occurs autonomously. Such a premise has been a major casualty of the critique of the modernization school by social scientists, for it is obvious to all but the most myopic of observers that the current changes have no respect for national boundaries; they are spread, albeit in different manifestations, across the whole face of the earth, with states almost powerless to influence them (Johnston, 1986). Hence we have global problems, a world recession, international stagnation – perhaps even a world in crisis.

Global thinking is not new, of course, and so the new perspectives are basically a shift of emphasis. The classification of countries into three types – First World (or developed/advanced/western/industrial), Second World (communist/socialist/command economy) and Third World (underdeveloped/developing) – has always indicated an awareness of circumstances affecting change beyond the bounds of individual countries, although as a typology it is rigid and inconsistent. An even more explicit formulation of such cross-boundary links was the early post-war work of Latin American economists, with their identification of core and periphery on a transnational scale. Such themes have been consolidated and expanded in recent years to produce what may prove to be a major qualitative shift in social science.

Currently, that shift is expressed in a variety of forms, and the simplest way to view them is to place studies on a continuum ranging from recognition to systematic. Close to the first pole are studies which are not far removed from their nation-state-oriented predecessors. The individual states remain the basic units of analysis and the only emphasis extending beyond national boundaries is one on interstate links. Change is viewed as a state-level process, but outside influences are allowed a stronger role than hitherto. This is explicitly demonstrated in political and welfare geography books of the 1970s (e.g. Coates et al., 1977; Smith, 1977; Cox, 1979) where the 'international scale' is added to studies at national and local scales, but with little effort at developing links among the three (see Taylor, 1981, 1982). In short, the importance of global analyses has been recognized but it has not altered the overall form of the work.

Moving away from that pole, more sophisticated studies emphasize the linkages between states and particular types of social relations. The best-known examples can be found in the dependency school of development studies, in which core–periphery linkages are the fulcrum of the process of change. Within geography, as a reaction to this emphasis on a single type of link, the concept of interdependence has been used to indicate a much more complex set of relations between countries (e.g. Brookfield, 1975; Reitsma, 1982), but change is still channelled through countries. These studies go beyond simple recognition of global processes but they fall far short of conceptualizing the world as a system.

At the other end of the continuum the focus is on some form of world-system (Thompson, 1983). Change is presented as a property of the system as a whole, and so what happens in any one place can only be understood within a holistic framework. In such a schema, individual countries are no more than parts of a much larger whole; it is impossible to understand economic decline in the United Kingdom, for example, by considering it alone outside the context of general global processes. There is a great variety of perspectives on the 'world-system'. Among them, a particularly relevant distinction is between those which present the current situation as a new global phase of capitalism, distinct from phases that have gone before, and those which argue that this is not a transient phase; to the proponents of this latter view, the capitalist world-system is by its nature transnational. The former group emphasizes the new features of the current situation (such as the industrialization of the Third World, presented as the 'new international division of labour') and builds an argument that the situation is a unique modern expression of capitalism (Ross, 1983). The other group, following Wallerstein (1974), views the 'modern world-system' as about 500 years old, as international throughout that period and as global since about 1900. According to this argument, the current situation involves a further set of structural changes that have been continual since the modern world-system was initiated in the sixteenth century. Such changes are manifested in different ways at different times and in different places: the present restructuring, for example, involves major changes in the ways that production is organized and also in its location (Lash and Urry, 1987). Both these perspectives, despite their important theoretical differences, emphasize a *holism* that is lacking from the studies of interconnections among states. For the 'interconnectionists' the world may be in crisis because of a failure in the connections – as in the distribution of food; for the 'holists' the world will be in crisis because of the contradictions involved in its operation as a system.

The holism of the world-systems approaches implies much more than just a change of geographical scale in the study of social change. World-system ideas can be traced back to the political economy perspective which pre-dates the specialization tendencies within the social sciences. For example, Wallerstein calls his project 'historical social science', so as to repudiate the separation of history from social inquiry, and he offers a 'unidisciplinary' approach rather than the ubiquitous call for multidisciplinary studies; 'historical social science', for Wallerstein, is a single discipline.

Geography shares some of the holistic heritage of the world-system analysts (Taylor, 1985). As an organized body of knowledge it pre-dates the specialization trends within the social sciences, although latterly it has largely succumbed to this general tendency of modern science. The 'subdisciplines' of human geography are mirror images of the basic trilogy of the social sciences, for example, with social, economic and political geography reflecting sociology, economics and political science respectively. Nevertheless, many of the extremes of myopic thinking in the social sciences (does politics matter?; there is no alternative to monetarism) appear to have been avoided in modern human geography. Despite the pressures for intradisciplinary specialization, human geographers have just about maintained a holistic viewpoint. Good geographical perspectives on the world in crisis are less narrow than those produced from other social sciences, as the chapters in the rest of this book demonstrate.

GEOGRAPHICAL PERSPECTIVES

The chapters in this book have not been written to a detailed brief and are not structured within any single approach to the study of the world in crisis; indeed, some of the authors are less certain than others that there is a crisis. Odell and Woods, for example, argue that the energy and population 'crises' will in the longer span of history be interpreted as 'generational blips', while both Blaikie and Bradley and Carter suggest that ecological and food crises are more localized than others sometimes argue (see also Watts, 1989). The selection of topics and authors reflected editorial decisions on the important issues that should be covered in such a book, and experts in the field were invited to contribute, interpreting the general brief as they believed the topic should be studied. All are geographers, by training if not by current affiliation, and they bring geographical perspectives to bear on crucial world issues.

For many people in the core of the world-economy, the clearest evidence of a crisis is the level of unemployment and the apparent inability of governments to produce policies that will stimulate sufficient employment growth. To some analysts, it is the policies adopted by the United States to produce jobs, and the consequent massive budget deficit and very high interest rates there, that prevent governments elsewhere from stimulating employment. By the late 1980s, some argued that the problems of the earlier years of the decade were over and that an inefficient system had finally, and painfully, been restructured. As yet, the restructuring was incomplete so that within the United Kingdom, for example, prosperity was currently greater in some regions than in others (Green, 1988; Martin, 1988). But the optimistic message was that, despite high levels of unemployment and unprecedented balance-of-trade deficits, the new system – termed flexible accumulation by some (e.g. Harvey, 1987) and disorganized capitalism by others (Lash and Urry, 1987) – was producing positive changes. Thus it is with the geography of those changes that the book starts, in Nigel Thrift's chapter on international economic disorder. In this wide-ranging survey, with much detail on the nature of the issue, he sets employment change in the context of capital restructuring on the international scale. As stressed already, the issue is one of the world-system as a whole, and Nigel Thrift identifies many of its ramifications in all parts of the world.

Perhaps the clearest statement of the onset of international economic disorder, at least in the popular mind, came in late 1973 with the OPEC actions to raise oil prices and use that vital natural resource as a political bargaining tool. These actions brought to an end the complacent optimistic years of cheap, apparently plentiful energy and stimulated much debate about and investigation into alternative sources. But is the world being drained of energy, so that a crisis is rapidly rising over the horizon? Peter Odell, an experienced analyst of the energy field and for long a trenchant critic of public and private sector energy policies, argues not. We have recently passed through a period of unparalleled growth in demand for energy, as a consequence of a particular conjuncture of causal factors, but that period is now over and we have returned to a manageable rate of growth. There are still problems of distribution and access, but his prognosis is much less pessimistic than many. If the crisis does come in the next few years it will not, it seems, be for a lack of energy supplies.

Through the late 1970s and the 1980s the focus of 'crisis attention' shifted away from energy and towards food. In particular, the problems in Africa, where several famines were linked to environmental deterioration (drought in the Sahel and the desertification of large tracts of land), were highlighted. These were forcibly emphasized in Western Europe and North America by television reports from Ethiopia, where famine on a very large scale was linked to a continuing civil war (see chapter 11). The spectre of a food crisis loomed large: not for the first time, it was argued by many that the world could not support its population at even a bare subsistence level, that it was raping its natural resource base and that the only solution – as practised by the Indian government in the mid-1970s – was a massive programme of birth control. Yet there were paradoxes in the situation, for the countries of the European Economic Community (EEC) were producing much more than their populations could eat or drink, US farmers were being paid not to produce and yields were being increased in many farming regions. So what is the real situation: is there a food crisis, an environmental crisis and a population crisis?

In chapter 4, Phil Bradley and Simon Carter tackle the issue of hunger, focusing particularly on Africa and Latin America. They contrast peasant subsistence agriculture with capitalist capital-intensive agriculture and show how the latter is replacing the former through much of the Third World. They argue that the problems of feeding peoples in those countries stem from the reorientation of local agriculture to the external markets of the core of the world-economy. This is a process initiated four centuries ago in the Caribbean, which has slowly spread through much of the rest of the periphery. (Indeed, it has only slowly penetrated much of the core, as recent work by Marsden et al., 1987, shows.) The condition of subjugation of periphery by core is now virtually complete, and the food problems of the former have become endemic. There is a crisis – not so much of productive capacity as of the orientation of production and the distribution of the output.

What of the pressures on the environment? Are we raping the earth of its resource base, and are there better ways of managing natural resources?

Piers Blaikie (chapter 5) finds much evidence of irresponsible use of resources, particularly but far from solely in the Third World where the infiltration of capitalist values – paralleling the agricultural changes outlined by Bradley and Carter – has led to rapid deterioration of the resource base in many areas. With regard to the future, Blaikie identifies many problems in developing more appropriate strategies. Those being applied in the Third World at present focus on the relative levels of technical expertise in the host countries and the multinational corporations which wish to exploit the resources there; political control allows the host government to exert some leverage on the potential users, but the users themselves have considerable leverage in situations where there are more potential suppliers than the market can currently support. Capitalist enterprises can contribute to responsible resource use, but will they? Would an alternative socialist mode of production lead to less destructive exploitation? From the evidence of the contemporary Second World, Blaikie identifies much environmental decline. Are human societies, whatever their mode of production, bound eventually to destroy that which sustains them?

Problems involved with the depletion of natural resources – especially those used in agriculture – and with the provision of food together suggest a population problem which, if insoluble within the present world-system, should lead inexorably to a population crisis. Is there such a crisis looming, with too many people in the world? This question has been asked many times before, and is reviewed in chapter 6 by Bob Woods. He focuses on the opposing views of Malthusians and Marxians on the resource–population equation. The former see the problem as stemming from the pressure on resources generated by rapid population growth, and identify a reduction (at least) in that rate of growth as the solution. The evidence presented by Bob Woods, both historical and contemporary, suggests that the required response does come, and that (in the same way that Peter Odell notes a return to previous rates of growth in energy use – chapter 3) in a longer time perspective the present century will be seen as a unique period of above-average growth. That is not to deny that there are places still experiencing the consequences of very rapid growth, and with little apparent ability to alleviate them in the short-term. For Marxians, however, there will always be a condition akin to overpopulation in the world under the capitalist mode of production, because of the need to ensure the existence of an industrial reserve army – a pool of underutilized labour power. Thus, according to the latter view, there will always be population problems, verging on crisis proportions at certain times and places, whereas for neo-Malthusians any potential problems can be countered, within the mode of production, by social action (incorporating individual action – perhaps coerced).

Social action implies decision-taking by society as a whole and by its individual members. What is the context for such action? The first chapters of the book focus on the economic base of society and pay little attention to the milieux within which economic decisions are made and implemented. It could be deduced from such an approach – and indeed has been in many simplistic analyses – that the capitalist mode of production is a singular form of social organization. But it is not; contemporary capitalism in the First and Third Worlds is characterized by a great variety of social formations – separate organizational forms within which the imperatives of capitalism are interpreted and reinterpreted. (The same is true, though to a lesser extent, in the self-ascribed socialist countries of the Second World.) The world is a mosaic of regional cultures, each the outcome of the interaction of past and present modes of production and each with its characteristic social and political forms (Massey and Allen, 1984).

This cultural variability is to some extent a hindrance to the spread of capitalism, and as a result powerful forces have sought to erode it and replace it by a 'world culture'. This process of erosion is the subject of chapter 7, where Dick Peet defines culture and investigates the central role of religion within it. Further, he looks at the commodification of culture, and the replacement of local modes by universal media-based forms. The result is what he terms *ultraculture*, elements of life 'launched on the market of minds in the pursuit of individual profit'.

The concept of ultraculture implies the development of the capitalist monolith – the replacement of local variability and variety by standard uniformity as all aspects of life become subject to the forces of the market-place and the core-dominated world-economy. What, then, is the position of the individual in such a situation? Is there a crisis of individuality? Capitalism is associated in much popular rhetoric with freedom to choose, to control one's own life and to influence the ways in which society makes its decisions. This is the basis of the concept of liberal democracy, a method of political organization which Ron Johnston (chapter 8) shows is closely associated with the core of the capitalist world-economy. Some models of economic and social development suggest that as the Third World periphery countries progress through the stages of economic growth (as outlined in Rostow's model; see pp. 6, 206) so their citizens will achieve greater freedoms; in this way capitalism is presented as a liberating mode of production. Ron Johnston's analyses suggest otherwise, however, for the freedom is tenuous, not only in the Third World countries but also in those of the First World. Liberal democracy has been created by states to legitimate the capitalist mode of production, which is an alienating form of economic and social life. But its ability to allow democracy to flourish is fragile, and the freedoms it permits are, at best, constrained. Further, it seems that there is a balance to be achieved. In almost all countries of the core of the world-economy, liberal democratic freedoms are granted and standards of life chances – as indexed by life expectancy – are high. In the periphery, liberal democratic freedoms are few and life chances are generally poor in the Third World; in the Second World, life expectancy is relatively high, even if liberal democratic freedoms are few.

Ron Johnston's chapter is the first in the book to pay explicit attention to the role of the state in the capitalist world-economy, though government policies are referred to in earlier chapters. The state is an increasingly important element in the operations of the capitalist world-economy, and much attention is now being paid by social scientists to the nature and role of the state apparatus (among geographers, see Johnston, 1982; Clark and Dear, 1984; Taylor, 1989). In those forms of economic organization that preceded the capitalist world-economy – the *world empires* in Wallerstein's terminology – political control by the state was the key element. With the transition to capitalism and the dominance of markets, the state's role changed; it was needed to support and to promote, but in a secondary (partly ideological) role. Nevertheless, states were necessary components of capitalism, providing the secure environment (including territorial definitions) within which it could prosper. Thus the creation of states was part of the creation of the geography of capitalism (Harvey, 1985).

A strong ideological link has developed between state-building (the process of creating territorially defined units) and nation-building (the creation of popular identification with the state), and the two processes were the focus of the classic work of the Norwegian social scientist Stein Rokkan (see Urwin and Rokkan, 1983). In some parts of the world the processes were initially more successful than they were in others, providing 'stable' springboards for some 'national' capitals in the competitive world-economy. But state and nation were not always spatially coincident – the newly defined territories were being overlaid on the pattern of pre-existing regional cultures, in some places with a better fit than in others. The consequence has been a tension between nation and state where the fits are poor, as illustrated by Colin Williams (chapter 9). Nationalism would seem to be a major problem today in many parts of the world. Is it a necessary concomitant of a growing economic crisis?

In the process of state-building, some states have become much more powerful than others, producing a sequence of hegemonies as Agnew and Corbridge (chapter 10) and O'Loughlin (chapter 11) illustrate. The powerful states seek to extend their political and economic influence, but their ability to do this declines as they face economic competition from others and have to invest more in the military support needed for their global organizations. At times, the tensions between the declining superpower and the burgeoning would-be replacements have led to major wars, with one superpower replacing another; John Agnew and Stuart Corbridge suggest that this is an unlikely consequence of the current crisis, however, because political and economic power are becoming increasingly separated. At other times, the tensions are played out in a cold war, with no major conflicts but with many smaller ones in the periphery as the hegemonic protagonists ally themselves with local forces – in some cases those fighting issues of national congruence, and in others linked to issues relating to economic organization. These are clearly illustrated in John O'Loughlin's chapter.

AND THE FUTURE?

So is there a crisis? Who knows – yet? Clearly there are problems, many of them – and all of them are intertwined in the operations of a capitalist world-economy which is hell-bent on annihilating space and place. These problems are severe now on a global scale, and life-destroying in some places. How, then, are they to be countered? Can they be used as the elements to create an intellectual crisis, as the stimuli to a turning-point, to a new set of decisions on how the world should be organized economically, socially and politically?

Change can be brought about in a variety of ways, one of which is a determination to do something new in the future because of the conclusions reached from analyses of the past and present. How should such analyses be conducted? The clear message of this book is that the approach must be holistic. It must not separate out the parts and treat their problems as curable independent of the other parts. We must take a global view, not only spatially but also through the adoption of a unified historical social science. This is Peter Taylor's conclusion in chapter 12, where he outlines in detail Wallerstein's world-systems project and locates the geographical perspective within it. This is not a call for a narrow orthodoxy, but for a realistic framework within which to analyse the problems of the world and work towards the removal of its many inequities. We are entering a period of ideological confusion in which the epistemological and historiographical premises of the social sciences are facing their severest challenge since their creation in the nineteenth century. This intellectual crisis is part of a wider set of political opportunities which, Taylor argues, we must grasp for our construction of a new world.

The task is urgent. Time may be running out, as we are reminded in Bill Bunge's epilogue (chapter 13). We possess the awesome power to destroy the world as we know it (and for all time, in all probability) through actions which may take only a few seconds. By such actions we would eradicate an environment in which, if we have the will, we could live in peace and universal prosperity. To make sure that this never happens, we must contribute to the creation of a better world order so that, to paraphrase one of Bunge's earlier (1973) and memorable phrases, the world will be full of happy regions.

References

Allen, J. and Massey, D. 1988: *Restructuring Britain: The Economy in Question*. London: Sage Publications.
Arrighi, G., Hopkins, T. K. and Wallerstein, I. 1988: *The Great Rehearsal*. Binghamton: State University of New York.
Brandt Commission 1980: *North–South: A Programme for Survival*. Cambridge, MA: The MIT Press.
Brookfield, H. C. 1975: *Interdependent Development*. London: Methuen.
Bunge, W. 1973: Ethics and logic in geography. In R. J. Chorley (ed.), *Directions in Geography*. London: Methuen, 317–31.
Clark, G. L. and Dear, M. J. 1984: *State Apparatus*. London: Allen & Unwin.
Coates, B. E., Johnston, R. J. and Knox, P. L. 1977: *Geography and Inequality*. Oxford: Oxford University Press.

Cox, K. R. 1979: *Location and Public Problems*. Chicago: Maaroufa.

Frank, A. G. 1982: Crisis of ideology and ideology of crisis. In S. Amin, G. Arrighi, A. G. Frank and I. Wallerstein (eds), *Dynamics of Global Crisis*. New York: Monthly Review Press, 109–66.

Galtung, J. 1980: *The True Worlds*. New York: Free Press.

Green, A. E. 1988: The north–south divide in Great Britain: an examination of the evidence. *Transactions, Institute of British Geographers*, NS13, 179–98.

Harvey, D. 1985: The geopolitics of capitalism. In D. Gregory and J. Urry (eds), *Social Relations and Spatial Structure*. London: Macmillan.

Harvey, D. 1987: Flexible accumulation through urbanisation. *Antipode*, 19, 260–86.

Johnston, R. J. 1982: *Geography and the State*. London: Macmillan.

Johnston, R. J. 1986: The state, the region and the division of labor. In M. J. Storper and A. J. Scott (eds), *Production, Work, Territory*. London: Allen & Unwin, 265–80.

Katsiaficas, G. 1987: *The Imagination of the New Left: A Global Analysis of 1968*. Boston: South End Press.

Kidron, M. and Segal, R. 1981: *The State of the World Atlas*. London: Pluto.

Lash, S. and Urry, J. 1987: *The End of Organized Capitalism*. Cambridge: Polity Press.

Marsden, T. K., Whatmore, S. J. and Munton, R. J. C. 1987: Uneven development and the restructuring process in British agriculture. *Journal of Rural Studies*, 3, 297–308.

Martin, R. L. 1988: The political economy of Britain's north–south divide. *Transactions, Institute of British Geographers*, NS13, 389–418.

Massey, D. and Allen J. (eds) 1984: *Geography Matters*. Cambridge: Cambridge University Press.

Peet, R. 1977: The development of radical geography in the United States. *Progress in Human Geography*, 1, 64–87.

Reitsma, H. J. A. 1982: Development geography, dependency relations and the capitalist scapegoat. *The Professional Geographer*, 34, 125–30.

Ross, R. S. J. 1983: Facing Leviathan: public policy and global capitalism. *Economic Geography*, 59, 144–60.

Rostow, W. W. 1971: *The Stages of Economic Growth: A Non-Communist Manifesto*, 2nd edn. Cambridge: Cambridge University Press.

Smith, D. M. 1977: *Human Geography: A Welfare Approach*, London: Edward Arnold.

Taylor, P. J. 1981: A world-system perspective on the social sciences. *Review*, 5, 3–11.

Taylor, P. J. 1982: A materialist framework for political geography. *Transactions, Institute of British Geographers*, NS7, 15–34.

Taylor, 1985: The value of a geographical perspective. In R. J. Johnston (ed.), *The Future of Geography*. London: Methuen.

Taylor, P. J. 1988: The error of developmentalism in human geography. In D. Gregory and R. Walford (eds), *New Horizons in Human Geography*. London: Macmillan.

Taylor, P. J. 1989: *Political Geography: World-Economy, Nation-State and Locality*. London: Longman.

Thompson, W. R. (ed.) 1983: *Contending Approaches to World-Systems Analysis*. London: Macmillan.

Urwin, D. and Rokkan, S. E. 1983: *Economy, Territory, Identity: Politics of West European Peripheries*. Beverly Hills, CA, and London: Sage Publications.

Wallerstein, I. 1974: The rise and future demise of the capitalist world-system. *Comparative Studies in Society and History*, 16, 387–418.

Wallerstein, I. 1982: Crisis as transition. In S. Amin, H. Arrighi, A. G. Frank and I. Wallerstein (eds), *Dynamics of Global Crisis*. New York: Monthly Review Press, 11–56.

Watts, M. 1989: The agrarian question in Africa: debating the crisis. *Progress in Human Geography*, 13, 1–41.

2

The Geography of International Economic Disorder

NIGEL THRIFT

The 1970s were a period of severe economic crisis worldwide. To some observers the 1980s have presented a much calmer prospect. The purpose of this chapter is to show that this prospect is an illusion. The economic crisis of the 1970s has not gone away: rather, in the 1980s it has been brushed under the carpet. Now it shows signs of emerging again but on a larger and much nastier scale.

The world-economy in the 1980s has consisted of a remarkable juggling act in which the three chief sets of actors on the world economic stage – multinational corporations, banks and the governments of countries – have all striven to gain maximum advantage whilst simultaneously promoting economic recovery. This act has produced a world-economic order which is more and more uncertain and yet which is also more and more inured to that uncertainty. It is a world-economic order which is addicted to the knife-edge. It is a world-economic order hooked on speed.

The first part of the chapter looks at the world-economy of the 1980s. The emphasis is very much upon the role of the multinational corporations and banks in bringing about world-economic order, although the power of the governments of countries is by no means forgotten. The second part looks at what has happened to people's jobs as a result of all the changes. It looks at patterns of employment and unemployment, and also at the rise of new kinds of jobs. Particular consideration is given to how far employment opportunities in some parts of the world have become more closely tied to employment opportunities in other parts through the rise of the so-called 'new international division of labour' (its growth facilitated by state action), through international migration, through new 'flexible' work practices and through the rise of 'world cities'. There is no need to stress that the chapter is geographical. All the changes described are inherently geographical. Indeed, their existence depends upon geography.

It is important first to clarify a number of issues. In this chapter the phrase 'world-economy' refers to the capitalist world-economy and the phrase 'world-economic order' refers to the capitalist world-economic order. No consideration is given to the socialist countries which some commentators (e.g. Lebkowski and Mankiewicz, 1986; McMillan, 1987; United Nations, 1988; Wallerstein, 1979, 1983) place within the orbit of the capitalist world-economy and the capitalist world-economic order. Such countries are certainly increasingly linked to it (see, for example, Bora, 1981; Kortus and Kaczerowski, 1981; Gutman and Arkwright, 1981), but the case has yet to be convincingly made that they are an integral part of it. Second, since what follows is such a compressed account, it is inevitable that changes in the world-economic order in the 1980s will be presented as remorseless, unitary movements forming an ordered and coherent whole. The reality, of course, is rather different. The changes taking place are still, even now, quite tentative processes which can conceivably be reversed – this is no surprise, for they are organized by humans. Also the world-economic order is not an ordered whole. Certainly, it has some semblance to order – otherwise it would not work – but it is not 'an order established *in order* to be coherent' (Lipietz, 1984b, p. 92). Third, and following from the second issue, it is important to stress that no attempt is made in what follows to give one determinant of the world-economic order of the 1980s absolute priority at the expense of any other. In particular, many writers have been guilty of emphasizing the role of one or two of the groups of actors on the world-economic stage to the detriment of others. For some writers, for example, the multinational corporations and the banks have fused into an overarching 'world capital' which transcends all national barriers (e.g. Harris, 1983, 1986). For others, what goes on inside national barriers must be given priority and multinational corporations are a sideshow (e.g. Aglietta, 1982; Lipietz, 1984b). Each of these reactions is equally incorrect. The world-economy is the outcome of a whole series of countervailing forces operating on a whole series of scales, no one of which makes sense without the others.

Finally, some questions of definition. In this chapter the phrase 'developed countries' generally refers to the 24 countries that belong to the Organization for Economic Co-operation and Development (OECD), namely Australia, Austria, Belgium, Canada, Denmark, Finland, France, West Germany, Greece, Iceland, Ireland, Italy, Japan, Luxemburg, the Netherlands, New Zealand, Norway, Portugal, Spain, Sweden, Switzerland, Turkey, the United Kingdom and the United States. The phrase 'developing countries' generally refers to those designated as low-income or middle-income developing countries by the World Bank in its annual *Development Report* (World Bank, 1988a). (In using these terms, there is no intention to imply the existence of naive developmental processes as was popular in some theories before the 1970s. Hence use of the term 'developed' does not suggest that the processes of change in the rich countries listed above have somehow finished. Nor does use of the term 'developing' indicate that such countries are actually experiencing significant levels of economic growth at the present time – for the data presented below clearly indicate that some are not. These two terms are firmly entrenched in the literature and are used in that context; the more discerning reader may wish to replace them by the alternative terms 'core' and 'noncore', which indicate countries' positions within the organizational structure of the capitalist world-economy.)

THE WORLD-ECONOMIC CRISIS AND A NEW WORLD-ECONOMIC ORDER

The 1970s and 1980s have not been an easy period in the history of the world-economy. The 1970s saw a decline in the rate of profits of companies which had been gathering pace since the 1960s, and the shock of rapid increases in the prices of primary products (most especially, the price of petroleum, which quadrupled in 1973/4) combining to produce slow economic growth and inflation – the phenomenon of 'stagflation' (Van der Wee, 1987). In the 1980s, inflation has been brought under control by many governments but the austere, supply-side policies which these governments have put in place in order to combat inflation, combined with the lack of serious international economic coordination by governments and high interest rates, have damped down any recovery. The result has been that *world* economic growth has continued to fall in the 1980s, although some *countries* (especially the United States) have enjoyed considerable prosperity (table 2.1). Indeed, for certain countries, most notably the developing countries of Africa and Latin America, the 1980s have been even more problematic than the 1970s.

How can such events be interpreted? Some commentators (Aglietta, 1979, 1982; Lipietz, 1984a, 1984b, 1986) have suggested that what we are seeing is the fall of one regime of capital accumulation and its replacement by another; in other words 'the crisis is calling into question a certain historical form of capitalism but not capitalism itself' (de Vroey, 1984, p. 64). In this depiction, the particular 'intensive' regime of capital accumulation (Aglietta, 1979) which had become typical of the economies of many of the industrialized market countries reached its peak in the 1960s. This regime was based on massive increases in productivity, brought about through widespread mechanization and sustained by equally massive increases in demand for the goods produced, generated by the linking of wages to productivity through the regulation of state or state-mediated institutions such as collective bargaining and systems of welfare. For all effects and purposes, this system of mass consumption – the 'powerhouse of demand' as it has often been called – excluded countries other than those of the industrialized core: 'Capitalism had temporarily resolved the question of markets on an internal basis. One could even say that the exports of manufactured goods to the periphery were only just covering the cost of raw materials' (Lipietz, 1984b, p. 99). Certainly, the 1950s and 1960s saw a large

Table 2.1   Growth rates of world output, by region/country, 1960–1986

Region/Country	1961–73	1974–80	1981–6
World output	5.5	3.6	2.7
Developed market economies	5.0	2.5	2.2
United States	4.0	2.2	2.4
Western Europe	4.8	2.4	1.5
Japan	9.8	3.8	3.6
Developing countries	6.3	5.1	1.5
Africa	5.5	4.4	− 0.9
Asia	5.1	6.0	4.8
Middle East	8.2	3.9	− 0.9
Developing Europe	5.5	6.4	2.9
Latin America and the Caribbean	6.7	5.2	1.0
Centrally planned economies of Europe	6.6	4.6	3.3
China	3.8	5.6	8.8

Source: United Nations, 1988, p. 17.

increase in the number of multinational corporations in the world, mainly those of North American extraction, but their attentions tended to be restricted to the other developed countries.

The changes we are now seeing coming about in the world-economy amount to the formative struggles of a new regime of accumulation, variously labelled neo-Fordism, or flexible accumulation (Harvey, 1987), or disorganized capitalism (Lash and Urry, 1987). This 'extensive' regime of accumulation has been born out of the reorganization of firms so that they could meet the heightened conditions of lower economic growth, and the reorganization of mechanisms of regulation by state or state-related institutions. It is identified by a whole set of associated indicators such as the automation of large parts of the labour process, increasing labour flexibility, the break-up of large integrated plants into smaller specialist units, and the greater coordination of products and distribution through the massive injection of information technology and telecommunications.

Whether the changes in the world-economy in the 1970s and 1980s can be interpreted as 'capitalism putting its books in order' (Margirier, 1983, p. 61) is a moot point; whether we are at the beginning of a new regime of accumulation or still in the death throes of the earlier regime is a clouded issue (Gordon, 1988; Massey, 1988); whether we have the theory rich enough to cope with all the changes seems unlikely (Cantwell, 1988; Casson, 1985). However, what is clear is that the 1970s and 1980s have been a time of experimentation by corporations and at the heart of this experimentation has been the process of the *internationalization of capital*.

The Internationalization of Capital

What does the 'internationalization of capital' mean? Above all, it means the export of capitalist relations of production, not just of money. *Capitalist relations* of production are created on a world scale through direct investment by foreign firms which create subsidiaries abroad, organized on capitalist lines. *Multinational corporations* are the main vehicles of this capital export and as they have increased in both number and size so they have taken on much greater importance in the world-economy than they had in the 1950s or 1960s.

At the heart of the world-economic order of the 1970s and 1980s has been the process of the internationalization of capital. In the 1970s, faced with falling rates of profit, firms were forced to 'automate, emigrate, or evaporate' (*New York Times*, cited by Frobel, 1983). Many firms chose the path of emigration (usually mixed with other strategies, such as automation), and for an obvious reason. Multinational corporations were more profitable than other enterprises (Andreff, 1984).

Thus the flow of foreign direct investment increased throughout the 1970s, in spite of economic crisis (Thrift, 1986). The pressures on competition in straitened economic circumstances meant that firms could not afford to let up in their search for profits on a worldwide basis. As more nationally-based firms decided to become multinationals, and as multinational corporations further expanded their network of subsidiaries, so many commentators concluded that the whole world would soon become a capitalist stage – Harris (1986) even argued that the end of the 'third world' was in sight.

Three main developments led commentators of the time to this conclusion. The first development that seemed to assist the continuing spread of capitalist social relations was the growth of the 'global corporation'. The largest multi-national corporations had become ever larger, adding on more and more subsidiaries. Indeed some of the larger multinational corporations now operated in the markets of virtually every country in which it was possible to make a profit. For them, the challenge was no longer expansion into the markets of new countries: rather, it was how to organize their world network of subsidiaries to turn the best possible profit. Until the late 1960s the challenge of putting into operation a global profit-making strategy would have been formidable. But technical developments, most notably in telecommunications and information technology, organizational developments, especially the setting up of 'regional headquarters offices' (Dunning and Norman, 1983; Grosse, 1982) and the rise of an international capital market enabled this challenge to be met by at least some of them. The result was the growth of a new type – the global corporation (Hunt, Parker, Rudden, 1982; Taylor and Thrift, 1982, 1986; *Economist*, 1984; Thrift and Taylor, 1988). Such corporations tend to promote global brand names, and production within them is organized on a regional or even on a global basis. The result of this integration of production and the ceaseless shifting of materials, components, money and information that it entails is the dramatic enlargement of markets internal to these corporations (Rugman, 1981, 1982;

Casson, 1986). Just how widespread these *internal markets* now are can be traced through figures on intra-firm trade across national boundaries. For example, for the United States in 1985, 40 per cent of total imports and 31 per cent of total exports could be identified as intra-firm. The proportion of intra-firm exports in the total exports of the United Kingdom increased from 29 per cent in 1976 to 30 per cent in 1981. Estimates suggest that between 30 and 40 per cent of world trade is now within corporations (United Nations, 1988).

The second development that seemed to show signs of capitalist social relations being exported by the multinational corporations to the rest of the world was clearest in certain manufacturing industries. These were the bell-wether industries of automobiles and electronics. In them could be found the beginnings of 'global factories', networks of production and associated facilities which were consciously coordinated on a worldwide basis, and even included Third World locations. The enormous multinational corporations of the automobile industry provided the stereotypical examples of this development. The example of Ford was often quoted (Bloomfield, 1981; Maxcy, 1981; Ballance and Sinclair, 1983; Cohen, 1983; Beynon, 1985; Dicken, 1986; Hill, 1987). Ford is currently the sixth largest company in the world, employing nearly 370,000 people worldwide, 53 per cent of them outside its home country of the United States. It has been a multinational corporation for some time. Even in 1930, the corporation was assembling cars in 20 countries. But until quite recently its structure of production consisted of quite separate manufacturing or assembly plants serving distinct national markets. However, as the example of Europe shows, in the 1960s its structure began to change. In 1967 Ford of Europe was created in an attempt to integrate the corporation's European operations. Among the factors involved in this decision were the desire to spread investment (especially because the British subsidiary was prone to labour unrest), the shortage of labour near the existing plant, the high cost of separate model development in a number of markets and the new market opportunities created by the founding of the EEC. By 1970 signs of this new kind of integration were already apparent (figure 2.1) and by 1978 integration of the European operations was becoming a reality. For example, the introduction of the Ford Fiesta in 1976 created new and complex patterns of movement within the corporation, integrating plants such as Belfast (carburettors and distributors) and Bordeaux (transmissions and axles) with assembly centres in the United Kingdom, Germany and Spain (Bloomfield, 1981). By 1988, Ford's European operations had gone a step further and were being integrated with Third World plants in Mexico and Brazil. Ford's European-made cars like the Escort and the Sierra were becoming world geographies on wheels. The size and scope of Ford's strategy to integrate its European operations can be gauged by the fact that in 1987 it had more than 12,000 tonnes of components in transit between its European plants at any one time. Ford even owns its own ships to transport components.

Similar kinds of events, although perhaps not on quite the same epic scale as in the highly concentrated automobile industry, were found in the electronics

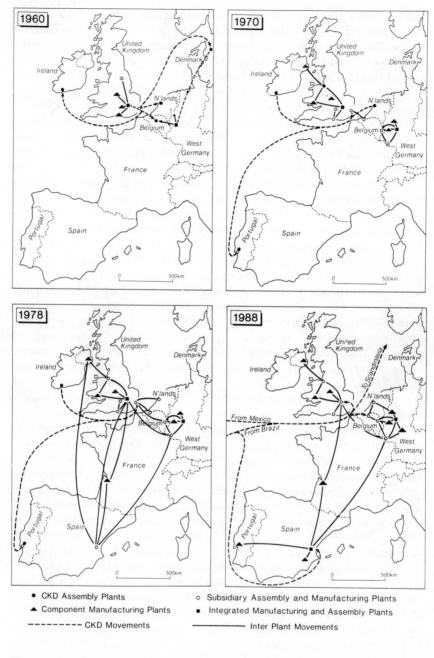

Figure 2.1   The changes in the organization of production of Ford vehicles in Europe from 1960 to 1988

Sources: Bloomfield, 1981, p. 284; Hamilton, 1984, p. 361

industry, especially in its semiconductors sector. There, the US-based corporations in particular were spreading out over the globe looking for new markets and low production cost assembly points (Ernst, 1985; Grunwald and Flamm, 1985; McGee, 1986; Scott, 1987, 1988). These corporations fixed particularly on Third World locations, giving rise to fears of the United States semiconductor industry becoming a 'runaway' from the United States (figure 2.2) (Morgan and Sayer, 1988).

The third development was the rise of the newly industrializing countries (the NICs). Through the 1970s, a number of countries were beginning to take off into economic prosperity which reached beyond the ruling elites. The four 'little tigers', Hong Kong, Singapore, South Korea and Taiwan, provided the most spectacular examples of export-led growth (Mitchell, 1988). Notwithstanding a number of problems (Hamilton, 1984), these four countries became so successful that they could no longer be classed as developing countries in any meaningful sense. Other countries like Argentina, Brazil and Mexico also seemed likely contenders to escape the more of poverty and dependency (Harris, 1986). They were producing more and more sophisticated products and their workforces were the recipients of steadily rising wages.

In the 1980s, the idea of a world-economy made up of global corporations, global factories and newly industrializing countries reaping the rewards of industrial progress began to look a little frayed around the edges. In particular, the rate of growth of foreign direct investment declined markedly (especially allowing for inflation). In addition, foreign direct investment was increasingly directed inwards to the core countries. The world-economy became inward-looking.

To begin with, the significance of the global corporation was questioned. Quite clearly the number of very large corporations has continued to grow. More and more multinational corporations joined the ranks of the billion dollar club – corporations with total sales of more than $1 billion – in the 1980s. By 1985 the largest 600 industrial companies accounted for between 20 and 25 per cent of the value added in the world's market economies. Their importance as world importers and exporters was even greater (United Nations, 1988).

But the 1980s saw the degree of multinationality of these very large corporations marking time. However, this slowdown concealed some important changes. The larger United States-based multinationals retreated, relatively speaking. Their greatest reach, measured by the ratio of foreign sales to total sales, was at the beginning of the 1980s. Since that time their degree of multinationality has declined (United Nations, 1988). Meanwhile the Japanese and Western European multinationals have continued to expand abroad. The larger Japanese multinational corporations continued to spread into the United States and Europe in the 1980s (Dicken, 1988), although their overseas presence is still relatively modest compared with Western European multinationals (or, indeed, those of the United States).

The degree of attention paid to the larger multinational corporations in the 1970s also deflected attention away from the importance of small- and

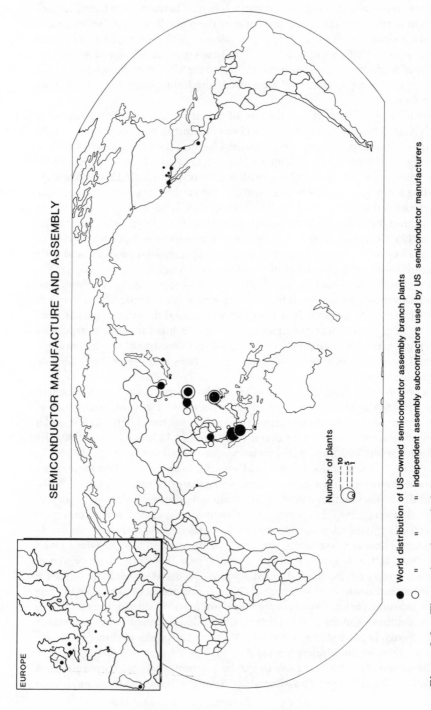

SEMICONDUCTOR MANUFACTURE AND ASSEMBLY

EUROPE

Number of plants

⊖ 10
○ 5
• 1

● = World distribution of US–owned semiconductor assembly branch plants

○ = independent assembly subcontractors used by US semiconductor manufacturers

Figure 2.2 The semiconductor industry overseas

Source: Scott and Angel, 1988, pp. 1058, 1059

medium-sized multinationals which offered the larger corporations considerable competition in the 1980s (United Nations, 1988). The evidence tends to suggest that, over time, these kinds of firms have become more important in the world-economy as the barriers to going overseas have become less substantive and the importance of serving overseas markets has become greater. Further, in the 1980s, they grew faster than their larger counterparts.

An important subset of these smaller multinational corporations consisted of firms from developing countries. Although there are a number of large multi-national corporations from developing countries – 17 have sales of $1 billion or more – most developing countries' multinationals tend to be small, relatively labour-intensive concerns concentrated in specific industrial sectors like textiles where their competitive advantages are clear. Most of them hail from a small number of developing countries such as Hong Kong, India, Brazil, Singapore, South Korea, Taiwan, Argentina, Mexico and Venezuela, and tend to invest first in other developing countries (Wells, 1983; Lall, 1984; Khan, 1986).

It was not only the idea of a world heading remorselessly into the thrall of global corporations which came under critical scrutiny. The idea of a move towards the global factory also began to be questioned. It was quite clearly based upon only a few industries and even within those industries it had become clear that running a global factory was no mean feat. In the automobile industry, the trend had seemed to be towards 'world-cars' built in 'world factories'. But the reality was rather different (Jones, 1985). An attempt at a world car, Ford's Escort, proved to be a success in Europe but failed to make an impact in the United States. Other world cars are now planned but although they will look alike and may be designed in different parts of the world from where they are produced (so that Ford Europe is now to be responsible for designing small cars, and Ford America large ones), they will not be *produced* globally. Rather, their production will be confined to regions like Western Europe and North America. In the electronics industry, there are also signs of limits having been reached. In the semiconductor industry, the growth of US-owned assembly plants in overseas locations has slowed, the result of a combination of an increase in automated assembly (which reduces the need for cheap labour), a decrease in the amount of assembly work as chip integration becomes greater, and new production techniques like 'just-in-time' which requires supplies to be close at hand (Sayer, 1986b; Morgan and Sayer, 1988). Since 1985, there have even been a few cases of relocation back to the United States (United Nations, 1988). However, it should also be noted that any withdrawal on a large scale is unlikely because overseas locations like Europe and Asia constitute important markets which cannot be served solely by exports and where competition has become much harsher (Scott, 1987).

The global factory has also become less important because multinational corporations have discovered other ways of achieving integration. Most particularly, they have all attempted to bring balance to their global networks of subsidiaries though a spate of mergers and acquisitions – Western European and Japanese multinational corporations have been particularly active in the

United States (United Nations, 1988) – and through a whole series of cooperative agreements with competitors. The automobile industry has been particularly keen to form strategic alliances. For example, Ford has bought into Mazda, has a joint venture with Volkswagen in South America and has growing links with Nissan worldwide. The same kind of alliances are being forged in electronics (Cooke, 1988).

These kinds of agreements provide corporate strategists with greater flexibility worldwide, since alliances can be forged and broken according to prevailing market conditions, risks (on research and development and other aspects) can be collectivized and latecomers can obtain access to new technologies and insights into new organizations of production. They stand alongside a number of interconnected innovations in the organization or production, pioneered by Japanese multinationals, which allow firms to integrate their networks of production in as flexible a manner as possible and also allow them to move towards volume production of small batches of goods. These innovations include: a greater but highly selective use of *subcontracting*; 'just-in-time' systems implying producing to order rather than to stock (or 'just-in-case') which require management to drive down stock inventories to minimal levels; *total quality control* (since there are no stock inventories to fall back on); and *automated manufacturing systems* (Sayer, 1986b, 1989). Ford has engaged in all these practices recently in its attempts to compete with Japanese automobile manufacturers (currently in Europe it is pruning its list of subcontractors so that it can set up more exacting supply contracts, and is moving towards a just-in-time system with a half-a-day lag). Some commentators (e.g. Leborgne and Lipietz, 1988; Scott, 1988) have suggested that these innovations in production have led to the formation of a whole series of new industrial spaces around the world because, jointly, they set in motion strong agglomerative tendencies.

Last but not least, the success of the newly industrialized countries has been questioned. The 1980s were a period of retrenchment in the world-economic order. It became clearer than ever that the world-economy has moved towards a tripolar regional structure based on North America, Western Europe, and the Asian Pacific Rim, focusing especially upon the Pacific Basin – the economic success of California in the 1980s has been nothing short of breathtaking (Dear, 1986). Most of the rise in the world's exports and imports in the 1980s has been soaked up by this triangle of wealth. The economic interaction between these three world regions has been increasing over time, and they now constitute an increasingly tightly knit system (figure 2.3). For the countries outside these three world regions, the winds have often blown exceedingly cold. The flow of foreign direct investment to developing countries decreased in the 1980s and in many of these countries there was no domestic growth to compensate. It is no surprise, then, to find that the newly industrialized countries outside Asia fell back in the 1980s (and even the Asian newly industrialized countries had their problems) (Van Liemt, 1988). They were hampered by slow world-economic growth, falls in the prices of raw materials and a combination of high interest rates and indebtedness (Peet, 1987). The pickings were even worse

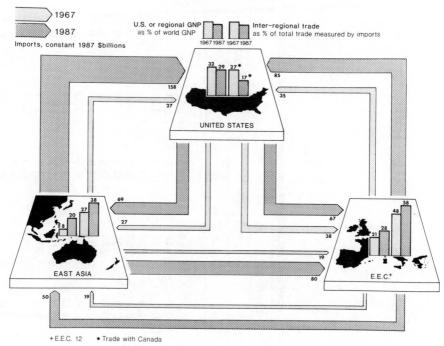

Figure 2.3   The tripolar world economy
Source: Economist, 24 December 1988, p. 41

Table 2.2 Annual average foreign direct investment by selected countries, 1975–1980 and 1981–1985

	Outflows				Inflows			
	1975–80		1981–85		1975–80		1981–85	
Region/Country	US $ million	%	US $ million	%	US $ million	%	US $ million	%
Developed market economies	37,774	98.7	44,454	98.1	24,642	76.6	36,593	75.1
Australia	296	0.7	1,146	2.6	1,271	3.9	1,858	3.8
Austria	74	0.2	131	0.3	139	0.4	253	0.5
Belgium	524	1.3	196	0.4	2,203	3.7	1,122	2.3
Canada	2,039	5.1	3,982	8.8	862	2.7	421	0.9
Denmark	89	0.2	95	0.2	75	0.2	62	0.1
Finland	74	0.2	274	0.6	44	0.1	28	0.1
France	1,825	4.5	2,692	5.9	2,127	6.6	2,145	4.4
West Germany	3,153	7.8	3,870	8.5	1,053	3.3	869	1.8
Italy	419	1.0	1,670	3.7	554	1.7	1,048	2.2
Japan	2,173	5.4	5,084	11.2	152	0.5	331	0.7
Netherlands	3,841	9.5	4,019	8.9	1,277	4.0	1,435	2.9
New Zealand	55	0.1	149	0.3	229	0.7	217	0.4
Norway	142	0.4	485	1.1	383	1.2	246	0.5
Portugal	5	0.0	16	0.0	90	0.3	182	0.4

South Africa	208	0.5	206	0.5	90	0.3	94	0.2
Spain	171	0.4	305	0.7	971	3.0	1,770	3.6
Sweden	571	1.4	1,033	2.3	100	0.3	167	0.3
Switzerland	–	–	–	–	–	–	–	–
United Kingdom	7,025	17.4	9,409	20.8	5,195	16.1	4,331	8.9
United States	17,092	42.4	8,640	19.1	7,895	24.5	19,156	39.3
Developing countries	501	1.2	859	1.9	7,539	23.4	12,142	24.9
East and Southeast Asia[a]	117	0.3	407	0.9	2,007	6.2	4,817	9.9
Latin America[b]	210	0.5	216	0.5	4,014	12.5	5,123	10.5
Total[c]	40,278	100.0	45,312	100.0	32,183	100.0	48,736	100.0

a Includes Oceania
b Includes the Caribbean
c Totals do not sum owing to the number of countries which have been omitted

Source: United Nations, 1988, pp. 504–10

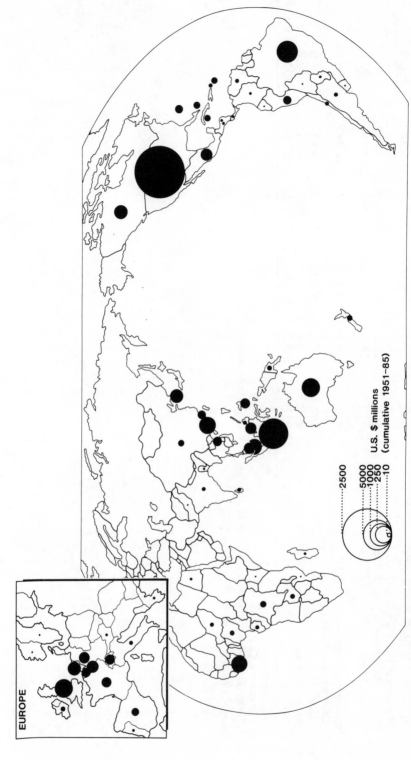

Figure 2.4 The location of Japanese overseas direct investment in 1985
Source: Dicken, 1988, p. 637

U.S. $ millions
(cumulative 1951–85)

2500
5000
1000
250
10

for many other developing countries whose position was often nothing short of catastrophic (table 2.2). The bald facts speak for themselves. In the 40 least developed countries, governments have reduced health spending by 50 per cent and education spending by 25 per cent since the beginning of the 1980s; average family incomes have fallen by between 10 to 25 per cent; and child malnutrition has increased dramatically (UNICEF, 1988).

Changes in the Origin and Destination of Foreign Direct Investment

These three changes are reflected in the pattern of foreign direct investment in the 1980s. The trends in the origin of direct investment are clear. The increasing relative importance of Japanese and West European (especially British, French and German) outward investment from year to year can be seen in table 2.2 and figure 2.4. Western Europe is now the world's largest source of foreign direct investment. The decreasing relative importance of United States outward investment is also apparent. The importance of US-based corporation investment peaked in the late 1960s or early 1970s (United Nations, 1988) and then declined from 42 per cent in the period 1975/80 to 19 per cent in the period 1981/5. But these trends should not be exaggerated, as they often are in the literature. The United States share of outward investment in 1981/5 was still nearly double that of Japan, and its stock of investment overseas in 1985 was nearly three times that of Japan (United Nations, 1988). Further, since 1985 there are signs of a US resurgence with US foreign direct investment rising faster than that of Japan (*Economist*, 1988a).

Turning to the destination of foreign direct investment, the most dramatic shift has been in the importance of the United States as a target for foreign investment. In the 1980s the United States has been besieged by foreign investors taking advantage of its growing market. Between 1981 and 1985 the United States accounted for nearly 40 per cent of foreign direct investment each year (most of the investment coming from Western Europe and Japan). The stock of direct investment accumulating in the country increased by over seven times between 1975 and 1986. The depreciation of the dollar since early 1985 has ensured that this level of investment has continued since. In the United States this level of investment has sometimes provoked a reaction surprisingly similar to the reaction of many Europeans in the 1960s to United States investment (Tolchin and Tolchin, 1988). Meanwhile, foreign direct investment into Western Europe declined relatively and absolutely in the early and mid 1980s (with the exception of countries like Spain), although it now shows distinct signs of taking off again in the run-up to the formation of a single European market in 1992 by the EEC.

In the 1970s the destination of foreign direct investment was neatly described as a double capital movement, towards the United States and the developing countries (Teulings, 1984). But in the 1980s it might more accurately be described as a circular capital movement between the United States, Western Europe and Asia. The developing countries have found the door to prosperity slammed shut.

The Internationalization of Services

Beginning in the 1970s, the sleeping giant of the world-economy woke and began to stalk the globe. Until then, service industries had tended to remain domestically bound. But in the 1970s and 1980s they began to internationalize, supported by a wide range of factors including the increased tradability of some services, occasioned by advances in telecommunications and information technology, as well as better packaging ('customization') and government deregulation (Enderwick, 1988; Marshall et al., 1988; Petit, 1986). By the 1980s service industries were growing faster than any other sector of the world-economy. By the mid-1980s, about 40 per cent of the world stock of foreign direct investment was in service industry (compared with 25 per cent at the beginning of the 1970s) and more than half of the annual flow of foreign direct investment was going into services (United Nations, 1988). Foreign direct investment in services tends to be more concentrated in the developed countries, in terms of both origin and destination, than other sectors of the world-economy, and its disproportionate growth has boosted the inward-looking nature of the world-economy in the 1980s.

Service industry covers a host of different products and activities. But there are four sectors which have internationalized particularly rapidly in the 1980s: tourism and travel (hotels, airlines, tour operators, car rentals); information services (data processing, software, telecommunications, information storage and retrieval); business services (accountancy, management consultancy, advertising, market research, public relations); and financial services (banking and finance, insurance). In each can be found large service multinationals like the accountancy firm Arthur Andersen, which employed more than 43,000 people worldwide in 1985 (figure 2.5). These four sectors are interlinked in various ways. Thus, the inputs of one sector are the outputs of another. For example, much of the increase in business travel which has taken place in the 1980s can be related to the internationalization of business and financial services. Again, many multinationals have diversified to the point where they are now engaged in activities in each of these sectors; Arthur Andersen's activities now span information, business and financial services. Even multinationals specializing in production have moved into these services. For example, Ford carved 18 per cent of its profits from financial services in 1987 and has plans to move further into finance in the future.

Of these four different service sectors, it has been financial services which have been the most prominent, not just in terms of their share of foreign direct investment in services, but also because of their wider effects on the world-economy. Indeed these effects are so significant that they deserve considerable elaboration.

The Internationalization of Finance

The expansion of production and, latterly, services overseas has needed to be financed. But the international expansion of financial services cannot be seen

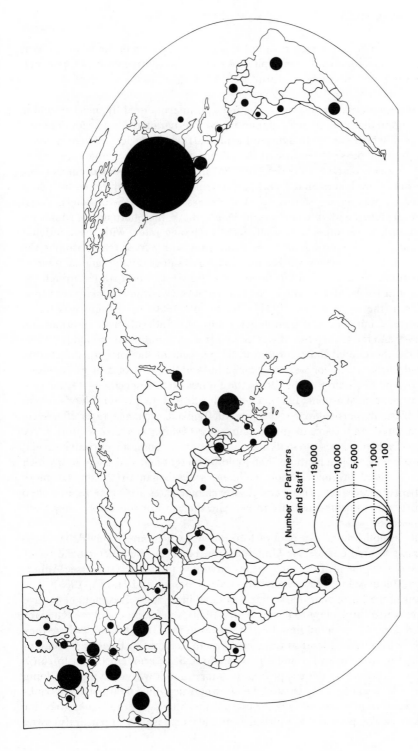

Figure 2.5 The distribution of employment in Arthur Andersen, in 1986
Source: Daniels, Thrift and Leyshon, 1988, p. 102

simply as an enabling factor: rather it is part and parcel of the whole process of the internationalization of capital. As such, it has three main components, which will be considered in turn.

The internationalization of money   The modern international financial system is based upon the creation of a number of new international markets for domestic currency which can be bought and sold, borrowed and lent. These markets have become possible because of the advent of floating exchange rates in the world money markets and the creation of a series of 'pseudo currencies', especially Eurodollars, in the world capital market. In 1944 the Bretton Woods agreement was signed, setting up what was essentially a US-run international financial system with three poles – the World Bank, the International Monetary Fund and, most important of all, fixed exchange rates, with the US dollar serving as the convertible medium of currency with a fixed relationship to the price of gold. This system worked only so long as sufficient international reserves of currency could be found to finance the growth of world trade through the 1950s and 1960s, and so long as the United States was the dominant economic power in the world (Daly, 1984). But by the 1960s the system was under pressure. Countries and companies could not find sufficient international reserves and the United States was no longer such a dominant economic power. In 1950 the United States produced 62 per cent of the total manufacturing output of the ten major western economies and 26 per cent of world exports. But by the beginning of the 1970s the figures were respectively 44 per cent and 19 per cent (Parboni, 1981). The result was that the Bretton Woods system crumbled. In particular, by the late 1960s fixed exchange rates effectively disappeared and every domestic currency became convertible into every other. Exchange rates 'floated' and, as a result, all domestic currencies could themselves become a medium that could be bought and sold and out of which a profit could be made. Soon exchange rates began to change far more frequently than other prices, and now exchange rates are changing in Tokyo when London businessmen are in bed and then in London before New York businessmen wake up.

   The establishment of a pool of Eurodollars, which are simply dollars held in banks located outside the United States, has been the other crucial factor in the development of the modern international financial system (see Aliber, 1979; Mendelsohn, 1980; Sampson, 1981; Coakley and Harris, 1983). The origins of the Eurodollar market are shrouded in mystery. It is thought to have started when, in the late 1940s and early 1950s, the Chinese and the Russians doubted the safety of holding dollar reserves in the United States (where they could be confiscated) and so transferred them to banks in Paris and London. Later, towards the end of the 1950s, as the US government began to run a balance-of-payments deficit, paying out more than it was receiving and doing so in dollars, so the newly created dollar owners deposited their dollars in banks in Europe rather than in New York. Because the European banks were far away from any potential US jurisdiction and (increasingly importantly) away from US control over interest rates they were able to pay higher interest rates on these dollar deposits than the American banks. Then three things happened which made the market in Eurodollars take off (see figure 2.5). First, during 1963 and 1964 President Kennedy, worried by the increasing flow of dollars abroad, announced an Interest Equalization Tax and a Voluntary Credit Restraint Programme, which were intended to reduce capital outflow. Instead, international borrowers looked to Europe and the Eurodollar market. (Prominent among these borrowers were US multinational corporations raising loans on the Eurodollar market so that they could continue to expand abroad). Second, from 1971 the United States government began to finance its budget deficit by paying in its own currency, flooding the world with dollars and helping to fuel the inflationary process worldwide. It is estimated that world monetary reserves increased *twelvefold* between 1970 and 1980 (see Parboni, 1981). Finally, there was the 'oil shock'. When the members of the Organization of Petroleum Exporting Countries (OPEC) simultaneously raised the price of oil in 1973 they also acquired huge reserves of dollars from their sales of oil, reserves which had to be invested. Many of these dollars were invested not in government bonds but in banks outside the United States, dramatically swelling the banks' Eurodollar deposits (figure 2.6).

The international banks, and other institutions, had to find somewhere to put all the money they suddenly found in their coffers. This was no easy task.

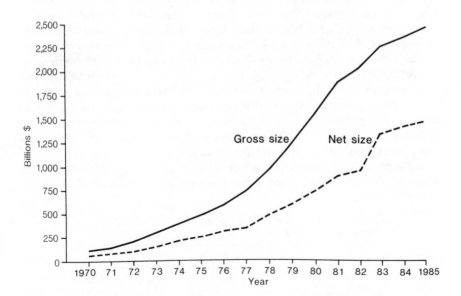

Figure 2.6 Eurocurrency market size, 1970–1985
Source: World Financial Markets, Morgan Guaranty, New York

It was difficult to find borrowers in the shape of existing governments, government-backed corporations and multinational corporations willing to take out new loans and soak up the surplus. One outlet was to create new markets. For example, the Eurodollar market became considerably more sophisticated in order to tempt a greater range of lenders and borrowers to participate. Eurodollars were joined by other Eurocurrencies held in banks located outside their country of issue: Japanese yen, British pounds sterling, West German marks, European ECUs and so on. The range of loans able to be made on the Eurodollar market expanded as well. For example, Eurobonds, relatively long-term IOUs issued by corporations and states and able to be sold on, became popular.

Another outlet the banks found to soak up the surplus was the better-off developing countries, especially those with some oil revenues. But a combination of falling oil prices and rising dollar interest rates (in which most of the loans were denominated) soon delivered a knock-out blow to many of the countries which had run up debts. They were unable to pay their debts off. In 1982 Mexico defaulted on its interest payments and was soon followed by Brazil, Argentina and others. The international debt crisis had begun (*Oxford Review of Economic Policy*, 1986; Corbridge, 1987; Strange, 1988). It has brought into existence a whole subset of the developing countries, the highly indebted countries (or HICs, including Argentina, Bolivia, Brazil, Chile, Colombia, Costa Rica, Ivory Coast, Ecuador, Jamaica, Mexico, Morocco, Nigeria, Peru, Philippines, Uruguay, Venezuela and Yugoslavia) which in 1987 had debts that were, on average, 60 per cent of their Gross National Product, and in the case of sub-Saharan Africa much worse: debts there reached 80 per cent of Gross National Product (table 2.3; figure 2.7). The crisis has actively led to transfers of funds back to the developed countries since 1983 (table 2.3). But in the end, these debts will probably never be wholly repaid. They will have to be 'forgiven'. In the meantime, the crisis has caused enormous damage to the development prospects of many countries and their populations as austerity programmes have been slotted into place. According to one estimate (UNICEF, 1988), half a million children died in 1987 as a result of the slow down or reversal of development programmes.

In the mid-1980s, it looked as though the debt crisis might bring the world's financial system tumbling down. But that has not happened. Instead, the world

Table 2.3   The level of debt of the highly indebted
countries (HICs) ($ bn)

	1982	1985	1988
Total external debt	391.0	454.0	529.0
Net flows to HICs	34.6	6.0	87.6
Net resource transfers	3.7	– 26.5	– 31.1

Source: World Bank, 1988

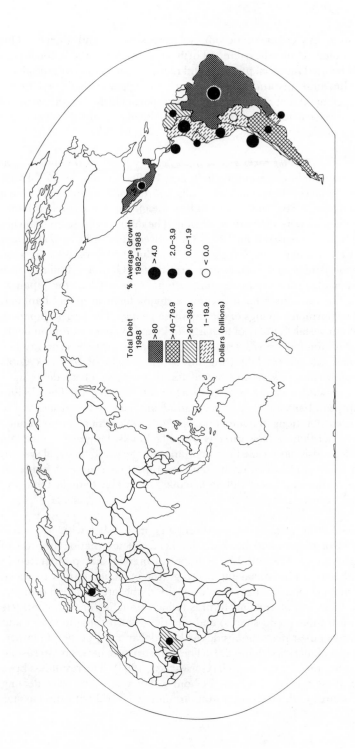

Figure 2.7 The highly indebted countries, 1988
Source: *Financial Times*, 1988

financial system has changed, becoming more sinuous and sinewy. This 'new international financial system' (Thrift and Leyshon, 1987) consists of electronically boosted financial products and markets spread out over the globe, in thrall to Japanese institutional investors' money, and dedicated to the constant recycling of debt and equity (most particularly, in the form of securities). It was even able to withstand the collapse of the chief growth market of the 1980s, the equity market, in the great crash of 1987.

The internationalization of the banks and securities houses   With all the money that was being pumped into them in the 1960s and 1970s, it comes as no surprise to learn that the larger banks and securities houses had become truly global corporations by the 1980s. In the race to become global, the US-based banks and securities houses were the most successful. The number of foreign branches of US-based banks increased from 124 to 173 between 1960 and 1976 and their assets grew by 1,816 per cent (Sampson, 1981). But the US banks were soon followed by the British and French and later by the German, Italian, Arab and Japanese banks. Like Japanese multinational corporations, the Japanese banks and securities houses have become a major force in global banking, backed by the enormous savings of the Japanese people. The largest Japanese banks have now overtaken many of the largest US and Western European banks (figure 2.8). A number of US and European banks (with exceptions like the Californian banks) are labouring under the weight of bad risk debts from developing countries arising out of the debt crisis. These debts have forced them to make enormous provisions against their reserves. The US and Western European banks have also been disadvantaged by the propensity of multinational corporations to issue bonds and other securities to raise money in the 1980s rather than borrow directly from the banks. However, in the late 1980s as the debt crisis has receded in importance (to the world financial system) and as bank borrowing has again become an important source of funds for multinational corporations, some of these banks will be able to make their way out of the doldrums.

The internationalization of the markets   Occurring in parallel with the internationalization of money and banking has been the internationalization of the financial markets. The number of financial markets has proliferated in the 1970s and 1980s as 'financial engineering' (the invention of new financial instruments) has become an important 'art form' of the late twentieth century. Markets like those in options and futures (which trade in forward contracts on commodities, money and shares) and equities (which trade in stocks and shares) have taken their place alongside the Eurodollar and Eurobond markets as important global markets (figure 2.9). The infrastructure necessary to sustain these markets has proliferated as well (Gorastraga, 1983). The result has been that many of these markets are now 24-hour-a-day businesses, with dealing in currency, shares and bonds migrating around the world with the passage of the sun.

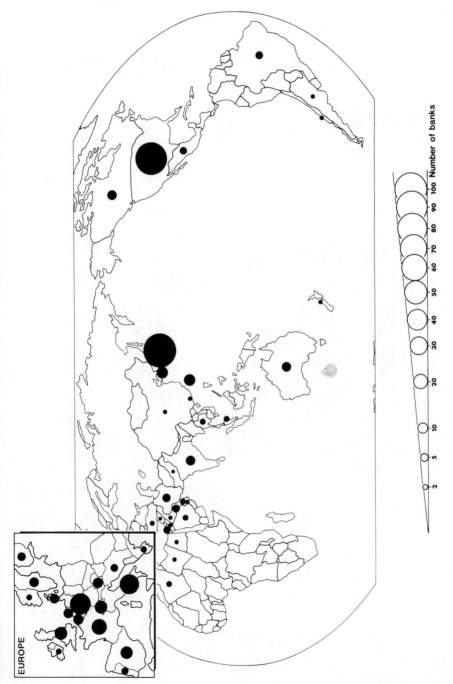

Figure 2.8 The country of origin of the top 500 banks
Source: The Banker, 1988

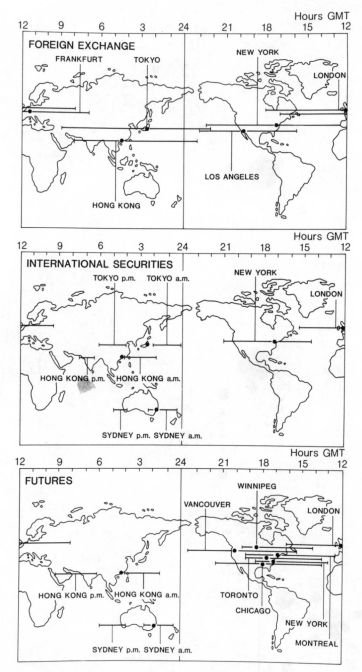

Figure 2.9 The rise of 24-hour markets
Source: Clarke, 1986, p. 68

The Internationalization of the State

The influence of the state is the final pole in an explanation of the world-economy of the 1970s and 1980s. It is, of course, quite possible to build up a theory which either gives the state an undue amount of power to resist capital or reduces the state to an agent of capital. The reality is rather different. States are the result of long-drawn-out conflict between different social groups to wrest different elements of power from one another. At different times and in different situations states play different roles. Four such roles can be identified, each of which has had some influence on the course of the world-economy in the 1970s and 1980s, and has both contributed to the internationalization of production and finance and hastened the integration of states into the world-economic order.

One role has been as a *market*. For example, a considerable part of the move of multinational corporations towards the United States in the 1980s, especially by corporations involved in the electronics industry, can be explained by the enormous military build-up sanctioned by President Reagan.

Another role has been *restrictive*. As economic growth has slowed, so a rising tide of protectionism has been touched off: the erection of barriers to trade like quotas, the insistence on local content in products, pollution regulations and so on. The increase in protectionism is one of the main reasons (along with the high yen) why Japanese multinational corporations went abroad in such numbers in the 1970s and 1980s (Dicken, 1988). It was becoming more and more difficult for the Japanese multinationals to export to particular countries. They had to produce from within them instead. For example, in the automobile industry the proportion of world trade in automobiles covered by 'voluntary' restrictions increased from less than 1 per cent in 1973 to about 50 per cent in 1983 and most of these restrictions were applied to Japanese imports. So Honda, Nissan, Toyota and all other Japanese automobile manufacturers had little choice but to become multinationals. The extreme case of restriction consists of the expropriation of foreign companies. This type of activity (practised chiefly in developing countries) has in fact declined precipitately in the 1980s (United Nations, 1988) as the balance of power has swung towards multinational corporations in the countries most likely to expropriate.

A third role of the state is an *enabling* one. Thus the 1980s have seen states adopt policies which are often very favourable to multinational corporations. The activities of foreign investment agencies in nearly every country of the world are legion. The battle in 1983 and 1984 to attract a Japanese (Nissan) automobile plant into Britain is only one notable example (Dicken, 1987). Then again, the state can act to help multinational corporations based in their own country to expand into other countries, either indirectly by providing expert advice and insurance, or directly through an explicit internationalization strategy, as in the case of the Ministry of International Trade and Industry (MITI) in Japan, or the efforts made by the South Korean government. Other enabling actions by states have become apparent in the 1980s. In particular, many states have adopted policies to 'liberalize' their economies. Most states have deregulated their financial markets. Some states have also deregulated their telecommunications markets. Again, a number of states have also privatized state-run industry, often allowing foreign influence on their economies to grow as a result.

Last but not least, the state can act as a *competitor*. In particular, it can compete for funds on the international capital markets. The United States has caused enormous problems by doing just this in the 1980s. It has allowed its budget deficit to grow to the point where in 1985 the United States became a debtor nation. There is nearly universal agreement that the United States budget deficit (combined with its growing trade deficit) has been the main destabilizing factor in the world-economy in recent years.

America's swollen demand for foreign capital caused a widening gap in the country's external payments, driving interest rates up and pushing the dollar first up and then down. So far . . . America has been able to borrow all it needs from the rest of the world without pushing interest rates so high that they cause a recession at home. The continuing risk is that foreign lenders – governments and private investors alike – will one day do so no more.

America's demands on the international capital market have meanwhile devilled economic policy-making in the rest of the world. Japan, Germany, and the other big industrial countries faced the risk of inflation as their currencies fell against the dollar, and later the strain of industrial dislocation as they rose. The developing countries fared worse. Soaring dollar interest rates helped to create the international debt crisis, leaving the debtor countries no choice but to cut deeply into their imports and their living standards. For good measure, in buying much less from abroad they made it all the harder for America – one of their principal suppliers – to keep its trade gap in check. (*Economist*, 1988b, p. 6)

It is important to remember that states do not have to be restricted in influence to their own boundaries. The number of international state organizations continues to multiply, all the way from the United Nations, through the IMF and the World Bank, to the 24-member OECD, the G7 countries, the EEC (with its monetary system), ASEAN and many other cross-border economic and political organizations (Nierop, 1989). These international organizations are not only used to coordinate the world-economy but also to extend a country's economic power abroad. Thus it is difficult not to see the IMF as an explicit arm of US economic and political policy, there to open up the economies of recalcitrant countries to US exports and to US multinational corporations, to discipline these countries to the laws of 'the market'. The IMF's role in the economic adjustment policies adopted by many of the highly indebted countries has been particularly transparent in this regard (Strange, 1988).

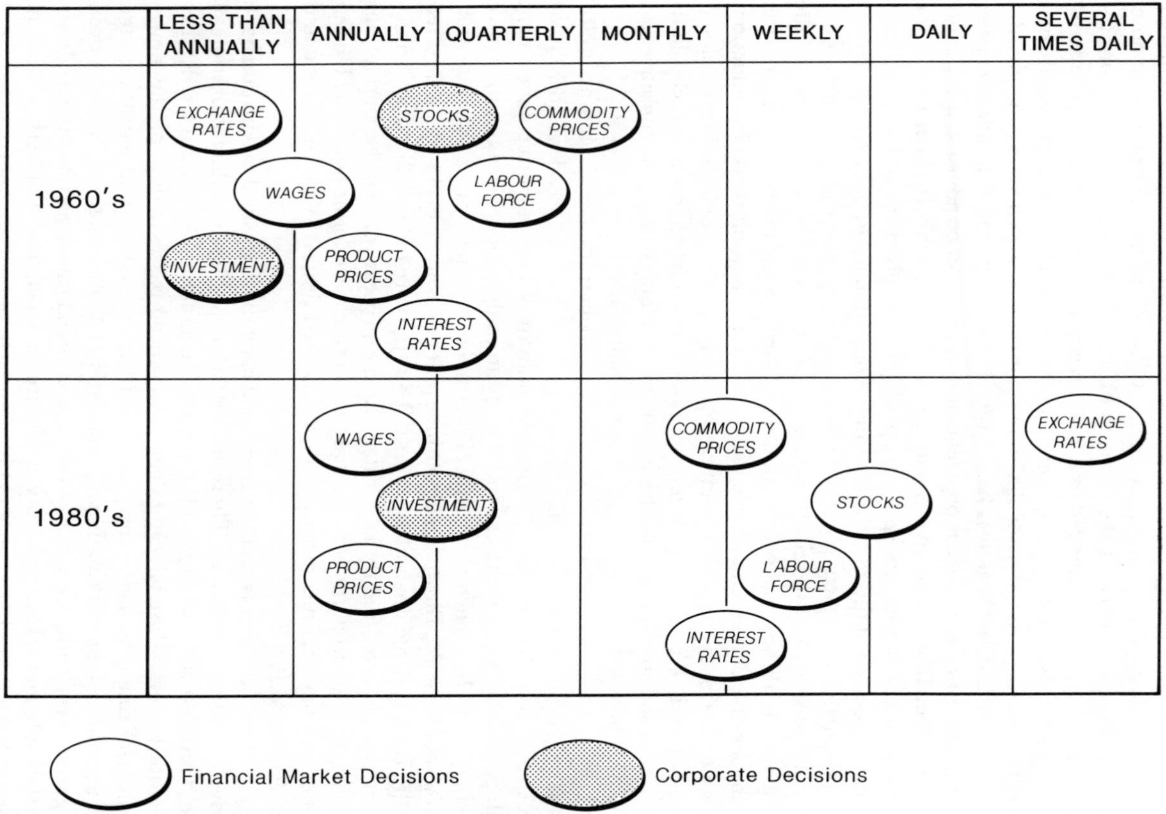

	LESS THAN ANNUALLY	ANNUALLY	QUARTERLY	MONTHLY	WEEKLY	DAILY	SEVERAL TIMES DAILY
1960's	EXCHANGE RATES, INVESTMENT	WAGES, PRODUCT PRICES	STOCKS, LABOUR FORCE, INTEREST RATES	COMMODITY PRICES			
1980's		WAGES, INVESTMENT, PRODUCT PRICES		COMMODITY PRICES	STOCKS, LABOUR FORCE, INTEREST RATES		EXCHANGE RATES

Financial Market Decisions

Corporate Decisions

Figure 2.10 The increasing pace of the world economy

Source: Economist, 23 September 1983, p. 11

Summary

Three main net effects of these changes will continue in the 1990s. The first is the acceleration of capital movement. Capital has become more footloose, both in time and space. Take the case of time first (figure 2.10). Before the 1970s, exchange rates changed only once every four years on average, interest rates moved perhaps twice a year, companies reviewed the prices of their products once a year and made decisions on investment even less frequently. But in the 1970s and 1980s all this has changed. Exchange rates change every few hours, interest rates change more frequently, prices can be adjusted much more swiftly, companies make investment decisions every year. In other words, the whole economic system has speeded up. Capital has also become more footloose in space. Because of the speed-up in time multinational corporations now review the productivity of their plants much more frequently and if plants do not produce the required level of profit they are likely to be shut down. At the same time multinational corporations can, through the medium of acquisition or cooperative agreements, move more swiftly into new markets or production, gaining control over existing plants or opening up new plants as they go along. Thus there is a constant process of strategic rationalization in which plants are set up and shut down more frequently than before, with all the consequences this has for the countries in which these plants are located.

The second net effect is the growing interpenetration of capital as multi-national corporations based in different countries have spread worldwide, as illustrated by two examples. The largest amount of Japanese direct investment still goes to the United States, and European multinational corporations have increasingly channelled their direct investment into the United States as well. On a smaller scale, European direct investment has increasingly been directed to the United Kingdom and vice versa. Thus the core industrialized economies have become more tightly integrated. In the process the world-economy has moved from an economy revolving around a single economic pole – the United States – to one that has three poles – the United States, Western Europe and the Asian Pacific Rim.

The third net effect is that the borders of capitalist industrial production have moved a little further out. There has been no wholesale industrialization of the Third World, as the dreadful poverty in much of Africa attests. Rather the border of industrial capitalism now encompasses a few of the newly industrializing countries, partly through their own efforts and partly through the attention of the multinational corporations. A combination of slow (or even negative) economic growth and the international debt crisis has blocked any further advance. This is hardly a picture of ascendant Southern industrial power. Rather local governments in the newly industrializing countries have attempted to maximize the possibilities for their indigenous capitalists, within a set of international economic and political constraints which makes preserving a space for indigenous capital a delicate business. This is not possible without permanent struggle. South Korea, Taiwan and Thailand have probably succeeded: Brazil, the Philippines, Argentina and Chile have not (Roddick, 1984, pp. 126-7).

THE WORLD MARKET FOR LABOUR

These changes in the structure of the world-economy in the 1970s and 1980s have brought prosperity to some people and agony to many others (literally so, in the case of some people working in developing countries with authoritarian regimes). The main cause of these mixed fortunes has been the changes in people's chances of getting a job.

There is still a very real difference between employment opportunities in the developed and the developing countries; there is little point in trying to play it down. Therefore, initially, the very different circumstances that can be found in each bloc will be enumerated. But it is a consequence of the form of internationalization of capital that has taken place in the 1970s and 1980s that some new links have now been forged between employment opportunities in the developed and the developing countries respectively: I will discuss five of these below.

The Developed Countries

The world-economic crisis of the 1970s and 1980s took place against the background of some quite important changes in the characteristics of the labour forces in the developed countries. The most important was the number of women (especially married women) joining the labour force. The participation of women increased in nearly all the OECD countries' labour forces over the period of the crisis. A second important factor was the 'baby boom' that lasted from the end of the Second World War until 1960. The generation born then entered the labour market at the very time that opportunities for employment started to plummet. In 1987 the size of the labour force in the OECD countries increased by 1-2 per cent, reflecting the rise in participation rates and the increase in the working population.

By far the most visible effect of the changes in the world-economy on the structure of opportunities for employment in the developed countries has been unemployment, which has increased dramatically. From 1960 to 1973, unemployment rates in North America varied between 4 and 7 per cent. In Europe and Japan they were never higher than 2 or 3 per cent. But after the first oil shock in 1973 unemployment rose quickly in nearly all countries until the end of 1975 (see figure 2.11). Then it remained relatively stable in most countries until 1979, when it again rose rapidly until 1983, as a result of the second oil shock, monetary tightening and recession. Since 1983 unemployment has levelled off in most countries and, more recently, declined slightly. In all, in the 24 OECD countries there were about 30 million persons out of work by 1983 (International Labour Organization, 1988b); about the same number of people remained jobless in 1987.

There are now wide variations between developed countries. While Spain's unemployment rate was more than 20 per cent in 1988, the United States has enjoyed a relatively impressive fall, from 7 per cent in 1985 to 5.5 per cent in 1988. During the same period, Japanese unemployment has remained below 3 per cent (figure 2.12).

Three trends in unemployment in developed countries have caused concern in the 1980s. First, the numbers of long-term unemployed have increased in a number of developed countries. Second, in many countries unemployment is not only underestimated, which is generally the case with official figures, but it is thought that the number of those who are unemployed but not counted in the official figures is increasing. Third, unemployment has had particularly severe effects on certain social groups. Those who are least likely to be unemployed are men between 24 and 54 years of age who have a good education or training. These people tend to fill what is often called the primary segment of the labour market, the segment with full-time jobs, promotion opportunities and incomes rising steadily with age. In contrast the young, the old, women and minorities are more likely to be members of the secondary labour market – the segment with unskilled, poorly paid, insecure jobs – that is, if they are employed at all. The young and the old are more likely to be unemployed in more developed countries. So are women. So are minorities. And any combination of these characteristics drastically reduces the person's chance of being employed. In short, on average in every developed country ten persons out of 100 are unemployed. Of these ten persons, five are young and three

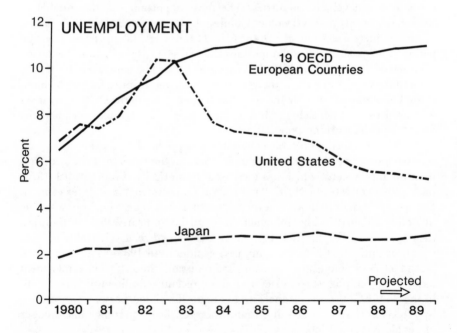

Figure 2.11   Unemployment in the developed countries
Source: *Financial Times*, 28 September 1988 Supplement, p. v

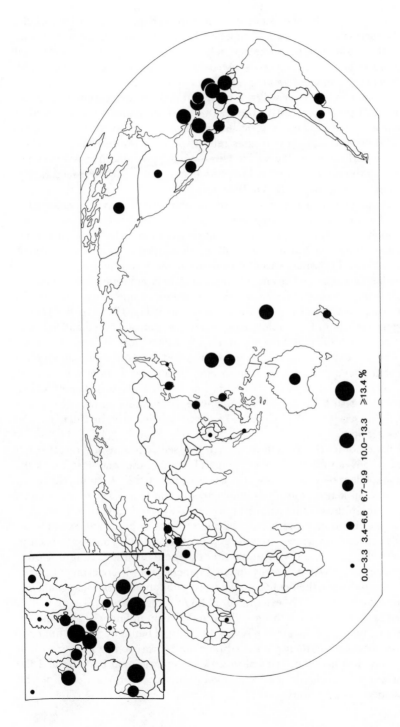

Figure 2.12 The rate of unemployment in various countries in 1987
Source: International Labour Organization

0.0–3.3 3.4–6.6 6.7–9.9 10.0–13.3 ≥13.4 %

of these are women. Further among the unemployed queueing up for unskilled jobs, the successful applicants will first be adult males (between 24 and 54 years of age), then women of the same age, followed by young persons; the last will be minorities and older workers (International Labour Organization, 1984, p. 46). *But* the chances of gaining employment are changing. The proportion of people aged 65 and over will rise in the OECD countries from 12 per cent in 1980 to 22 per cent in 2040, the result of a combination of falling fertility and mortality rates. The sharp decline in birth rates from the 'baby boom' years of the 1960s that these figures reflect will alter the size of the working population and will probably lead to falls in unemployment in most developed countries (although increases in the participation rate and other factors may cancel out some gains) (OECD, 1988).

If unemployment is the most clearly visible effect of recent changes in the world-economy this does not mean that these changes have had no other effects on the structure of opportunities for employment. Thus, manufacturing employment has declined in nearly all the developed countries since the oil shock of 1973. The main exception has been the electronics sector, where demand for computers, office and telecommunications equipment and consumer electronics has kept employment up. Meanwhile employment in services has grown in nearly all the developed countries since 1973. In the OECD countries, for example, employment in services, on average, had risen to 63 per cent of total employment in 1981. Employment in services in the United States and Canada is now more than two-thirds of total employment and is well above 50 per cent in most other countries. Part of this increase is, of course, a relative effect of the decline in manufacturing employment but there has also been an absolute increase in numbers employed in some parts of the service sector since 1973, most particularly in financial and business services.

Of course not all the growth in the services sector has been a direct result of recent changes in the world-economy. The role of the service sector in the economies of developed countries was increasing before the 1970s. But the recent changes have hastened its growth – employment in the services sector grew much more rapidly in the OECD countries between 1974 and 1986 than from 1960 to 1973, and in some cases it has had more direct impact. For example, in many countries the greatest growth in government employment has been in social services concerned with promotion of employment: 'Owing to slower economic growth and a rise in unemployment, many governments have expanded activities in the field of employment promotion, vocational guidance, training and security' (International Labour Organization, 1984, p. 49). Similarly, although much of the growth in employment in financial and business services during the 1970s and 1980s was the result of banks and other financial institutions expanding their domestic branch networks, some of the growth in employment was also the result of the rise of international banking and the increase in the information services and other producer services demanded by multinational corporations.

It is important to note that much of the increase in service employment consisted of an increase in the number of part-time workers. By 1987, part-time work accounted for more than a quarter of all persons in employment in Sweden, 20 per cent in the UK and more than 10 per cent in most of the other OECD countries. Most of this part-time work is concentrated in service-sector industries which need flexibility from their employees to meet their peaked patterns of business activity during the day (for example, in banks and restaurants) or during the week (for example, in shops). The bulk of this part-time work is done by women (Petit, 1986). Nowhere is the figure for female part-time employment below 60 per cent and in a number of the developed countries it is well above this figure. In the United States, for example, by 1987 the figure was 70 per cent and in the United Kingdom it was over 90 per cent for the same year. Further, the proportion of women in part-time work is increasing.

The second effect which recent changes in the world-economy have had on the state of employment opportunities has been to hasten the rise of an informal economy. According to Pahl (1980) the informal economy has two major components. The first is the household economy. Production in the household has increased enormously in the last 20 years in the developed countries as households have switched from buying services to producing their own. More precisely, households now increasingly buy goods, which are in effect capital equipment, to which they then add their own labour. So there has been a rise in do-it-yourself, in vegetable gardening, and so on. We no longer go to the laundry, or employ servants; instead we use washing-machines and vacuum cleaners which we operate ourselves. This is the rise of the so-called 'self-service' economy (Gershuny, 1978). The second component is the underground or clandestine economy (de Grazia, 1984; Gershuny and Miles, 1983; Redclift and Mingione, 1985; Pahl, 1988) which consists of the production of goods and services that evade systems of public regulation and taxation. It takes three main forms. The first is the undeclared employment of workers, especially illegal immigrants (to be found among outworkers in the clothing industry and workers in building, agriculture, the hotel and catering trades, housework, etc.). The second is undeclared self-employment (to be found among workers in dressmaking, in car repairs, in household repairs, among those running a market stall, etc.). The final form is undeclared multiple jobholding (to be found especially now among teachers, businessmen and policemen). The numbers employed in each of these forms of employment are all on the increase.

This might not matter if the underground economy were small, but it is not. Figure 2.13 shows estimates of the percentage of GNP taken up by the underground economy for selected countries. The proportions are large, and numbers employed in the underground economy are large. Clandestine employment is, of course, the only job that many workers have, especially for the unemployed, migrant workers, pensioners and housewives. There are, it must be remembered, over 500,000 clandestine immigrant workers in Europe and more than 5 million in the United States. Estimates vary concerning the

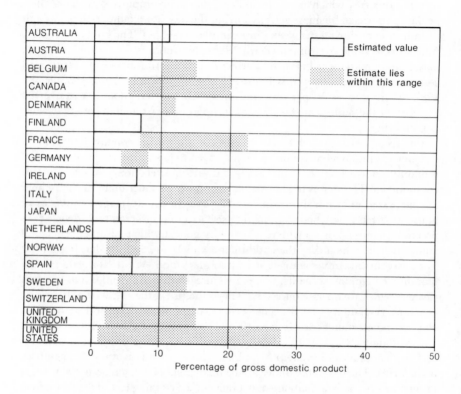

Figure 2.13 Estimates of the size of the informal economy in selected countries
Source: Tanzi, 1982; Petit, 1986

number of the unemployed who have jobs, from 80 per cent in some *départements* of France to very few in parts of the United Kingdom (see Pahl, 1984). If the figures of those who have a clandestine second job are added in to the total then the numbers become very large indeed. In West Germany it is reckoned that 8 per cent of the workforce have a second undeclared job. In France, Sweden and Belgium the proportion of the workforce so employed varies from 5 to 15 per cent. In the United States and Canada some estimates are as high as 25 per cent (International Labour Organization, 1984).

It is difficult to know precisely how the world-economic crisis has affected the informal economy but it has almost certainly fuelled its growth. Faced with competition, many firms have resorted to clandestine employment to reduce labour costs, and there are many more unemployed who are prepared to work at what they can find in order to eke out a living.

The Developing Countries

In the developing countries employment problems are far more serious because of the massive expansion of the labour base in most countries which makes the 'baby boom' in the developed countries seem insignificant. Demography reigns supreme.

Undoubtedly changes in the world-economy have had some effect on employment opportunities but in the face of the massive rates of growth of many labour forces these are rather difficult to identify. Certainly, it seems clear that problems of the 1980s have had a severe impact on the labour forces of certain of the newly industrializing countries, which are more tightly linked to the world-economy. In these countries the manufacturing sector has transmitted adverse circumstances in the shape of labour demand for manufactured goods from the developed countries.

The truth is that figures on unemployment in the developing countries are hard to come by, difficult to compare because of differing definitions, and applicable only to the wage labour force. These figures are summarized in figure 2.12. In general, unemployment rates do not seem particularly high by comparison with rates in the developed countries. However, the *underemployment* rate (those who are in employment of less than normal duration) is very high. In the Asian economies, for example, it is estimated that 40 per cent of rural workers are underemployed, as are 23 per cent of urban workers.

As in the developed countries, so in the developing countries, changes in the world-economy have had some effects on the structure of employment opportunities, as well as provoking the simple lack of employment opportunity through unemployment and underemployment. The first important effect has a close parallel with the developed countries; the growth of the so-called *urban informal sector* of employment (Bromley and Gerry, 1979; Sethuraman, 1981; Redclift and Mingione, 1985). It seems certain that changes in the world-economy have speeded the expansion of this sector.

The urban informal sector consists of what is often the largest part of the urban labour force in the developing countries, workers who carry on their livelihoods outside the domain of public regulation and taxation. The activities that are included in this sector are too diverse to itemize one by one, but they include small-scale commodity production (for example, of bicycles) and repair shops, street vending (the selling of food, cigarettes and drink), the operation of low-income forms of transport (such as tricycles, scooter-based three-wheelers and jeeps), and a whole range of activities that are often illegal, such as prostitution and running drinking dens. The urban informal sector has come about, in part at least, because the activities that it comprises are the only ones that many rural–urban migrants – who stream into the cities of so many developing countries – can gain access to. Since the urban population of the developing countries is expected to double over the next ten years, in great measure owing to rural–urban migration, it seems highly unlikely that the sector will diminish in size and highly likely that it will continue to grow. Recent changes in the world-economy, such as the international debt crisis, with their effects on many developing countries will certainly exacerbate the problem.

The second way in which the world-economic crisis has had effects on the structure of employment opportunities has been through manufacturing employment. Until the 1980s manufacturing grew as a proportion of total employment in nearly all developing countries. This growth came about in part through the expansion of indigenous industries trying to satisfy new levels of consumer demand, in part through the demands of a thriving local construction industry, in part through aggressive (and usually state-led) export strategies, and in part through decisions made by some foreign-based multinational corporations to relocate some of their unskilled assembly production to the dozen or so newly industrializing countries, this apparently providing the beginnings of a new world spatial division of labour often called the 'new international division of labour'.

The Global Effects of Cycles in the World Economy on Employment Opportunities

This new spatial division of labour is the first of five direct effects on employment of the economic restructuring that has taken place around the globe in the 1970s and 1980s. The other four effects are: the growth of free trade zones; the growth of international migration; the growth of labour 'flexibility'; the growth of world cities. Each of these effects will be considered in turn.

The new international division of labour   By the mid-1980s, 65 million workers were directly employed in jobs provided by multinational corporations. Probably the same number were employed indirectly (United Nations, 1988). Most of this employment was provided by multinationals from only twelve countries (table 2.4), of which the United States, the United Kingdom, West Germany, Japan and the Netherlands provided the bulk of employment abroad (figure 2.14).

Table 2.4   Direct employment by multinational corporations in home and host country operations, by country of origin of parent corporation, latest year ('000 workers)

Country of origin	Year	Total	% employment abroad
Austria	1983	400	25
Belgium	1975	345	53
Canada	1984	1,764	40
France	1981	3,930	20
West Germany	1983	9,632	25
Italy	1981	1,000	25
Japan	1985	4,630	20
Netherlands	1980	1,454	74
Sweden	1984	950	30
Switzerland	1986	744	78
United Kingdom	1981	5,250	40
United States	1984	24,560	26

Source: United Nations, 198, p. 212

A relatively new aspect of this pattern of employment has been that so much of it is in developing countries. Multinationals now employ about 5 million workers, or 8 per cent of their world labour force, in the developing countries. Until the 1980s, it looked as though the trend towards growth of employment in developing countries would continue. This increasing presence led a number of writers to posit the existence of a 'new international division of labour' (Frobel, Heinrichs and Kreye, 1980; Tharakan, 1980) in which the parts of the production process which require cheap, unskilled labour were increasingly relocated to the Third World. One reason why this new international division of labour took root was the breakdown of traditional economic and social structures in many developing countries, leading to an inexhaustible supply of cheap labour. Another reason was that the production process had become more fragmented and more homogenized, making it possible for many sub-processes to be spatially separated and carried out by unskilled workers after very short training periods. A third reason was that as transport and communications technology has developed so it had become possible to carry out complete or partial production processes at many new sites around the world without prohibitive technical, organizational or cost problems.

The trouble with this simple explanation of the new international division of labour was that it was difficult to sustain the argument that it existed as a significant *global* tendency (Thrift, 1980). The explanation was not just simple, it was simplistic. Thus, only limited evidence could be found for the actual relocation of employment, that is for the direct physical movement of plant from a location in a developed country to another in a developing country (Gaffikin and Nickson, 1984; Flynn and Taylor, 1986). Again, there were real difficulties in interpreting the new international division of labour as simply the result of a search for low-cost wage locations by multinational corporations

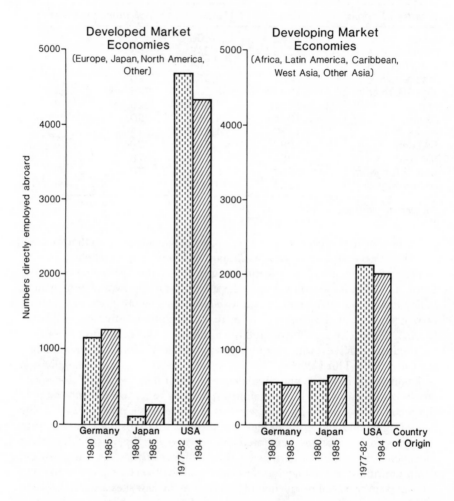

Figure 2.14 Direct employment abroad by the multinational corporations of West Germany, Japan and the United States, by region, 1980 and 1985
Source: United Nations, 1988, p. 217

(Jenkins, 1984; Corbridge, 1986). Certainly, wage levels are a determinant of location in certain cases but they are hardly the be-all and end-all. First of all, there is the technical problem of making simple comparisons of wage costs from country to country. As Gordon (1988) points out, these comparisons ignore the degree to which labour in different countries has different levels of skill and technological back-up which can produce enormous differences in labour productivity per unit cost. Then there is the fact that labour costs are a relatively small proportion of the total cost of many commodities. More than this, setting up a new plant overseas requires a relatively long period of operation in order to recoup the initial investment. This means that the stability of the country in which it is located (including factors such as volatility of exchange rates, levels of inflation and degree of labour discipline) is likely to be very important. Most important of all, investments in many developing countries (especially in Asia) seem to have been prompted chiefly by the degree to which these countries have large and growing markets (Jenkins, 1984).

It is clear that, in any case, the new international division of labour is limited in its impact, by both country and industry. The redeployment of manufacturing employment by multinational corporations in the developing countries has generally been restricted to a very few countries, mainly the newly industrializing ones. Even in the latter, the part played by multinational corporations in stimulating employment varies enormously. For example, the share of employment in both foreign-owned and joint-venture firms is high in Singapore. But in South Korea and Hong Kong the proportions are much smaller (see International Labour Organization, 1984; Jenkins, 1984; United Nations, 1988).

Moreover, the new international division of labour is restricted to a few industries, especially textiles and clothing (Clairmonte and Cavanagh, 1981) and electronics (Morgan and Sayer, 1988; Economist Intelligence Unit, 1984), where the economic and technological considerations of making particular products can often make it easier to relocate production abroad. The difference between the worldwide patterns of employment in the two industries, shown in figures 2.15 and 2.16, is that in the textile and clothing industry there is strong indigenous industry in the newly industrializing countries. Multinational corporations' involvement in production abroad in textiles and clothing is still considerable, but in the electronics industry – where the newly industrializing countries have severe problems in terms of gaining access to technology and sufficient capital for research – multinational corporations' involvement has, until recently, been all but total. But even here, there are signs of indigenous capacity increasing (Scott, 1987).

So, seen in the narrow sense, as the relocation of elements of the production process to developing countries by multinational corporations searching for low wages, the new international division of labour is clearly of less importance than originally thought. Indeed, as pointed out above, there are signs that increasing automation and other factors may be negating some of the developing countries' advantages.

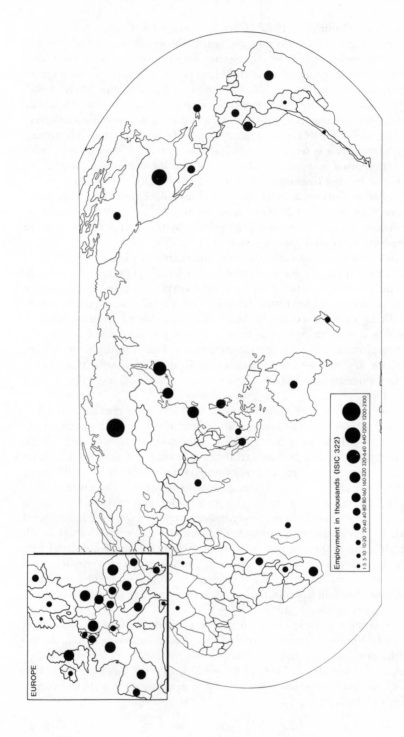

EUROPE

Employment in thousands (ISIC 322)

1·5 5 10 10·20 20·40 40·80 80·160 160·320 320·640 640·1200 1200·2100

Figure 2.15 Worldwide employment in the clothing industry, 1983

Source: International Labour Organization, 1985

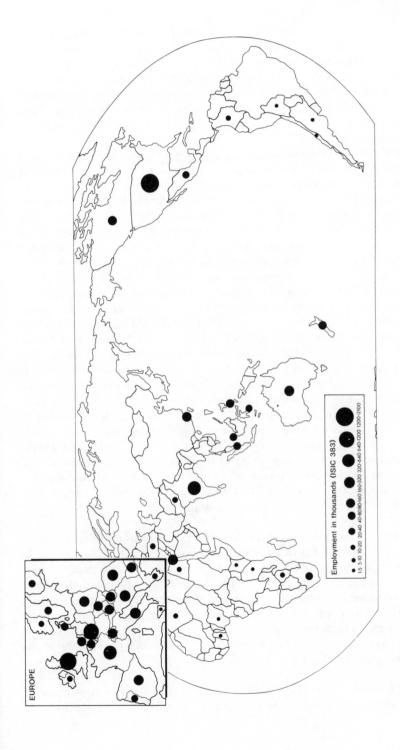

Figure 2.16 Worldwide employment in the electronics industry, 1983
Source: International Labour Organization, 1985

EUROPE

Employment in thousands (ISIC 383)

1-5 5-10 10-20 20-40 40-80 80-160 160-320 320-640 640-1200 1200-2100

However, the new international division of labour is of much greater significance if to the direct shift in employment from developed to developing countries is added the relative shift in employment between the countries of the world. Seen in this light, the concept has very considerable merit in describing the employment outcomes of the strategies pursued by multinational corporations as they have restructured during the world-economic crisis. Then the spatial pattern of employment more or less follows the spatial pattern of direct investment outlined above.

In particular, there seems to have been a general fall-off in the labour force of most multinational corporations in the 1980s as a response to the various restructurings noted above. But within this trend of a general decline in employment there is a shift in where the remaining employment is concentrated because employment falls more in certain countries and falls less, if at all, in others. Specifically, employment decreases more in the home countries of multinational corporations, and decreases less, or even expands, in other countries. For example, among European multinational corporations there has been a greater tendency for employment to fall in the home country's plants and offices, while staying static elsewhere in Europe and expanding in North America and some of the developing countries. A number of corporate studies have made the picture clearer, for example the study by Teulings (1984) of the operations of the Dutch-based multinational corporation Philips, and the study by Clarke (1982, 1985) of the British-based multinational corporation, ICI.

Reinterpreted in this more general light the new international division of labour is a tendency which has undoubtedly had serious repercussions on the employment prospects of workers in developed countries, particularly in those countries with 'open' economies. These are countries like the Netherlands, Sweden and the United Kingdom whose corporations have historically carried out much of their production overseas (Dunning, 1981; Taylor and Thrift, 1981) and which are themselves extensively penetrated by multinational corporations. In these countries the behaviour of just their 'own' multinational corporations led to extensive job loss in the late 1970s and the early 1980s, although the position has now recovered somewhat.

Export processing zones   Governments have not stood idly by as the refurbishing of the world-economy has taken place. They have been active participants in all the ways mentioned in the first part of this chapter. But one relatively new phenomenon has been the export-processing zone, an adaptation of the free trade zone which has been common around many ports for some time (Currie, 1985; Marsden, 1980; Salita and Juanico, 1983; Wong and Chu, 1984), which is partly a response by governments, especially the governments of the developing countries, to the new international division of labour. In the words of Frobel, Heinrichs and Kreye (1980, p. 283):

> World market production cannot be undertaken at every location in the developing countries where there happens to be an unemployed workforce.

Profitable industrial production for the world market requires an adequate provision of industrial inputs and a sophisticated infrastructure as well as a labour force. These factors are not necessarily available at those sites where there is an abundant supply of unemployed labour. In addition, profitable industrial production requires the lifting of the national restrictions on international transfers which exist in most developing countries as a result of their chronic balance of payments deficit. In fact, it is the function of [export processing zones] to fulfil the requirements for profitable world market-orientated industrial production in those places in the developing countries where unemployed labour is available and suitable for industrial utilisation.

The United Nations Industrial Development Organization (UNIDO) defines export-processing zones as 'small closely definable areas within countries in which favourable investment and trade conditions are created to attract export-oriented industries, usually foreign-owned'. Within export-processing zones four conditions are usually found. First, import provisions are made for goods used in the production of items for duty-free export, and export duties are waived. There is no foreign exchange control and there is generally freedom to repatriate profits. Second, infrastructure, utilities, factory space and warehousing are usually provided at subsidized rates. Third, tax holidays, usually of five years, are offered. Finally, abundant, disciplined labour is provided at low wage-rates.

The first successful implementation of an export-processing zone was in 1956 at Ireland's Shannon International Airport. Puerto Rico (in 1962) and India (in 1965) were the first two developing countries to follow suit, followed by Taiwan, the Philippines, the Dominican Republic, Mexico, Panama and Brazil in the period from 1966 to 1970 (Wong and Chu, 1984). The greatest expansion in the number of zones came after 1970. By 1975 there were 31 zones in 18 countries. By 1987, on one count, 260 zones were in operation or planned in about 40 (principally developing) countries but the bulk of the zones were concentrated into only a few countries (United Nations, 1988), including some of the lowest income countries like Bangladesh (figure 2.17). The major world regions in which export-processing zones are found are in the Caribbean, Central and Latin America and Asia, although there are also a few in Africa and the Middle East. In the rest of this section, particular attention will be given to the cases of Mexico and Asia.

About three-quarters of the industry in most export-processing zones is owned by foreign-based multinational corporations and, not surprisingly, most employment in export-processing zones is in industries that have expanded strongly in developing countries (partly as a result of multinational corporations' intervention), especially electrical and electronic goods, textiles and clothing and motor vehicle parts. In Mexico in 1985, 60 per cent of the *maquiladoras*, the assembly plant factories located in special zones near the US border with Mexico, were concerned in some way with electronics and electrical assembly.

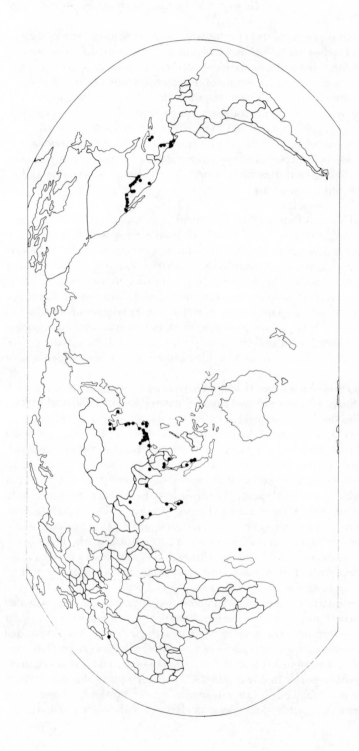

Figure 2.17 Export processing zones, 1987
Source: Rubin, 1988

Another 8 per cent were in textiles and clothing. In Asia again approximately 60 per cent of employment is in the electronics industry with the clothing and footwear industries second. The advent of electronics companies is fairly recent, which makes the build-up even more impressive. For example, in 1975 the Philippines' export-processing zones were almost entirely given over to textiles and clothing, but then in 1976 multinational electronics companies started to move in.

About 1.3 million people are employed in the export-processing zones (International Labour Organization, 1988b) with 95 per cent of employment concentrated in just 14 countries. The numbers are modest, with few exceptions. The case of Mexico is the best known (Grunwald and Flamm, 1985). In 1987 *maquiladora* employment numbered 307,866 in 1132 plants, including those of major US corporations such as RCA, Ford, General Motors, Chrysler and General Electric, as well as Japanese companies (figure 2.18). Some of the zones in Asia have also generated considerable employment. There are about 500,000 workers directly employed in export-processing zones in Asia, but of this number nearly 50 per cent are employed in the zones of Singapore (105,000) and South Korea (121,700). Zone employment is also comparatively high in Malaysia (101,110) and Hong Kong (70,000) and is growing rapidly in China.

Most employment in the zones, usually over 60 per cent, is of young, unmarried women aged between 16 and 25; in Mexico 66 per cent of *maquiladora* manual employment is of young women (Fernandez-Kelly, 1983). In some of the zones in Asia the figures are even higher (Lin, 1987). In Sri Lanka young women account for 88 per cent of zone employment, in Malaysia and Taiwan 85 per cent, in South Korea 75 per cent and in the Philippines 74 per cent (Morello, 1983).

Young women are employed because their wages can be set at rates 50 to 75 per cent lower than men's because they are considered 'dexterous', because they are easier to hire and lay-off, and because they are 'more able to cope with repetitive work'. Very little skill is required of them, certainly, and wages are very low, although usually higher than in local industries outside the zone. Overtime and incentive payments account for a high proportion of wages. Working weeks range from 45 to 55 hours. What is more, in a few zones the iniquitous trainee system is used, in which 'trainees' are paid only 60 to 75 per cent of the local minimum wages and are constantly fired and rehired so as to obtain a permanent reduction in the wages bill.

Weighing up the advantages and disadvantages of the export-processing zones to the countries in which they are located is no simple task. There is the matter of employment creation. In some areas, the employment advantages have been substantial. For example, since their establishment export-processing zones have accounted for at least 60 per cent of manufacturing employment expansion in Malaysia and Singapore and for about 10 per cent in Hong Kong, the Philippines and South Korea. But except in these and a few other cases such as Mexico, the impact of the zones on joblessness has been marginal and even

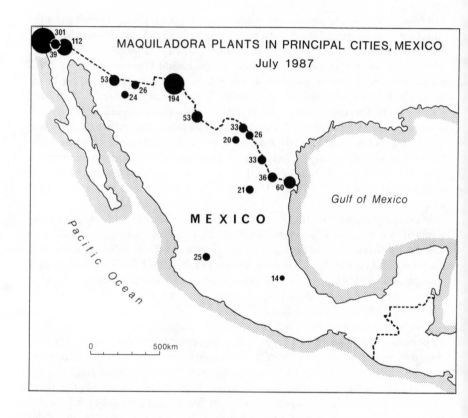

Figure 2.18 Distribution of *maquiladora* employment in Mexico, 1987
Source: Rubin, 1988

in the successful cases most of the employment created has been temporary and low paid.

Then there is the matter of the generation of foreign exchange. This can be very high: the foreign exchange generated by the *maquiladora* industries in Mexico is fourth after oil revenues, remittances and tourism. But the figures can also be deceptive: in 1981 the value of the Philippines' three export-processing zones' export total was US $236.2 million but the figure for exports minus the imports into the zones was only US $62.5 million, and this was *before* interest payments and repatriation of profits was taken into account (Morello, 1983).

It is arguable that export-processing zones, through their linkages to local firms, might create extra employment outside. But in fact links to local economies are generally minimal, usually only about 3 per cent of components and supplies are locally sourced. The only real linkages tend to be through service sector activities such as banking, transport, electricity and maintenance. Finally, other advantages have been mooted such as transfer of technology, demonstration effects and upgrading of workers' skills. The evidence here is ambiguous.

To sum up, export-processing zones are marginal in most, although not all, cases. For example, one study of the Indonesian export-processing zone on Batam Island, located so as to take advantage of Singapore's need for cheap labour, found that benefits only outweighed costs when bribes were taken into account. Further, competition between zones can now undermine what benefits there are. Competitive bidding between prospective sites for firms has resulted in more and more incentives having to be offered. So export-processing zones are, on the whole, a much commented upon but relatively unimportant outgrowth of the new international division of labour.

International migration   The export-processing zone represents an attempt to bring work to the workers. Another solution is for workers to go to the work. International migration has been an important part of the world-economic system since before the turn of the century, but changes in the world-economy have influenced international migration in new ways (Portes and Walton, 1981; Petras and McLean, 1981; Sassen, 1988; Cohen, 1987).

The International Labour Organization (1984) estimates the stock of economically active migrants in the world today as between 19.7 and 21.7 million (plus a similar number of dependants living with them). In general the flow of migrants is from the less to the most economically successful countries. Thus the United States has some 5 million legal immigrants, half of whom were in the labour force and two-thirds from developing countries, with a further 2.5 to 4 million illegal immigrants, most in the labour force and nearly all from Mexico. Canada also has a large immigrant workforce. Western Europe has numerous economically active immigrants, perhaps as many as 6.3 million (table 2.5). The Arab countries employ some 2.8 million foreigners, nearly all of whom are from developing countries (table 2.6). But many

Table 2.5 Recorded number of migrant workers in selected western European countries, 1980 ('000s)

Out-migration country	In-migration country									
	Austria	Belgium	France	W. Germany	Luxembourg	Netherlands	Sweden	Switzerland	UK	Total
Algeria	–	3.2	322.7	1.6	–	–	–	–	–	327.5
Finland	–	–	–	3.7	–	–	108.0	–	1.0	112.7
France	–	38.5	–	54.0	8.5	2.0	–	–	14.0	117.0
Greece	–	10.8	3.0	138.4	–	1.2	7.5	–	6.0	166.9
Italy	–	90.5	146.4	324.3	11.2	12.0	–	301.0	73.0	958.4
Morocco	–	37.3	116.1	16.6	–	33.7	–	–	–	203.7
Portugal	–	6.3	430.6	59.9	13.7	4.2	–	–	5.0	519.7
Spain	–	32.0	157.7	89.3	2.3	10.4	–	85.7	17.0	394.4
Tunisia	–	4.7	65.3	–	–	1.1	–	–	–	71.1
Turkey	28.2	23.0	20.6	623.9	–	53.2	–	20.1	4.0	773.0
Yugoslavia	115.2	3.1	32.2	367.0	0.6	6.6	24.0	62.5	5.0	616.2
Other	31.3	83.2	192.4	490.1	15.6	70.2	94.6	237.0	804.0	2,018.4
Total	174.7	332.6	1,487.0	2,168.8	51.9	194.6	234.1	706.3	929.0	6,279.0

No figure given if no migrants recorded or estimated, if magnitude less than 500, or if not applicable.
Source: International Labour Organization, 1984, p. 100

developing countries also have substantial migrant-worker labour forces; those of South America have 3.5 to 4 million migrant workers and those of West Africa have 1.3 million (International Labour Organization, 1984).

This migration has considerable economic importance for all the countries concerned, increasing the interdependence of the world-economy. For example:

In 1987, the flow of migrant workers' savings (remittances) to major migrant-sending countries amounted to US $24,000 million. This provided a notable element of international financial flexibility and much-needed hard currency for many developing countries. On the part of migrant-receiving countries, for example, one in six cars made in the Federal Republic of Germany in 1980 can be attributed to the work of Mediterranean migrants and the ratio would be higher if only production-line workers were counted. On the part of migrant-sending countries, for example, the inflow of remittances in the Yemen Arab Republic since the early 1970s has covered the country's growing trade gap or – which amounts to the same thing – migrants' transfer payments have enabled the country to increase its imports correspondingly, amounting to about 600 times the level of exports at the end of the 1970s (International Labour Organization, 1984, pp. 102–3).

Conventionally, international migration is split into four categories; settlement migration, irregular migration, contract migration and official and business migration. In each of these cases there has been some response to changes in the world-economy. However, it is among contract migration and official and business migration that the most important responses can be found (see Petras and McLean, 1981).

Contract migration has grown swiftly during the 1980s, involving migration for a period of time which is dependent upon the issue of a contract by an employer in the country concerned (Findlay, 1989). Indeed quite often a group of workers is admitted under a collective contract for the span of a particular project or part of a project. Thus a group of workers may be brought into a country on contract to build factories and infrastructure and they are then replaced by another group who specialize in operations and maintenance. The contract for Riyadh University was won by a consortium – Bouyges–Blount – dominated by the French Bouyges corporation. The joint venture acted as a kind of broker, sub-contracting each part of the project. So a consortium of companies from South Korea held the general construction contract and brought in Korean workers to do most of the work. An Italian group had the electricity contract and brought in Italian workers, and so on. The number of subcontract workers reached 10,000 at the height of the project (International Labour Organization, 1984). This kind of project, offering a complete package, is becoming more common and as it does so it is possible to see how flows of migrants and the objectives of multinational corporations can become much more closely integrated.

Table 2.6 Estimated number of migrant workers in Arab countries, 1980 (actual figures)

Out-migration country	In-migration country										
	Bahrain	Jordan	Iraq	Kuwait	Libya	Oman	Qatar	Saudi Arabia	UAR	Yemen	Total
Democratic Yemen	1,125	–	–	9,500	–	120	1,500	65,000	6,600	–	83,845
Egypt	2,800	68,500	100,000	85,000	250,000	6,300	5,750	155,100	18,200	4,000	695,650
India	12,300	500	2,000	45,000	32,000	35,600	11,850	29,700	109,500	2,000	280,450
Iraq	310	–	–	40,000	–	–	–	3,250	1,200	–	44,760
Jordan (incl. Palestine)	1,400	–	7,500	55,000	15,000	2,250	7,800	140,000	19,400	2,000	250,350
Lebanon	300	–	4,500	8,000	5,700	1,500	750	33,200	6,600	500	61,050
Oman	900	–	–	2,000	–	–	1,150	10,000	19,400	–	33,450
Pakistan	26,160	4,000	7,500	34,000	65,000	44,500	20,770	29,700	137,000	3,000	371,630
Somalia	–	–	–	500	5,000	400	–	8,300	5,000	500	19,700
Sudan	900	–	500	5,500	21,000	620	750	55,600	2,100	2,250	89,220
Syrian Arab Republic	150	–	–	35,000	15,000	600	1,000	24,600	5,800	1,000	83,150
Yemen	1,125	–	–	3,000	–	120	1,500	325,000	5,400	–	336,145
Other Arab	–	–	–	300	65,600	120	–	500	–	–	66,520
Other Asian	10,000	1,000	1,500	10,000	27,000	–	4,500	93,500	20,700	300	168,500
Other	10,250	2,000	2,000	45,900	44,200	4,670	22,930	49,800	54,100	1,450	237,300
Total	67,720	76,000	125,500	378,700	545,500	96,800	80,250	1,023,250	411,000	17,000	2,821,720

No figure given if migrants recorded or estimated, if magnitude less than 500, or if not applicable.
Source: Birks and Sinclair, 1980

Official and business migration has also become more important in the period of the crisis as a result of increasing activity by multinational corporations, and especially of their 'globalization'. This category of migrants includes many types of temporary migrant, in particular the subcategory of business people and intra-company transferees. As business people have been forced to go from country to country more frequently, and as intra-company transfers of personnel have had to be made more often as multinational companies have grown larger and established themselves in more countries, so these subcategories have increased in proportion. Here patterns of migration closely mirror flows of foreign direct investment. The internationalization of the service industries has accentuated the growth of business migration still further since it is in the nature of many services that people have to be transported very frequently (figure 2.19).

However, international migration has also produced problems. The potential number of international migrants is increasing in the poor countries of Asia, Central and Latin America and Africa, but the opportunities for employment in the developed countries are falling away. It is no surprise, then, that many developed countries are now trying to rid themselves of the 1950s' and 1960s' migrants who originally formed a useful substratum of the working class but in the recession have formed the leading ranks of the unemployed (Castles and Kozack, 1985).

Flexible work practices   The rise of Japanese industry has brought with it considerable imitation of Japanese work practices. Thoughout the other developed countries there is a shift towards work practices based upon the Japanese model (although with many local innovations, adjustments and compromises). The shift, which is very complex, intertwines with the moves towards flexible methods of production outlined in the first section of this chapter and can be found in both manufacturing and services industries (Scott and Cooke, 1988). It includes three particular kinds of flexible work practices, each of which will be considered in turn.

The first kind of flexibility is *numerical* and consists of ways of enabling frequent changes to be made to the numbers in a firm's labour force, allowing rapid adjustment to any change in business conditions. Examples include the greater systematic use of temporary workers, which means that firms can hire and fire according to all kinds of exigencies, and more use of part-time workers, a shift particularly noticeable in the service industries. The second kind of flexibility is *functional* and is mainly concerned with changing the way workers do their work. It includes the introduction of practices such as: 'task flexibility' (for example, the breakdown of shopfloor demarcation barriers between workers, workers being taught a variety of skills and workers being taught to do routine maintenance tasks which were formerly the responsibility of special maintenance engineers); changes to shift patterns; the use of work teams, quality controls and other attempts to reconnect what mass production disconnected, the use of workers' intellects; and an associated use of various kinds of new

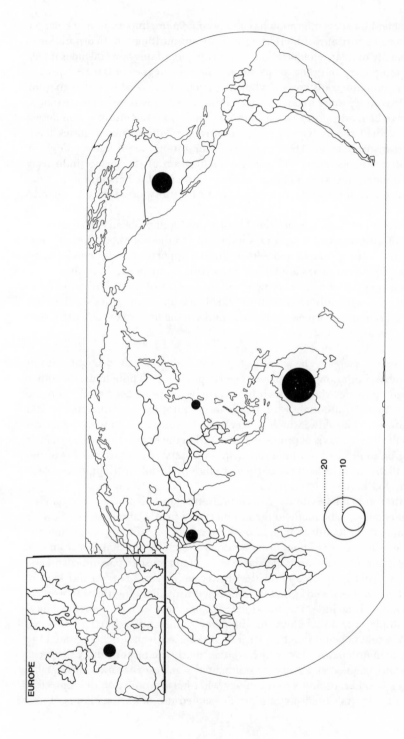

Figure 2.19 Location of overseas secondments from the London office of Arthur Andersen, 1986

Source: research by J. Beaverstock, Department of Geography, University of Bristol

technology. The third kind of flexibility is *organizational* and consists of the way that all the different new work practices are amalgamated. Atkinson (1985) suggests that a core-periphery model of the firm is emerging in which a core of functionally flexible workers is surrounded by a group of semi-peripheral workers who are flexible and a group of peripheral workers in subcontracting firms whose use offers extreme numerical flexibility. Certain firms, such as Ford, certainly seem to be heading in this direction. For example, Ford now uses many more workers on temporary contracts, and has made many attempts at functional flexibility (Sayer, 1986b).

The degree to which these flexible work practices have been or are likely to be adopted around the world remains open to debate. The evidence is often ambiguous and depends critically on what definitions of flexibility are adopted. Even so, it seems clear that flexible work practices are a new and growing component of labour organization in the developed countries. In turn, these new practices may well have impacts on the developing countries. In particular, they are likely to produce a further weakening of interest in developing countries as suppliers of low-cost labour, both because the new kinds of flexible work practice require mainly skilled workers which are less likely to be found in developing countries (Schoenberger, 1988) and because, to a degree, the developed countries are developing low-cost labour businesses in their own backyard, through the greater use of numerical flexibility.

World cities   The final global economic effect on labour has come from the growth of the international services economy, especially that part of it based upon financial and corporate activities and allied producer services. This growth can be traced to the internationalization of production and finance. The expansion in the plants and offices of global corporations and the subsequent growth of their international markets has produced an increase in the administrative load. The growth of the international financial system has similarly led to considerable increases in administration as banking and finance offices have carpeted the globe and market capacities have expanded. Indeed the administrative load increased to the point where it could have constrained future world-economic expansion. But three events have allowed expansion to continue. The first of these is the invention of the regional headquarters office (Heenan, 1977; Grosse, 1982; Dunning and Norman, 1987), which provide the intermediate tier of management between the administration, much of it strategic, carried out by the head office of a corporation or bank, and the administration, much of it routine, carried out by regional branch plants and offices. They are responsible for definable regions of the world (such as Asia and the Pacific) and are intended to reduce the internal costs of many information and marketing decisions. The second event has consisted of the growth of information technology and telecommunications which has allowed information to be analysed and transmitted at much greater speeds (Langdale, 1984; Moss, 1987). Finally, the growth of so-called producer services has occurred. These activities could, no doubt, be produced within the corporation but, for various reasons, especially because they can be obtained more cheaply outside, are externalized and bought in. Producer services, then, are 'activities that assist user firms in carrying out administrative, developmental (research and development, strategic planning) and financial functions, banking, insurance, real estate, accounting, legal services, consulting, advertising and so forth' (Noyelle, 1983a, pp. 117–18). All these services have internationalized too, sometimes following in the wake of the multinational corporations, but sometimes expanding independently of them as well (see Thrift, 1985). For example, accountants have internationalized to meet the needs of their corporate customers but also in order to search for new customers (Daniels, Thrift and Leyshon, 1988).

The most important result of the international expansion of production and finance, telecommunications and producer services has been the development of a complex of international corporate activities in particular cities. For various reasons to do with the continuing need for face-to-face contacts in business, access to office space, sophisticated telecommunications, airports and government, and economies of scale, multinational corporations and allied producer services tend to cluster together in key urban centres around the world in which they can come to dominate economic, social and cultural life. These 'world cities' (Cohen, 1981; Friedmann and Wolff, 1982; Noyelle, 1983a, 1983b; Soja, Morales and Wolff, 1983; Thrift, 1985, 1987; Dogan and Kasarda, 1987) occupy the upper layers of the hierarchy and can be divided into three main strata. At the apex of the hierarchy there are truly global centres – New York, London and Tokyo. These contain many head offices, branch offices and regional headquarters offices of the large corporations, and include the head offices or representative offices of many banks. They account for most large business dealings on a world scale. Second, there are the *zonal* centres – cities like Paris, Singapore, Hong Kong and Los Angeles. These also have many corporate offices of various types and serve as important links in the international financial system but they are responsible for particular zones rather than for business on a world scale. Finally, there are the *regional* centres – cities like Sydney, Chicago, Dallas, Miami, Honolulu and San Francisco. These host many corporate offices and foreign financial outlets but they are not essential links in the international financial system. Some specialize in providing space for corporate regional headquarters serving particular regions. Thus Miami is a regional headquarters node for US-based multinational corporations operating in Latin America (with at least 150 such offices) and Honolulu is a regional headquarters node for US-based multinational corporations operating in Asia (with at least 50 such offices).

These world cities tend to become more and more oriented towards the world-economy over time and as they internationalize they form distinctive enclaves which increasingly tend to detach themselves from their host countries. New York is a perfect example of a modern world city (Drennan, 1988). The New York region, with nearly 20 million people living within its boundaries, had a gross regional product in 1985 which was equivalent to 11 per cent of US output, more than the national output of Canada, Brazil or China. This economic powerhouse has been expanding its role as a headquarters for international business, international banking and international producer services throughout the 1980s. Although a number of headquarters offices of large US corporations moved out of the city in the 1970s and 1980s, those that remain are all very international in character. Their foreign revenues as a percentage of total revenues were 41 per cent, compared with 24 per cent for corporations headquartered elsewhere in the United States. In commercial banking, New York is similarly international in character. The number of foreign bank offices in the city has grown from 47 in 1970 through 84 in 1975 to 191 in 1985. In producer services, the city acts as a major international node. For example, half of all US law firms with more than one foreign office are headquartered in New York. As a result of this concentration of international activity, New York dominates international telecommunications in the United States. In 1982, 20 per cent of the US total of telephone overseas message units emanated from New York. Not surprisingly, the city is also a hub of international transport. In 1986 62 per cent of the international passengers flying into or out of the United States flew into or out of the three major New York airports. Given these indices of New York's eminence as a global service centre, it is no surprise to find that jobs in service industries there grew rapidly in the 1970s and 1980s (while manufacturing jobs declined). For example, a third of overall job growth between 1978 and 1986 was in financial services. Producer services jobs doubled in number between 1978 and 1986, accounting for about 40 per cent of overall job growth. Nor is it any surprise to find that the city's office market has boomed as the expansion of jobs has taken place, especially in Manhattan (Bateman, 1985). As if to underline the city's global character, about one sixth of Manhattan's office space is now foreign-owned, mainly by Canadian, Japanese, British and Dutch investors.

The importance of the expansion of international activity in world cities is not just economic. It is also social and cultural. In the world cities can be found a growing and privileged class of professionals and managers working in the international sector who set the city's social and cultural tone through their buying power in markets for housing, consumer goods and so on. New York is a good example of this process in action. There the growing class of professionals and managers has been responsible for the gentrification of much of New York (Zukin, 1988), for the growth in conspicuous consumption (Wolfe, 1988) and for many of the myths and realities of yuppiedom. But it is also important to remember that New York, like many other world cities, is also a place in which many poor people live, often Third World immigrants trapped in jobs outside the mainstream services industries (Sassen, 1988).

CONCLUSIONS

Perhaps the most important conclusion to draw from this chapter is that, as a result of the world-economic crisis, the world-economy has become more integrated than ever before. There are new links between multinational corporations, banks and countries and the old links have been strengthened.

Take, first, the interconnections between the multinational corporations and the banks. The internationalization of production and finance has forged much closer links between the two. Company finance and profits now depend on interest rate and currency swaps, on Eurobond issues, on financial futures and on foreign exchange options, all usually arranged by the banks. Banks advise on corporate finance, on mergers, on acquisitions. And banks are increasingly the shareholders of the multinational corporations (see Fennema, 1982; Grou, 1983; Andreff, 1984). Second, the ties between multinational corporations and countries have been strengthened. In nearly every nation of the world the level of foreign direct investment has increased and national economies have consequently become more 'open'. Indeed, country action to prevent import penetration by the use of tariffs and other barriers has only hastened this process. In a few countries like the United Kingdom and the Netherlands, which have particularly high levels of foreign direct investment and which conduct much of their production abroad, it is possible to ask whether a coherent national economy still exists (Radice, 1984). Finally, there are much greater connections now between countries and banks. The present international debt crisis is simply the most extreme illustration of this fact (Lipietz, 1984a, 1986).

The greater level of integration of the world-economy should not be overemphasized. The rise of a new world-economic order does not give overwhelming power to the multinational corporation, it does not give the banks total control, and it does not mean the end of national sovereignty. The internationalization of capital, understood in its broadest sense, is a very messy business. It is, as Michalet (1976) has pointed out, a process which produces partial incomes; it is not an economic apocalypse. But it is still the most important economic event taking place in the world at present and it is having crucial reverberations – economic, social and cultural – on many countries. Yet, still only a few human geographers seem willing to come out of their national shells and take the wider view which would enable them to understand what is going on *within their own countries*. They seem quite willing to document the national effects of international processes. This insular view of the world cannot continue, because the world is no longer like that.

References

Aglietta, M. 1979: *A Theory of Capitalist Regulation: The U.S. Experience*. London: New Left Books.

Aglietta, M. 1982: World capitalism in the eighties. *New Left Review*, 136, 25–36.

Aliber, R. Z. 1979: *The International Money Game*. New York: Basic Books.

Andreff, W. 1984: The international centralization of capital and the re-ordering of world capitalism. *Capital and Class*, 22, 58–80.

Atkinson, J. 1985: The changing corporation. In D. Clutterbuck (ed.), *New Patterns of Work*. Aldershot: Gower, 17–34.

Ballance, R. J. and Sinclair, S. W. 1983: *Collapse and Survival: Industry Strategies in a Changing World*. London: Allen & Unwin.

Bateman, M. 1985: *Office Development: A Geographical Analysis*. London: Croom Helm.

Beynon, H. 1985: *Working for Ford* (2nd edn). Harmondsworth: Penguin.

Bloomfield, G. T. 1981: The changing spatial organisation of multinational corporations in the world automotive industry. In F. E. I. Hamilton and G. J. R. Linge (eds), *Spatial Analysis, Industry and the Industrial Environment*, vol. 2: *International Industrial Systems*. Chichester: Wiley, 357–94.

Bora, G. 1981: International division of labour and the national industrial system: the case of Hungary. In F. E. I. Hamilton and G. J. R. Linge (eds), *Spatial Analysis, Industry and the Industrial Environment*, vol. 2: *International Industrial Systems*. Chichester: Wiley, 155–84.

Bromley, R. and Gerry, C. (eds) 1979: *Casual Work and Poverty in Third World Cities*. Chichester: Wiley.

Cantwell, J. 1988: Theories of international production. *University of Reading Discussion Papers in International Investment and Business*, 125.

Casson, M. 1985: Recent trends in international business: A new analysis. *University of Reading Discussion Papers in International Investment and Business*, 112.

Casson, M. 1986: *Multinationals and World Trade: Vertical Integration and the Division of Labour in World Industries*. London: Allen & Unwin.

Castles, S. and Kozack, G. 1985: *Immigrant Workers and the Class Structure of Western Europe*. Oxford: Oxford University Press.

Clairmonte, F. and Cavanagh, J. 1981: *The World in their Web: Dynamics of Textile Multinationals*. London: Zed Press.

Clarke, I. M. 1982: The changing international division of labour within I.C.I. In M. J. Taylor and N. J. Thrift (eds), *The Geography of Multinationals*. London: Croom Helm, 90–116.

Clarke, I. M. 1985: *The Spatial Organisation of Multinational Corporations*. Beckenham: Croom Helm.

Coakley, J. and Harris, L. 1983: *The City of Capital*. Oxford: Basil Blackwell.

Cohen, R. B. 1981: The new international division of labour, multinational corporations and urban hierarchy. In M. J. Dear and A. J. Scott (eds), *Urbanisation and Urban Planning in Capitalist Society*. London: Methuen, 287–315.

Cohen, R. B. 1983: The new spatial organisation of the European and American automotive industries. In F. W. Moulaert and P. B. Salinas (eds), *Regional Analysis and the New International Division of Labour*. Boston, MA: Kluwer Nijhoff, 135–43.

Cohen, R. B. 1987: *The New Helots: Migrants in the International Division of Labour*. Aldershot: Gower.

Cooke, P. 1988: Flexible integration, scope economies and strategic alliances: social and spatial mediations. *Environment and Planning D. Society and Space*, 6, 281–300.

Corbridge, S. 1986: *Capitalist World Development: A Critique of Radical Development Geography*. Basingstoke: Macmillan.

Corbridge, S. (ed.) 1987: Special issue on the international debt crisis. *Geoforum*, 18/3.

Currie, J. 1985: *Export Processing Zones in the 1980s. Customs Free Manufacturing*. London: Economist Intelligence Unit.

Daly, M. T. 1984: The revolution in international capital markets: urban growth and Australian cities. *Environment and Planning A*, 16, 1003–20.

Daniels, P. W., Thrift, N. J. and Leyshon, A. 1988: Internationalization of Professional Producer Services: Accountancy Conglomerates. In D. Endewick (ed.), *Multinational Service Firms*, London: Routledge, 79–106.

Dear, M. (ed.) 1986: Special issue on Los Angeles. *Environment and Planning D. Society and Space*, 3/2.

Dicken, P. 1986: *Global Shift: Industrial Change in a Turbulent World*. London: Paul Chapman.

Dicken, P. 1987: Japanese penetration of the European automobile industry: the arrival of Nissan in the United Kingdom. *Tijdschrift voor Economische en Sociale Geografie*, 78, 59–72.

Dicken, P. 1988: The changing geography of Japanese foreign direct investment in manufacturing industry. A global perspective. *Environment and Planning A*, 20, 633–54.

Dogan, M. and Kasarda, J. (eds) 1987: *The Metropolis Era* 2 vols. Beverly Hills: Sage.

Drennan, M. P. 1988: Local economy and local revenues. In C. Brecher and R. D. Horton (eds), *Setting Municipal Priorities, 1988*. New York: New York University Press.

Dunning, J. H. 1981: *International Production and the Multinational Enterprise*. London: Allen & Unwin.

Dunning, J. H. and Norman, G. 1983: The theory of the multinational enterprise: an application to multinational office location. *Environment and Planning A*, 15, 675–92.

Dunning, J. H. and Norman, G. 1987: The location choices of offices of international corporations. *Environment and Planning A*, 19, 613–32.

Economist 1984: Multinationals vs. globals. *Economist*, 5 May, 67.

Economist, 1988a: America still buys the world. *Economist*, 17 September, 99–100.

Economist, 1988b: One-armed policy maker. A survey of the world economy. *Economist*, 24 September, 1–72.

Economist Intelligence Unit 1984: How to make offshore manufacturing pay. *Economist Intelligence Unit Special Report* 171. London: Economist Intelligence Unit.

Enderwick, P. (ed.) 1988: *Multinational Service Firms*. London: Routledge.

Ernst, D. 1985: Automation and worldwide restructuring of the electronics industry. *World Development*, 13, 333–52.

Fennema, M. 1982: *International Networks of Banks and Industry*. The Hague: Martinus Nijhoff.

Fernandez-Kelly, M. P. 1983: *For We Are Sold, I and My People: Women and Industry in Mexico's Frontier*. Albany, NY: State University of New York Press.

Findlay, A. 1989: The impact of international migration on British localities. In P. Cooke and N. J. Thrift (eds), *Captive Britain?*. Cambridge: Cambridge University Press.

Flynn, N. and Taylor, A. P. 1986: Inside the rust belt: an analysis of the decline of the West Midland economy. 1. International and national economic conditions. *Environment and Planning A*, 18, 865–900.

Friedmann, G. and Wolff, G. 1982: World city formation: an agenda for research and action. *International Journal of Urban and Regional Research*, 6, 309–44.

Frobel, F., Heinrichs, J. and Kreye, O. 1980: *The New International Division of Labour*. Cambridge: Cambridge University Press.

Gaffikin, F. and Nickson, A. 1984: *Jobs Crisis and the Multinationals: The Case of the West Midlands*. Nottingham: Russel Press.

Gershuny, J. I. 1978: *The Self-Service Economy*. London: Frances Pinter.

Gershuny, J. I. and Miles, I. 1983: *The New Service Economy*. London: Frances Pinter.

Gorastraga, X. 1983: *International Financial Centres in Underdeveloped Countries*. London: Croom Helm.

Gordon, D. M. 1988: The global economy: new edifice or crumbling foundations. *New Left Review*, 168, 24–64.

de Grazia, R. 1984: *Clandestine Employment*. Geneva: International Labour Office.

Grosse, R. E. 1982: Regional offices in multinational firms. In A. M. Rugman (ed.), *New Theories of the Multinational Enterprise*. London: Croom Helm, 107–32.

Grou, P. 1983: *La Structure financière du capitalisme multinational*. Paris: Presses de la Fondation Nationale des Sciences Politiques.

Grunwald, J. and Flamm, K. 1985: *The Global Factory, Foreign Assembly in International Trade*. Washington, DC: Brookings Institute.

Gutman, P. and Arkwright, F. 1981: Tripartite industrial cooperation between East, West and South. In F. E. I. Hamilton and G. J. R. Linge (eds), *Spatial Analysis, Industry and the Industrial Environment*, vol. 2: *International Industrial Systems*. Chichester: Wiley, 185–214.

Hamilton, F. E. I. 1984: Industrial restructuring: an international problem. *Geoforum*, 15, 349–64.

Harris, N. 1983: *Of Bread and Guns: The World in Economic Crisis*. Harmondsworth: Penguin.

Harris, N. 1986: *The End of the Third World*. London: Penguin.

Harvey, D. 1987: Flexible accumulation through urbanisation: reflections on postmodernism in the American city. *Antipode*, 19, 260–86.

Heenan, D. A. 1977: Global cities of tomorrow. *Harvard Business Review*, 55, 79–92.

Hill, R. C. 1987: Global factory and company town: the changing division of labour in the international automobile industries. In J. Henderson and M. Castells (eds), *Global Restructuring and Territorial Development*. London: Sage.

Hunt, T., Parker, M. E. and Rudden, E. 1982: How global companies win out. *Harvard Business Review*, 60, 98–108.

International Labour Organization 1984: *World Labour Report 1. Employment, Incomes, Social Protection, New Information Technology*. Geneva: International Labour Office.

International Labour Organization 1988a: *Yearbook of Labour Statistics 1987*. Geneva: International Labour Office.

International Labour Organization 1988b: *The Economic and Social Effects of Multinational Enterprises in Export Processing Zones*. Geneva: International Labour Office.

Jenkins, R. 1984: Divisions over the international division of labour. *Capital and Class*, 22, 28–57.

Jones, D. T. 1985: The internationalisation of the world automobile industry. *Journal of General Management*, 10, 23–44.

Khan, K. M. (ed.) 1986: *Multinationals of the South: New Actors in the International Economy*. London: Frances Pinter.

Kortus, B. and Kaczerowski, W. 1981: Polish industry forges external links. In F. E. I. Hamilton and G. J. R. Linge (eds), *Spatial Analysis, Industry and the Industrial Environment*, vol. 2: *International Industrial Systems*. Chichester: Wiley, 119–54.

Lall, S. 1984: *The New Multinationals: The Spread of Third World Enterprises*. Chichester: Wiley.

Langdale, J. 1984: Electronic funds, transfer and the internationalisation of the banking and finance industry. *Geoforum*, 16, 1–13.

Lash, S. and Urry, J. 1987: *The End of Organized Capitalism*. Cambridge: Polity Press.

Lebkowski, M. and Mankiewicz, J. 1986: Western direct investments in centrally planned economies. *Journal of World Trade Law* 20.

Leborgne, D. and Lipietz, A. 1988: New technologies, new modes of regulation: some spatial implications. *Environment and Planning D. Society and Space*, 6, 281–300.

Lin, V. 1987: Women in the semiconductor industry in Southeast Asia. In J. Henderson and M. Castells (eds) *Global Restructuring and Territorial Development* London: Sage.

Lipietz, A. 1984a: How monetarism has choked Third World industrialisation. *New Left Review*, 145, 71–87.

Lipietz, A. 1984b: Imperialism or the beast of the apocalypse. *Capital and Class*, 22, 81–109.

Lipietz, A. 1986: *Mirages and Miracles: The Crisis of Global Fordism*. London: Verso.

McGee, T. 1986: Joining the global assembly line: Malaysia's role in the international semiconductor industry. In T. McGee (ed.), *Industrialisation and Labour Force Processes: A Case Study of Peninsular Malaysia*. Australian National University, 35–67.

McMillan, C. H. 1987: *Multinationals from the Second World: Growth of Foreign Investment by Soviet and East European State Enterprises*. London: Macmillan.

Margirier, G. 1983: The eighties: a second phase of crisis? *Capital and Class*, 21, 61–86.

Marsden, J. 1980: Export processing zones in developing countries. *UNIDO Working Papers on Structural Change* 19.

Marshall, J. N. et al. 1988: *Services and Uneven Development*. Oxford: Oxford University Press.

Massey, D. 1988: Uneven development: social change and spatial divisions of labour. In D. Massey and J. Allen (eds), *Uneven Re-development*. London: Hodder & Stoughton, 250–76.

Maxcy, G. 1981: *The Multinational Motor Industry*. London: Croom Helm.

Mendelsohn, M. S. 1980: *Money on the Move*. New York: McGraw Hill.

Michalet, C. A. 1976: *Le Capitalisme mondiale*. Paris: Presses Universitaires de France.

Mitchell, T. 1988: *From a Developing Country to a Newly Industrialised Country: The Republic of Korea 1961–82*. Geneva: International Labour Office.

Morello, T. 1983: Sweatshops in the sun? *Far Eastern Economic Review*, 15 September, 88–9.

Morgan, K. and Sayer, A. 1988: *Microcircuits of Capital*. Cambridge: Polity Press.

Moss, M. 1987: Telecommunications, world cities and urban policy. *Urban Studies*, 24, 534–46.

Nierop, T. 1989: Macro-regions and the global institutional network 1950–1980. *Political Ecography Quarterly*, 8, 43–66.

Noyelle, T. J. 1983a: The implications of industry restructuring for spatial organisation in the United States. In F. Moulaert and P. W. Salinas (eds), *Regional Analysis and the New International Division of Labour*, Boston: Kluwer Nijhoff, 113–33.

Noyelle, T. J. 1983b: The rise of advanced services. Some implications for economic development in U.S. cities. *Journal of the American Planning Association*, 25, 280–90.

OECD 1988: *Ageing Populations: The Social Policy Implications*. Paris: OECD.

Oxford Review of Economic Policy 1986: Special issue on The International Debt Crisis. *Oxford Review of Economic Policy*, 2/1.

Pahl, R. E. 1980: Employment, work and the domestic division of labour. *International Journal of Urban and Regional Research*, 4, 1–19.

Pahl, R. E. 1984: *Divisions of Labour*. Oxford: Basil Blackwell.

Pahl, R. E. (ed.) 1988: *On Work*. Oxford: Basil Blackwell.

Parboni, R. 1981: *The Dollar and its Rivals*. London: Verso.

Peet, J. R. (ed.), 1987: *International Capitalism and Industrial Restructuring*, Boston: Allen & Unwin.

Petit, P. 1986: *Slow Growth and the Service Economy*. London: Frances Pinter.

Petras, E. and McLean, M. 1981: The global labour market in the world economy. In M. M. Kritz, C. B. Keely and S. M. Tomasi (eds), *Global Trends in Migration*. New York: Centre for Migration Studies, 44–63.

Portes, A. and Walton, J. 1981: *Labour, Class and the International System*. New York: Academic Press.

Radice, H. 1984: The national economy – a Keynesian myth? *Capital and Class*, 22, 111–40.

Redclift, N. and Mingione, E. 1985: *Beyond Employment: Household, Gender and Subsistence*. Oxford: Basil Blackwell.

Roddick, J. 1984: Crisis, Seignorage and the modern world system; rising Third World power or declining US hegemony. *Capital and Class*, 23, 121–34.

Rugman, A. M. 1981: *Inside the Multinationals*. London: Croom Helm.

Rugman, A. M. (ed.) 1982: *New Theories of the Multinational Enterprise*. London: Croom Helm.

Salita, D. C. and Juanico, M. B. 1983: Export processing zones: new catalysts for economic development. In F. E. I. Hamilton and G. J. R. Linge (eds), *Spatial Analysis, Industry and the Industrial Environment*, vol. 3: Regional Economies and Industrial Systems. Chichester: Wiley, 441–61.

Sampson, A. 1981: *The Money Lenders*. London: Hodder & Stoughton.

Sassen, S. 1988: *The Mobility of Labour and Capital: A Study in International Investment and Labour Flow*. Cambridge: Cambridge University Press.

Sayer, A. 1986a: Industrial location on a world scale: the case of the semiconductor industry. In A. J. Scott and M. Storper (eds), *Production, Work and Territory: The Geographical Anatomy of Industrial Capitalism*. London: Allen & Unwin.

Sayer, A. 1986b: New developments in manufacturing: the just-in-time system. *Capital and Class*, 30, 43–72.

Sayer, A. 1989: *International Journal of Urban and Regional Research*, forthcoming.

Schoenberger, E. 1988: From Fordism to flexible accumulation: technology, competitive strategies and international location. *Environment and Planning D. Society and Space*, 6, 245–62.

Scott, A. J. 1987: The semiconductor industry in Southeast Asia: organisation, location and the international division of labour. *Regional Studies*, 21, 143–160.

Scott, A. J. 1988: *New Industrial Spaces*. London: Pion.

Scott, A. J. and Cooke, P. 1988: Special issue on the new geography and sociology of production. *Environment and Planning D. Society and Space*, 6/3.

Sethuraman, S. V. (ed.) 1981: *The Urban Informal Sector in Developing Countries*. Geneva: International Labour Office.

Soja, E., Morales, R. and Wolff, G. 1983: Urban restructuring: an analysis of social and spatial change in Los Angeles. *Economic Geography*, 59, 195–230.

Strange, S. 1988: *Till Debt us do Part*. Harmondsworth: Penguin.

Taylor, M. J. and Thrift, N. J. 1981: British capital overseas: direct investment and firm development in Australia. *Regional Studies*, 15, 183–212.

Taylor, M. J. and Thrift, N. J. (eds) 1982: *The Geography of Multinationals*. London: Croom Helm.

Taylor, M. J. and Thrift, N. J. 1986: *Multinationals and the Restructuring of the World Economy*. Beckenham: Croom Helm.

Teulings, A. W. M. 1984: The internationalisation squeeze: double capital movement and job transfer within Philips worldwide. *Environment and Planning A*, 16, 565–706.

Tharakan, P. M. 1980: *The New International Division of Labour and Multinational Companies*. Farnborough: Saxon House.

Thrift, N. J. 1980: Frobel and the new international division of labour. In J. R. Peet (ed.), *An Introduction to Marxist Theories of Underdevelopment*. Australian National University, Department of Human Geography, HG14, 181–9.

Thrift, N. J. 1985: The internationalisation of producer services and the integration of the Pacific Basin property market. In M. J. Taylor and N. J. Thrift (eds), *Multinationals and the Restructuring of the World Economy*. London: Croom Helm.

Thrift, N. J. 1986: The geography of international economic disorder. In R. J. Johnston and P. J. Taylor (eds), *A World in Crisis? Geographical Perspectives* (1st edn). Oxford: Basil Blackwell.

Thrift, N. J. 1987: The fixers: The urban geography of international commercial capital. In J. Henderson and M. Castells (eds), *Global Restructuring and Territorial Development*. London: Sage, 219–47.

Thrift, N. J. and Leyshon, A. 1987: The gambling propensity: Banks, developing country debt exposures and the new international financial system. *Geoforum* 18.

Thrift, N. J. and Taylor, M. J. 1988: What could we do? The new geography of multinational corporations. In D. Gregory and R. Walford (eds), *New Horizons in Human Geography*. London: Macmillan.

Tolchin, M. and Tolchin, S. 1988: *Buying into America. How Foreign Money is Changing the Face of our Nation*. New York: Times Books.

UNICEF 1988: *State of the World's Children*. Oxford: Oxford University Press.

United Nations 1988: *Transnational Corporations in World Development: Trends and Prospects*. New York: United Nations Centre on Transnational Corporations.

Van der Wee, H. 1987: *Prosperity and Upheaval: The World Economy 1945–1980*. Harmondsworth: Penguin.

Van Liemt, G. 1988: *Bridging the Gap: Four Newly Industrialising Centres and the Changing International Division of Labour*. Geneva: International Labour Office.

de Vroey, M. 1984: A regulation approach interpretation of contemporary crisis. *Capital and Class*, 23, 45–66.

Wallerstein, I. 1979: *The Capitalist World Economy*. Cambridge: Cambridge University Press.

Wells, L. T., Jr 1983: *Third World Multinationals: The Rise of Foreign Investment from Developing Countries*. Cambridge, MA: MIT Press.

Wolfe, T. 1988: *The Bonfire of the Vanities*. New York: Basic Books.

Wong, K. and Chu, D. K. Y. 1984: Export processing zones and special economic zones as generators of economic development: the Asian experience. *Geografiska Annaler*, Series B, 66, 1–16.

World Bank 1988a: *World Development Report 1988*. Oxford: World Bank/Oxford University Press.

World Bank 1988b: *World Debt Tables*. Washington, DC: World Bank.

Zukin, S. 1988: *Loft Living* (2nd edn). London: Radius.

3

Draining the World of Energy

PETER R. ODELL

Each and every society's capability of development, or indeed its ability to survive, depends on continuing access to energy in appropriate forms and quantities and at acceptable levels of cost (Cook, 1976). There is a relationship, which changes over time, between the degree of development in a society (economy) and the use of energy. This is illustrated in figure 3.1. The relationship has, until very recently (when a small number of wealthy and highly industrialized countries started to use less energy), been one of an increasingly intensive use of energy as development occurred. Thus the gradual transformation over the last 200-plus years from a world in the mid-eighteenth century made up of primitive and peasant societies which were largely subsistence in their structure and organization to a world consisting mainly of post-industrial, industrial and industrializing economies has led to a global use of energy which is now over 20 times greater than it is estimated to have been in 1860 – the earliest date for which world energy use can be estimated with any degree of confidence (Marchetti and Nakicenovic, 1979). Even by 1860, however, world energy use was probably about twice what it had been at the beginning of the industrial revolution in the third quarter of the eighteenth century, so that in 1988 the world is using energy at a rate which is more than 40 times greater than it was in 1788. Figure 3.2 shows the evolution of energy use over the 128 years back to 1860. It demonstrates what a remarkable consistent rate of growth there has been over the whole of the period since then (at just over 2 per cent per annum), except for the years between 1950 and 1973 when the rate more than doubled to almost 5 per cent per annum. It is, however, the very long-term consistent 2-plus per cent per annum growth rate in energy use against which the availability and potential availability of the world's energy supplies need to be measured when looking at the energy prospects on a global scale for the medium- to long-term future.

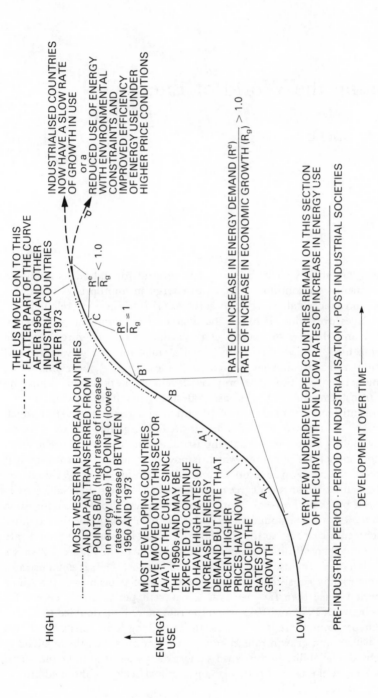

Figure 3.1 The relationship between energy use and economic development: the rate of increase (decrease) in energy use as a function of the stage of economic development

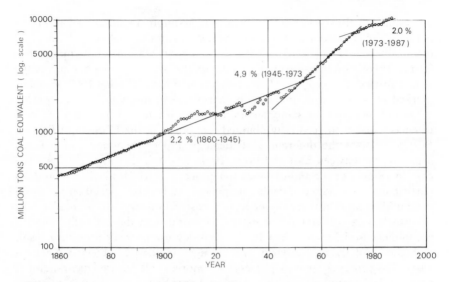

Figure 3.2   Trends in the evolution of world energy use since 1860
©EGI 144/83 rev 88

THE DEMAND FOR ENERGY

The near 5 per cent per annum energy-use growth rate of the years between 1945 and 1973 is important because attitudes to the world energy situation and its outlook remain heavily dependent on the idea that this exceptionally high rate of growth in energy use represents the 'norm' – the 'required' rate of increase to match a continuation of population increase and desirable economic progress. This is unlikely to be the case, however, as the rate of increase for the period 1945–73 can be shown to be the result of a temporally unique set of factors. At that time the world's nations were almost all on the steepest part of the curve shown in figure 3.1 – in a situation, that is, in which their economies were going through the most heavily energy-intensive period of development.

First, the small number of rich countries in the western economic system were then in the later stages of the traditional industrialization process – a stage marked by an emphasis on products with high energy inputs, such as motor vehicles, household durable goods and petrochemical products (Schurr and Netschert, 1977). As a result, the use of energy on the production side of such economies greatly increased. At the same time, the increase in the per-capita use of energy on the consumption side of these countries' economies was even more dramatic. This arose from the suburbanization of cities and the increasing length of the journey to work, the switch from public to private transport and the mass use of motor cars, the expanded availability of leisure time and the 'annihilation of space' in the public's use of such time, the mechanization of households by the use of electrically powered equipment and the achievement of much higher standards of comfort (by heating and/or cooling) in homes and other buildings (Leach et al., 1979). Cheap energy was one of the main bases for the 'revolution of rising expectations' on the part of the populations of the rich countries. The progress of that revolution in the 1950s and 1960s was thus marked by a very rapid increase in energy use in the industrialized world.

Second, many of the same factors positively influenced the rate of growth in energy use in the centrally planned economies of the USSR and Eastern Europe, despite the differences of ideology and of economic and political organization between East and West (Park, 1979). This was particularly the case in respect of the industrialization process in which all these countries participated as a matter of deliberate policy – and with a special emphasis on the rapid expansion of heavy energy-intensive industry. To a smaller but, nevertheless, a still significant extent, consumers in the centrally planned economies also increased their levels of energy use in this period as a result of higher real living standards and changes in lifestyle. Electrification, a particularly energy-intensive process, was moreover a declared central aim of such planned economies – not least because Lenin had specified it as a necessary part of the evolution to communism (Lenin 1966). Thus all these countries were also moving up the steepest part of the curve shown in figure 3.1 during this period.

Meanwhile, most of the poorer countries in the Third World moved off the lowest part of that curve as these (for the large part) newly independent nations pursued policies of deliberate industrialization (viewed as the panacea for and the *sine qua non* of economic progress) and found that such policies were necessarily accompanied by rapid urbanization with its much enhanced energy requirements. The types of industry which were established were, moreover, either heavy industries which were very energy intensive – such as iron and steel, metal fabricating, vehicles and cement – or industries such as textiles and household goods which were relatively energy intensive. The concurrent urbanization process meant that peasants and landless agricultural labourers were transferred from their low-energy ways of living (in which situation, moreover, most of the energy required was collected rather than purchased) to lifestyles in the city or urban environment which, no matter how poor the living standards achieved turned out to be, were nevertheless much more demanding in their requirement for energy generally and in their use of electricity and petroleum products in particular (Dunkerley, 1981).

It was essentially the temporal coincidence of these fundamental societal developments in most countries of the world in the 1950s and 1960s which caused the abnormally high rate of growth in global energy use over that period. It is not without significance that the United States moved off the steepest part of the curve in the late 1960s and early 1970s as its rate of growth in energy use started to fall away under the impact of structural changes in its economy when the service sector grew more quickly than manufacturing. Because of the importance of the United States in total world energy use, this also started

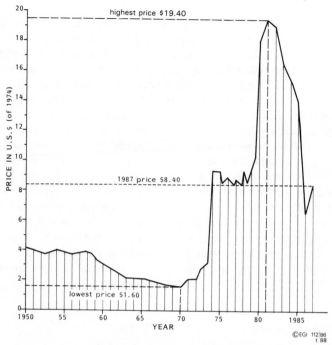

Figure 3.3 The price of oil, 1950–1987 (Saudi Arabian light crude in 1974 US$)
©EGI 112.86 r. 88

to exercise a downward pull on global rates at that time – some years, that is, prior to the first oil price shock of 1973. There was, however, another powerful factor at work in influencing the rate in increase in energy use in the more than 20 years before that price shock. This was the long-continuing decline in the real price of energy after 1950 (Adelman, 1972). This is shown in the left-hand part of the curve in figure 3.3 in which the evolution of the (real) price of Saudi Arabian light crude oil over the years from 1950 is illustrated. It shows how the market value of a barrel of this oil fell between 1950 and 1970 by over 60 per cent from $4.25 to $1.60 under the impact of the technologically efficient and politically powerful oil companies which, at that time, were able both to ignore the interests of the oil-producing countries in the exploitation of their resources and to persuade the importing countries of the continuity of these low-cost oil supplies (Odell, 1986).

This decline in the oil price brought about a falling market price for all other sources of energy throughout the world during this period, though not to the same degree as the fall in the price of crude oil when calculated in terms of the reduction in price to the final consumer of energy. This fall in energy prices was especially important in the western industrial countries with their open – or relatively open – economies under their post-war liberal trading regimes. In these countries local energy production – such as coal in Western Europe and Japan, and oil and natural gas in the United States – either had to be reduced in price to enable it to compete with falling-price imported oil, or else the industries concerned had to be cut back or eliminated (Manners, 1971).

Thus both the actual decline in the cost of energy over this period and the perception among energy users that energy was cheap and getting cheaper – so hardly worth worrying about in terms of the care and efficiency with which it was used – created conditions in which the careless and wasteful consumption of energy became a hallmark of both the technological and behavioural aspects of societal and economic developments. There was a consequent emphasis on systems of production, transport and consumption which were quite unnecessarily energy intensive (Odell, 1975) and against which a reaction of significant proportions could be expected once the real price curve started to move up. This phenomenon dates from 1970, but in 1973–4, as shown in figure 3.3, it took a massive upward leap with the so-called 'first oil price shock'. By the time of the second price shock in 1979–81 structural changes in the energy-careless and wasteful systems described above were already well under way, and they have since both intensified and become geographically much more diffuse.

STRUCTURAL CHANGES IN THE ENERGY SYSTEM SINCE THE MID-1970s

In other words, the order-of-magnitude jump in oil and other energy prices in the 1970s has terminated the perception of energy as a near-costless input to economic and societal developments in most parts of the world. Thus, as shown in figure 3.2, the rate of increase in energy use has now been brought back to its historic long-term figure of about 2 per cent per annum. To date, the most important element in the change has been the quite dramatic decline in the energy intensity of economic activities in the western industrialized countries, where the motivation to achieve energy savings has been greater than elsewhere and the ability to do so has been higher (Fritsch, 1982). Both behavioural and technological components have contributed to this development.

Behaviourally, users have responded to the higher prices by taking steps which save energy (such as turning down thermostats and 'trip combinations' in the use of motor vehicles). They have, moreover, been subjected to 'Save It'-style energy-conservationist campaigns which have imparted knowledge whereby energy-using behaviour could be adjusted appropriately. The technological component has been much more effective, however, largely because there was so much 'fat' to work out of the system from the energy-careless and thoughtless technology of the pre-1970 period. More efficient energy processes in factories, more efficient lighting in offices, better insulation in buildings, the development of motor vehicles, planes and ships which give more kilometres per litre (or ton) of fuel and the expansion of systems of electricity production which are inherently more energy efficient have jointly combined to produce significantly lower rates of energy use per unit of output of goods and services. Overall in Western Europe, for example, the energy intensity of energy growth from 1973 to 1980 was less than half that of the preceding 10 years (International Energy Agency, 1984), and since 1980 the ratio has continued to fall, albeit at a slower rate especially since energy prices declined sharply in real terms after 1985 (see figure 3.3).

Despite the already considerable reductions in the intensity of energy use over the last 15 years, both the behavioural and technological components involved in reducing energy use still have a long way to go before the energy-inefficient systems which were the norm in the period of low- and decreasing-cost energy are finally replaced. This is true even in the world's rich countries where the required investment funds and the other necessary inputs of knowledge, managerial and technical expertise and know-how are available, so rendering the changes simply a matter of time (Darmstadter et al., 1977, 1983; Owen, 1986).

The diffusion of more efficient energy-using systems and infrastructure to the modern sectors of most Third World countries will be a slower process because of the relative scarcity in such countries of the inputs that are required to implement the changes – notably the lack of investment capital and of expertise. Such diffusion is, however, taking place under the powerful stimulus of the very high foreign exchange costs of energy imports (particularly oil) on which Third World countries depend to a larger degree even than most industrialized countries (Dunkerley, 1981; Pachauri, 1988). The alternative to more efficient energy use in such cases is, indeed, too little energy to keep the systems going because a limit has to be placed on the amount of oil that can be imported. Thus enhanced levels of energy-utilization efficiency will eventually be achieved in the Third World. This is an important consideration in relation to the longer-term evolution of global energy demand, given that the percentage of world energy used in these countries must increase steadily under the joint impact of their high rates of population growth and of changes in the structure of their economies which are going through their most energy-intensive period of development, as illustrated in figure 3.1 (Desai et al., 1987). The countries of the North, which already use much more energy per capita, will meanwhile experience a slow or even a negative growth rate in energy use (Hoffman and Johnson, 1981; Guilmot, 1986).

The countries with centrally planned economies lie somewhere between the industrialized and the Third World nations in respect of their energy-use pattern and the prospects for their development (Dienes and Shabad, 1979; Park, 1979; Hoffman, 1985). To date they have done significantly less well than the market economies in saving energy – partly for economic (price) reasons, and partly for technical and organizational reasons. The importance of energy conservation has now, however, become much more generally recognized by the governments concerned, so that, given the opportunities which exist in such 'command' economies for the speedy implementation of

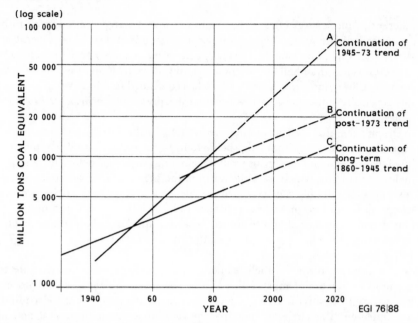

Figure 3.4   World energy use in 2020 given the extrapolation of different historical trends in energy use
©EGI 76 88

policies, measures to save energy are now being accorded a much higher priority than hitherto.

Given these structural changes in attitudes and policies towards the energy sectors of the economies of countries in all parts of the world, there are now long-term prospects for a slow rate of increase in energy demand – even with a renewed higher rate of economic growth in the western system. The huge differences in the quantities of energy which will be used as a result of these changes are illustrated in figure 3.4. Thus, for example, a continuation of the abnormally high growth rate in energy use of the years between the late 1940s and the early 1970s would, by 2020, have required an annual supply of energy of almost 80,000 million tons of coal equivalent (mtce). With a continued growth rate of 2.0 per cent per annum from the 1973 base, the use of energy in 2020 will be only about 21,000 million tons, and if the high 1945–73 rate of growth is compensated by a rate of growth in the meantime so that by 2020 the curve is brought back to the value as defined by an extrapolation of the 1860–1945 growth trend, then energy use in 2020 will be of the order of 12,000 mtce. As the latter development would imply an average annual growth rate of only a little over 1 per cent from 1973 it would seem overly optimistic to reckon on such a 'good' result from increasing efficiency of energy use. This sort of increase would, moreover, imply a probable failure of the world-economy to expand at a high enough rate to improve living standards in the Third World, and would thus indicate a situation of increasing economic conflict between North and South. If this is to be avoided then enough energy will have to be used to sustain an adequate development of the world's economy. If it is assumed that this can be achieved, then my best guess of global energy demand in 2000 is for 15,000–16,000 mtce (compared with a use in 1988 of about 11,000 mtce), and in 2020 for a use of between 25,000 and 26,000 mtce.

These are formidable totals of energy to be supplied – compared with what has been achieved to date – but they are very much lower expectations than those which emerge from the earlier near-automatic extrapolation of the high rates of growth in energy use between 1945 and 1973, despite the experience over the past 15 years in the achievement of greater efficiencies in energy use and the impact of much higher energy prices. The now much more modest expectations of the demand for energy over the next critical 30 years or so (after which the world seems likely to be in a position to turn increasingly to the utilization of renewable and benign energy sources – notably the direct use of the sun's energy) are highly significant when viewed in relation to the potential for increasing the supply of energy. In particular, it adds to the significance of the world's available fossil fuel supplies because it is these which offer the best prospects for a relatively safe supply at the lowest possible cost for the world's energy-using systems over the next 30–40 years. This eliminates both the potential dangers from a too rapid rate of development of relatively untested and still unproven nuclear energy (Lovins, 1977; Hansen et al., 1988), and the need that would otherwise arise for the expenditure of too high a proportion of the world's wealth and its available scientific and technical expertise on a 'crash programme' of development of non-conventional energy sources (Pryde, 1983).

THE SUPPLY AND PRICE OF FOSSIL FUELS

Oil and Natural Gas

In the 1945–73 period of low and declining energy prices the geography of energy production gradually became dominated (outside the United States and the Soviet Bloc) by a rapidly increasing flow of oil from the Middle East and a small number of countries elsewhere in the world as shown in figure 3.5. The prolific oil resources with extremely low production costs of this small group of countries undermined the economic production of most other sources of energy in most of the rest of the world (Adelman, 1972). Indeed, the survival of energy production elsewhere, in the face of low-cost oil from the Middle East and a few other countries where it was efficiently produced and from where it was equally efficiently transported by the seven major international oil corporations and a number of other somewhat smaller, but still large, oil companies (Odell, 1986; Sampson, 1975; Blair, 1976), depended largely on protectionist legislation. Despite such protectionism many pre-existing energy supply industries were closed down or were severely cut back, as in the case

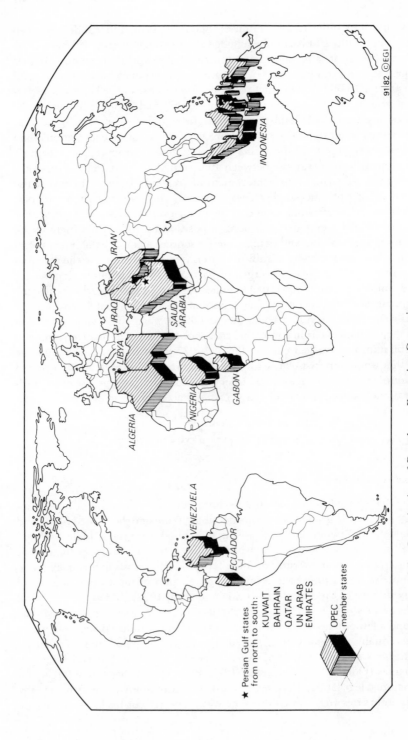

Figure 3.5 Members of the Organization of Petroleum Exporting Countries
©EGI 91|82

of the coal industry in Britain. Thus the exploitation of the extensive resources of somewhat higher-cost energy was undermined by cheap Middle East oil. This was particularly important for the industrializing countries of the Third World where, generally, local energy sources had not hitherto been greatly developed because of the lack of demand. Now they could not be developed because capital could not be attracted to enterprises unable to compete with low-cost imported oil (Odell and Vallenilla, 1978).

Those countries, such as the United States, which tried to protect their own indigenous energy production (of coal, oil and natural gas) and so had to pass on the higher prices to the consumers of energy, found that they then suffered adverse economic consequences because they could not compete for markets for industrial goods with countries such as Japan and many countries in Western Europe which enjoyed the cost advantages given to them by their use of low-cost oil imports from the Middle East. Thus in the early 1970s even the United States had to open up its markets to international oil (Blair, 1976), although by then the price of such oil was already beginning to rise (as shown in figure 3.3) under the impact of the initial successes of the organization which the main oil exporting countries had formed, i.e. the Organization of Petroleum Exporting Countries (OPEC), and of its collusion at that time with the major international oil companies (Odell, 1986). The latter were then also seeking higher prices in order to enhance their profitability, and their ability to go on finding the increasing volumes of new oil required to satisfy a market which was expanding at about 7.5 per cent per annum (Anderson, 1984).

By 1973, as shown in table 3.1, oil exports from the OPEC countries accounted for 36.5 per cent of the total energy supply in the non-communist world outside the OPEC countries themselves. The production of indigenous oil, natural gas and coal were then each only a little over half as important as imports of OPEC oil. Twenty-five years earlier (in 1948) oil exports from the countries

Table 3.1 Sources of energy used in the non-communist world (excluding the OPEC countries) in 1973 and 1986

	mtoe[a]	Percentage of total	mtoe[a]	Percentage of total
Total energy use	4,045	100	4,172	100
Imports of OPEC oil	1,480	36.5	674	16.2
Other energy imports[b]	100	2.5	207	5.0
Indigenous production	2,465	60.9	3,291	78.8
Oil	760	18.8	1,191	28.1
Natural gas	765	18.9	741	18.0
Coal	805	19.9	1,076	25.7
Other	135	3.3	283	6.7

[a] million tons oil equivalent
[b] oil, natural gas and coal from the centrally planned economies

that were later to become the members of OPEC had satisfied less than 7 per cent of the non-communist world's demand for energy.

The order-of-magnitude increase in oil prices between 1973 and 1981 and the increasing uncertainty over the willingness of OPEC countries to continue to produce sufficient oil to meet the world's demand for energy produced a general reappraisal of the prospects for the indigenous production of energy in most countries. The prospects for such indigenous production were significantly enhanced in both economic and political terms and, as a result, there was a marked fall in the contribution of OPEC oil to world energy supplies. This can be seen in table 3.1 in the contrast between the 1973 and the 1986 divisions of the non-communist world energy market between oil imported from the OPEC countries on the one hand and the use of other sources of energy supply on the other hand. It shows how dramatically OPEC's contribution to the non-communist world's energy supply has fallen compared with that provided by energy production elsewhere in the non-communist world and, on a smaller scale, with imports from the centrally planned economies. In particular, oil production in the western world excluding OPEC now comfortably exceeds OPEC's oil exports, whereas in 1973 the latter were almost twice as important as the former. OPEC's oil exports in 1986 were, moreover, little more than gas production in the western world and well below the contribution from indigenous coal.

In other words a geographically much more diffuse pattern of world energy production developed rapidly under post-1973 conditions. This diffusion of energy production is set to continue as long as the price of OPEC oil remains above the cost of producing alternatives – and as long as OPEC oil is perceived to be an unreliable source of supply. This remains a strong perception, given not only the interruptions from time to time in the flow of oil from OPEC countries (for example, as a result of the Iran–Iraq war) but also OPEC's policy of deliberately restraining supplies through the quota system which it first established in 1983 as a means of maintaining prices (Odell, 1986).

Oil and/or natural gas are already being produced in significant quantities in 75 countries – compared with under 60 in 1973 – and the number will continue to increase as exploration becomes more widespread and more successful. Most countries are, indeed, aiming for self-sufficiency, or better, in terms of oil and gas. This includes the countries of the industrialized world, where a number, such as Norway, Britain, the Netherlands, Denmark and Australia, have already been successful and will continue to build up their production potential. There are also good prospects geologically in many Third World countries, but these are often thwarted by financial and other problems in their exploration and exploitation efforts so that the process of increasing production in most of these countries has been gradual rather than spectacular (United Nations, 1982). Nevertheless, by 1986 non-OPEC developing countries had, as a group, become a small net exporter of oil rather than a group with large net imports of oil from the OPEC countries as in 1973. Figures 3.6 and 3.7 show, however, how many opportunities remain for finding and developing

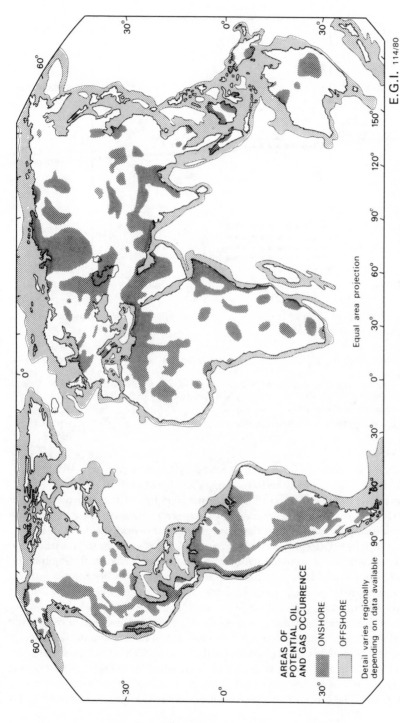

Figure 3.6 The world's potentially petroliferous regions (excluding Antarctica and the deep oceans)
©EGI 114/80

AREAS OF
POTENTIAL OIL
AND GAS OCCURRENCE

ONSHORE

OFFSHORE

Detail varies regionally
depending on data available

Equal area projection

E.G.I. 114/80

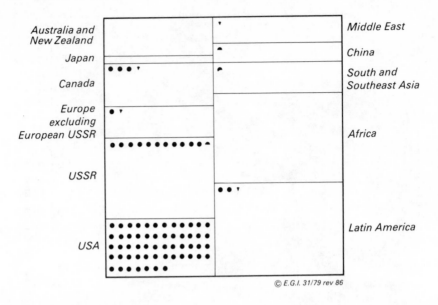

Figure 3.7   The regional distribution of potentially petroliferous areas shown in proportion to the world total
©EGI 31/79 rev 86

Within each region the number of exploration and development wells which have been drilled is shown: each full circle represents 50,000 wells drilled by the end of 1987. Segments of circles are included only for regions in which the total number of drilled wells to 1987 is less than 50,000. Relative to the United States, all other parts of the world, but especially the regions of the Third World, are very little drilled for their oil.

Source: Grossling, 1976, p. 83 (updated by author)

the oil and gas resources of Latin America, Africa and Asia (Grossling, 1976; World Bank, 1983; Odell and Rosing, 1983; Khan, 1988).

Thus, over at least the rest of the century, the importance that the member countries of OPEC hitherto managed to achieve in terms of their domination of the oil supply pattern of the non-communist world, seems unlikely to return. Moreover, despite their very considerable resources, their natural gas production will be constrained by the difficulties that they have in finding markets in the rest of the world where indigenous production is, and will generally remain, sufficient to meet expanding needs. Such indigenous gas will certainly be the preferred source of supply.

There is, however, a penalty involved in this changing global pattern of world oil and gas production – with a greater emphasis on the exploitation of reserves within the consuming countries and a reduced emphasis on the exploitation of oil and gas from areas with relatively little demand for energy, so that export markets are essential. Success in achieving exports was, as shown in figure 3.7, a function of the low-cost production of oil in the OPEC countries. In so far as much of the use of this low-cost production is being replaced by the use of higher-cost energy production from other parts of the world, the inevitable consequence will be a reduction in the overall potential for economic growth in the western world's economic system, thus ensuring that fewer of the world's people achieve an adequate standard of living. The ultimate penalty for not producing and using the world's lowest-cost energy resources is a lesser degree of economic development measured at the global level.

However, the much reduced rate of growth in demand for oil and gas since 1973, and the enhanced supply potential which has been created by the diffusion of exploration and production activities, are together having the effect of postponing for several decades the occurrence of peak volumes of oil and gas output. Thus conventional oil and gas now seem likely to be available in the quantities demanded until well into the second quarter of the twenty-first century. Thereafter, oil from non-conventional habitats (such as tar sands and oil shales, as well as heavy oil from relatively shallow horizons) will be developed and thus ensure the continued slow expansion of the oil industry until some time in the second half of the twenty-first century (Odell and Rosing, 1983).

A similar, but even more emphatic argument holds for natural gas: the expansion of conventional supplies from an increasing number of locations will enable this industry to increase its contribution to world energy needs over many decades into the future – as has, for example, happened in Western Europe over the last 20 years and which is now beginning to take place in many countries of Latin America, in South and South-east Asia, and even in parts of Africa (Halbouty, 1983; Davison et al., 1988). Moreover, the prospects for additional gas from non-conventional habitats are even brighter than those for oil. For example, deep gas dissolved in a saline solution offers a vast potential on which work is already under way in the United States (Hodgson, 1978), deep ocean and Arctic gas hydrates offer a longer-term prospect and beyond these is the possibility of a near-infinite source of natural gas arising from the hypothesis of an abiogenic primeval origin for most of the world's methane (Gold and Soter, 1980; Gold, 1987). Based on resources which are already known and exploitable, plus a 50-year period during which one or other of the sources of non-conventional gas can be proven and commercialized, natural gas could easily provide a steadily increasing proportion of the world's slowly increasing energy needs throughout the twenty-first century. The prospects for natural gas are, indeed, more a function of the geography of resources' development, relative to the geography of demand, than of the global aggregate of the resources available or potentially available. The internationalization at best, or at least the regionalization, of the natural gas industry is required for an effective use of the world's resources of this minimum-polluting source of fossil energy, and this, of course, is an essentially geopolitical question as, for example, in the degree to which Soviet gas will be exported to western Europe, or Mexican and Canadian gas to the United States.

Coal

Meanwhile, the high oil prices between 1973 and 1985 stimulated the exploitation of the world's low-cost coal reserves so that, as shown in table 3.1, production of coal in the non-communist world increased by about 30 per cent between 1973 and 1986. Coal's share of total energy use over the decade grew from under 20 per cent to almost 26 per cent. In the centrally planned economies over the same period coal production rose by a somewhat higher percentage – about 35 per cent – and in 1987 these countries used rather more coal than all the rest of the world (about 2,000 million tons compared with 1,600 million tons). In the aftermath of the oil price increases – and the fears at that time of the inadequacy of oil reserves – there was a great deal of enthusiasm and optimism over the prospects for developing the world's production and use of coal (WOCOL Report, 1980). However, rapid expansion has, to date, proved to be neither possible nor necessary (because of the dramatic decline in the rate of increase in energy use), so that the net global expansion of the coal industry has been relatively modest (Gordon, 1987).

Moreover, it should be noted that the increase in world coal production is a net effect emerging from a rapid growth in some areas offset, in part, by continuing stagnation or decline in the traditional deep-mined coal industries of western Europe, Japan and a few other countries. Though there are plenty of remaining coal reserves in countries such as Britain, the Federal Republic of Germany, Belgium and France, their exploitation has proved to be much too costly, largely because of the high price of the main component in the overall production costs, i.e. labour costs which account for between 50 and 60 per cent of the total costs of traditional deep-mined coal. The prospects for any significant change for this high-cost coal are poor, particularly in circumstances in which coal can be produced elsewhere at much lower cost from both deep mines in countries with lower wage costs (such as South Africa, India, Taiwan and Poland) and open-cast mines in countries with vast reserves of good quality coal at or near the surface. This is possible in, for example, western Canada, the American Mid-West and Australia, in all of which major developments have already taken place and more are planned (International Energy Agency, 1984; Gordon, 1987). In addition, however, large new surface mines are under development or are planned in a number of other countries such as Colombia, Botswana and Indonesia.

These, and other developments elsewhere in the Third World (as well as in China), will provide much of the increasing supply of coal for international trade over the next 20 years. This is currently dominated by Australia, the United States and South Africa, but the new suppliers will gradually add to the complexity of such trade. The importance of internationally traded coal in total world energy supplies is likely to increase slowly – in part at the expense of traditional deep-mined production from traditional areas, but in greater part as a result of the increasing use of coal, particularly for power generation as a lower-cost alternative to both the continued use of oil and the expansion of capital-intensive nuclear power. From the viewpoint of supply–demand relationships there is no reason why the price of coal in real terms should increase above its present levels for at least the rest of the century. Thus, given its current price advantage over oil (and over natural gas where, as in western Europe, its price is related to that of oil), it should be able to find expanding markets to serve, provided that the air pollution problems associated with its use can be controlled and reduced (Flohn, 1980; James, 1982).

ALTERNATIVE ENERGY SOURCES

The high oil/energy prices and the supply problems from time to time arising from political and military events since 1973, together with the increasing concern for environmental problems caused by the use of fossil fuels (European Environment Bureau, 1981) have created enthusiastic lobbies and pressures for the rapid expansion of benign energy systems based on solar, wind, water, waste and biomass energy potential (Foley, 1976). To date, however, there has been only a modest official response to this enthusiasm in most parts of the world. This is largely a result of the difficulties which are involved in incorporating such dispersed energy-producing systems into the highly centralized and bureaucratic energy systems which were developed in most parts of the world in the period before 1973 and which most governments have assumed, for the purposes of energy policy planning, must be the basis for organizing future energy supplies. In addition, of course, the technology of producing benign energy remained underdeveloped for nearly three decades after 1945 as there was little motivation for governments or private firms to invest in the necessary research at a time when fossil fuels (plus nuclear power) were thought to be more than sufficient to sustain future energy needs. There thus remain formidable problems to overcome before the production of benign energy sources can be much increased. The relative contribution of such sources of energy to world energy thus remains small at about 4 per cent of the total (even including the long-developed hydroelectricity component) and its share seems unlikely to increase very much at all until after the turn of the century. Moreover, even these longer-term prospects still depend on the investment of increased amounts of research and development funds. This is a development which seems likely to depend more on the private sector in respect of improving prospects for commercial solar power (through so-called solar ponds and as a result of developments in photovoltaics) than on governments, most of which have decided to concentrate their investments in an alternative to fossil fuels on the nuclear power industry (Pryde, 1983).

Nuclear power, indeed, secured the support of most governments, and of all intergovernmental energy organizations, as the best means of reducing dependence on fossil fuels in general and on oil in particular. This expansion has thus been generously, even extravagantly, funded – in part, at least, because it is linked to the development of the major powers' military interests and their military–industrial establishments. Nevertheless, nuclear power remains less important than hydroelectricity in global terms and it still contributes only about 5 per cent to the world's total use of energy. (This is measured, as by the United Nations in its Energy Statistics, in terms of the heat value of the electricity produced by nuclear power. In contrast the pro-nuclear governments and intergovernmental agencies measure nuclear power's contribution to total energy supply in terms of the amount of fossil fuels which would have been required to produce an equal amount of electricity. This approach has the effect of more than doubling the apparent importance of nuclear power.) Officialdom generally remains convinced of the potential importance of nuclear power (with Sweden, Denmark and a few other countries as notable exceptions), but the combination of a number of negative factors, i.e. cost escalation in building nuclear power stations, public concern for safety (whether justified or not is irrelevant) and the lack of sufficiently fast growth in base-load electricity demand to justify large expansions of nuclear generating capacity (which has to be used in base-load configuration to stand any chance of being economic), has severely restricted its rate of growth in most of the non-communist world and has even more severely undermined its prospects for rapid development (Hansen et al., 1988). Therefore it, too, seems destined to make only a very limited additional contribution to world energy supplies over the rest of the century – and beyond.

Figures 3.8 and 3.9 summarize the outlook for world energy developments (except for so-called non-commercial energy which is dealt with in the final section of this chapter) over the period of the next 23 years to 2010. Growth in use will be modest as the historic trend of about 2 per cent growth in global energy use continues to reassert itself, after the short abnormal period of a much higher growth rate between 1948 and 1973. The contribution of fossil fuels will remain dominant in the world energy picture, though (as shown) oil, which enhanced its position very markedly in the 25 years prior to 1973, will lose part of its share of the total market to natural gas and coal. Alternative energy sources, taken together, seem unlikely to be very much more important by 2010 (relative to the increased amount of energy used) than they are at the present time with a combined contribution of about 16 per cent. The expansion of the share of nuclear power is particularly at risk. This is because in the aftermath of the incident at Chernobyl in the Soviet Union, another accident of serious dimensions to a nuclear power station anywhere in the world will very likely lead to the closure of similar nuclear power stations everywhere. This would dramatically reduce the total contribution of nuclear power to the world's total use of energy for at least a generation – and possibly for ever.

THE POOR WORLD'S ENERGY PROBLEM

One important element in the world's energy situation and prospects has so far been virtually excluded from the content of the description and analysis as presented in this chapter. Nevertheless, it is of central importance for very

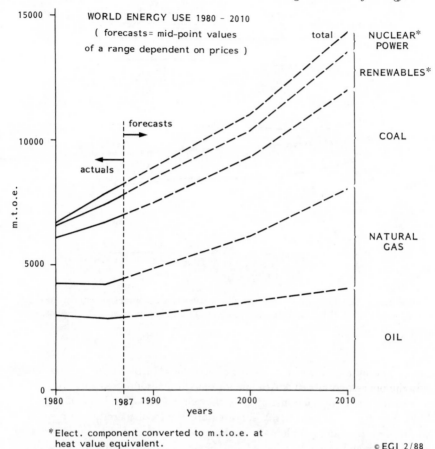

Figure 3.8   The prospects for supplies of the main sources of world energy to 2010 (forecasts are the midpoint values of a range dependent on prices) ©EGI 2/88

* For nuclear power and renewables the electricity component is converted to mtoe at the heat value equivalent

large parts of the world with even larger populations – the supply and use of locally available energy in societies which remain largely or mainly subsistence in their economic and social organization.

The per-capita energy use in such societies is small (it is measured in *kilograms* of oil equivalent rather than in *tons* of oil equivalent as in the world's most industrialized countries), and it depends essentially on the locally available supply of combustible materials for cooking and lighting (and sometimes heating) needs. At the global level, however, this form of energy supply is still large, simply because of the very large numbers of people involved in such societies. It is estimated that it still accounts for about 20 per cent of total world

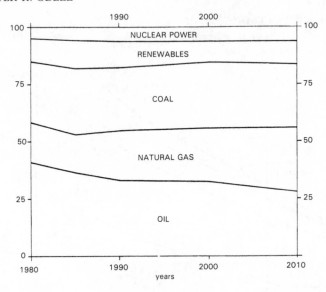

Figure 3.9   The percentage contribution of the main sources of energy to world energy use, 1980–2010 (based on midpoint values for forecasts of world energy use – see figure 3.8)
©EGI 1/88

energy supply overall, but in some parts of the Third World (most notably in the poorest countries of Africa and Asia) its contribution reaches over 90 per cent of the total energy consumed by the populations (Smil and Knowland, 1980; Stevens, 1987). In many such regions the local scarcity of firewood is becoming a pressing problem – a problem which is related to the increasing distances which have to be covered, and hence to the increasing time and effort required, to collect the volume of wood required for basic survival purposes. A solution to this basic development problem requires the implementation of effective 'energy-farming' practices whereby wood becomes an annual crop and new, albeit simple and low-cost, technological developments for enhancing efficiencies in the utilization of wood are developed etc. This is a world energy problem about which little is known (even though it has its parallels in seventeenth and eighteenth century Europe when there was a similar wood scarcity problem) and about which even less is being done, compared, that is, with the attention that is being given to the problems of oil and the other energy sources required for intensive use – often for non-essential purposes – in the developed world, both capitalist and communist, and in the modern sectors of the economies of the developing countries.

References

Adelman, M. A. 1972: *The World Petroleum Market*. Baltimore, MD: Johns Hopkins University Press.

Anderson, J. 1984: *Oil: The Real Story behind the Energy Crisis.* London: Sidgwick & Jackson.

Blair, J. M. 1976: *The Control of Oil.* London, Macmillan.

Cook, E. 1976: *Man, Energy and Society.* San Francisco, CA: Freeman.

Darmstadter, J., Dunkerley, J. and Alterman, J. 1977: *How Industrial Societies Use Energy.* Baltimore, MD: Johns Hopkins University Press.

Darmstadter, J., Landsberg, H. H., Morton, H. C. and Lodu, M. 1983: *Energy Today and Tomorrow: Living with Uncertainty.* Englewood Cliffs, NJ: Prentice-Hall.

Davison, A. et al. 1988: *Natural Gas: Governments and Oil Companies in the Third World.* Oxford: Oxford University Press.

Desai, H. et al. (eds) 1987: Special LDC Issue. *Energy Journal*, 8.

Dienes, L. and Shabad, T. 1979: *The Soviet Energy System*, Washington, DC: Winston.

Dunkerley, J. 1981: *Energy Strategies for Developing Nations.* Baltimore, MD: Johns Hopkins University Press.

European Environment Bureau 1981: *The Milano Declaration on Energy, Economy and Society.* Brussels: European Environment Bureau.

Flohn, J. 1980: *Possible Climatic Consequences of a Man-Made Global Warming.* Laxenburg: International Institute for Applied Systems Analysis.

Foley, G. 1976: *The Energy Question.* Harmondsworth: Penguin.

Fritsch, B. 1982: *The Energy Demand of Industrialized and Developing Countries until 1990.* Zurich: Institute of Technology.

Gold, T. 1987: *Power from the Earth.* London: Dent.

Gold, T. and Soter, S. 1980: The deep earth gas hypothesis. *Scientific American*, June, 154–61.

Gordon, R. L. 1987: *World Coal: Economics, Policies and Prospects.* Cambridge University Press.

Grossling, B. 1976: *Window on Oil: A Survey of World Petroleum Sources.* London: Financial Times.

Guilmot, J. F. 1986: *Energy 2000.* Cambridge: Cambridge University Press.

Halbouty, M. T. 1983: Reserves of natural gas outside the communist bloc countries. *11th World Petroleum Congress.* Winchester: Wiley.

Hansen, U. et al. 1988: Nuclear energy after Chernobyl. *Energy Journal*, 9 (1).

Hodgson, B. 1978: Natural gas: the search goes on. *National Geographic*, November, 132–51.

Hoffman, G. W. 1985: *The European Energy Challenge: East and West.* Durham, NC: Duke University Press.

Hoffman, T., and Johnson, B. 1981: *The World Energy Triangle.* Cambridge, MA: Ballinger.

International Energy Agency 1982: *The World Energy Outlook to 2020.* Paris: OECD.

International Energy Agency 1984: *Coal Prospects and Policies in I.E.A. Countries.* Paris: OECD.

James, P. 1982: *The Future of Coal.* London: Macmillan.

Khan, K. (ed.) 1988: *Petroleum Resources and Development: Economic, Legal and Policy Issues for Developing Countries.* London, New York: Bellhaven.

Leach, G. et al. 1979: *A Low Energy Strategy for the United Kingdom.* London: IIED.

Lenin, V. I. 1966: Report of the work of the Council of People's Commissars and the Eighth All-Russia Congress of Soviets – December 22 1920. In *Collected Works of V. I. Lenin, 31, April–December 1920.* Moscow: Progress Publishers.

Lovins, A. B. 1977: *Soft Energy Paths.* Harmondsworth: Penguin.

Manners, G. 1971: *The Geography of Energy.* London: Hutchinson.

Marchetti, C. and Nakicenovic, N. 1979: *The Dynamics of Energy Systems*. Laxenburg: ILASA.

Odell, P. R. 1975: *The Western European Energy Economy, Challenges and Opportunities*. London: Athlone Press.

Odell, P. R. 1984: Energy issues. In A. M. Kirby and J. R. Short (eds), *The Human Geography of Contemporary Britain*. Harmondsworth: Penguin.

Odell, P. R. 1986: *Oil and World Power*, 8th edn. Harmondsworth: Penguin.

Odell, P. R. and Rosing, K. E. 1983: *The Future of Oil: World Resources and Use*, 2nd edn. London: Kogan Page.

Odell, P. R. and Vallenilla, L. 1978: *The Pressures of Oil: A Strategy for Economic Revival*. London: Harper & Row.

Owen, S. 1986: *Energy, Planning and Urban Form*. London: Pion.

Pachauri, R. K. 1988: Energy and growth: beyond the myths and myopia. *The Energy Journal*, 10(1), 1–20.

Park, D. 1979: *Oil and Gas in Comecon Countries*. London: Kogan Page.

Pryde, P. R. 1983: *Non-Conventional Energy Resources*. New York: Wiley.

Sampson, A. 1975: *The Seven Sisters: The Great Oil Companies and the World They Made*. London: Hodder & Stoughton.

Schurr, S. and Netschert, B. 1977: *Energy in the American Economy, 1850–1975*. Baltimore, MD: Johns Hopkins University Press.

Smil, V. and Knowland, W. E. 1980: *Energy in the Developing World: The Real Energy Crisis*. New York: Oxford University Press.

Stevens, P. (ed.) 1987: *Energy Demand: Prospects and Trends*. London: Macmillan.

United Nations 1982: *Petroleum Exploration Strategies in Developing Countries*. London: Graham & Trotman.

WOCOL Report 1980: *Coal, Bridge to the Future*. Cambridge, MA: Ballinger.

World Bank 1983: *The Energy Transition in Developing Countries*. Washington, DC: World Bank.

4

Food Production and Distribution – and Hunger

P. N. BRADLEY and S. E. CARTER

INTRODUCTION – THE MAGNITUDE OF THE FOOD CRISIS

One of the gravest aspects of the current 'crisis', the subject of this book, is the concentration of hunger in the poorest nations of the world, affecting hundreds of millions of people. The so-called 'food crisis' has been a major issue in discussions about development and the future of the Third World since the early 1970s. Despite increasing attention from governments of the industrialized nations and from the global techno-bureaucracies such as the World Bank and the Food and Agriculture Organization of the United Nations (FAO), little has been achieved to prevent a continuation and even a deepening of the crisis. There are more hungry people in the world at the end of the 1980s than ever before in the history of humankind, despite the fact that enough food is currently produced to feed a global population equivalent to that projected for the year 2000 (Bennett, 1987, p. 18). In this chapter we seek to show why this is so, and why hunger will not be eradicated within the present world economic system.

Global Food Production

World food production is increasing steadily and at a higher rate than world population, so that more food is being produced per capita every year. Figure 4.1 illustrates the upward trends in global production of the three major cereal crops, wheat, paddy rice and maize, between 1970 and 1986, as reported by the FAO. Production of these three crops rose by 70 per cent, 55 per cent and 80 per cent respectively over the period in question (FAO, 1975, 1986).

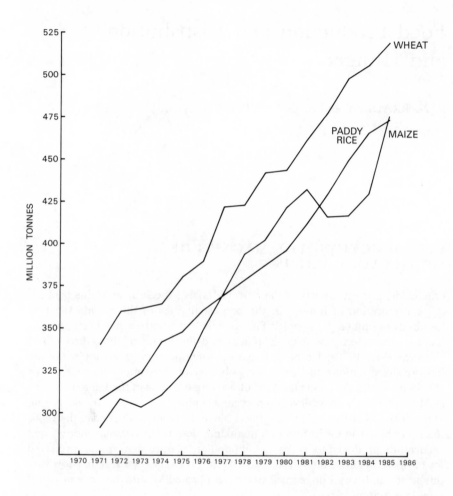

Figure 4.1   World production of major cereals, 1970–1986 (three-year moving means)

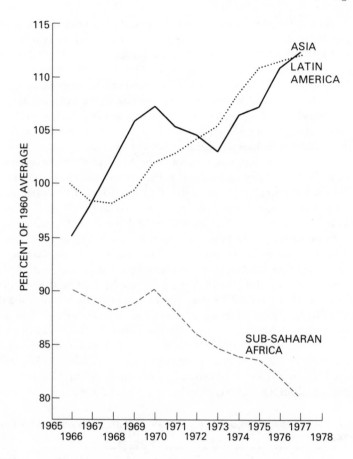

Figure 4.2   Index of per-capita food production for sub-Saharan Africa, Asia and Latin America, 1965–1978 (three-year moving means based on a 1960 index of 100)

Similar trends were observed for other crops and for livestock (although see p. 129 on the reliability of the data used). There is a considerable reserve capacity for further expansion, not only in North America, Europe and Australia but also in South America and a number of African countries (FAO, 1984). Yield levels continue to increase. At the global level, then, there would seem to be adequate food production.

Global Hunger

Trends in food production have been accompanied by increasing hunger and malnutrition. A World Bank study reported that there were 730 million people without enough to eat by the end of the 1970s, and that both the absolute number and the proportion of the world's population who were hungry had risen since 1970. A fifth of these people were in Africa, whilst two-thirds were in South Asia (World Bank, 1986; cited by Lappé and Collins, 1988, p. 43).

At the aggregate level, Africa appears to be the worst affected area (see figure 4.2). Attention has been strongly focused on this region, as a result of the wave of disasters which have affected the continent since the late 1960s: Biafra, the Sahelian drought, the famines in Ethiopa and Sudan, and the wars in the Horn, Angola and Mozambique. To some extent media coverage has led to the problems of hunger elsewhere being overshadowed. Undoubtedly Africa has fared worse than Latin America or Asia in terms of aggregate food production. In 1961 Africa's food self-sufficiency ratio stood at 98 per cent. By 1971 it had declined to 89 per cent, and by 1978 it had fallen as low as 78 per cent. The corollary was that, between 1970 and 1980, food imports for the continent showed an annual increase of 8.4 per cent. Cereal imports for human consumption cost $5 billion per annum. Excluding South Africa, only seven of the remaining 47 states increased per-capita food production during that period, and these states represented only 5 per cent of the continent's total population (again excluding South Africa). Per-capita food production declined throughout the 1960s and 1970s (figure 4.2), a trend which continued during the 1980s (FAO, 1986) and which is in marked contrast with the situation in Latin America and Asia. Aggregate statistics such as those shown in figure 4.2 hide the variations that occur both between and within individual states. Table 4.1 illustrates the different degrees to which a number of African states are affected by the current food crisis. In turn this hides the differences, both spatial and social, in the occurrence of hunger within these states. By extension,

Table 4.1   Selected indices of the food crisis as it affects Zambia, Mali, Nigeria and Tanzania

	Zambia	Mali	Nigeria	Tanzania
Increase in volume of wheat imports 1971–81 (%)	260	508	495	1,004
Increase in value of wheat imports 1971–81 (%)	790	2,133	2,130	4,300
Increase in consumer food price index 1968–78 (%)	n.a.[a]	65	436	163
Change in per capita food production index 1969–71 to 1977–9 (%)	−1	−12	−13	−6
Population with calorie intake below 1.2 basic metabolic rate (%)	34	49	n.a.[a]	35

[a] Data not available.

Sources: FAO, 1981; ILO, 1978; United Nations, 1980; World Bank, 1981

we begin to see how hunger can also exist on a massive scale in the other regions, which are apparently doing so much better than Africa. The forces which have generated Africa's food crisis operate throughout the periphery of the world economy; they are the fundamental processes that govern the global distribution of food consumption.

Hunger in Asia or Latin America is less easily associable with factors such as war, drought or declining per-capita production, evidence that these factors are not the real causes of the African food crisis. India has a third of the world's hungry. In 1985 'surpluses' of wheat and rice in India totalled more than 24 million tons (Lappé and Collins, 1988, p. 8). This is not at all strange for a Third World country, and most are net exporters of agricultural produce. In Latin America there are no signs that hunger is becoming any less common: 40 per cent of Brazil's population is malnourished, and the corresponding figure in rural Mexico has been as high as 80 per cent in recent years (Burbach and Flynn, 1980, p. 105). Most countries in Latin America, and to a lesser extent Asia, have large proportions of their populations in cities, so that hunger is more of an urban phenomenon than in Africa. As we shall see, the forces which generate hunger are the same in both rural and urban areas, and although famine may be less likely in the more urbanized countries, the end result is essentially the same.

SOME MYTHS DISPELLED

A number of unsatisfactory explanations have been put forward to account for the continued increase of hunger and poverty in the world. Such myths need to be overturned if the underlying causes of the problem are to be revealed. As they are currently presented, the 'liberal' explanations of the global food crisis do little more than rephrase the questions that they seek to answer. Far from offering a genuine explanation, they draw attention away from the causes towards the symptoms.

Overpopulation is the Problem

A popular explanation for the food crisis is overpopulation. As we have seen, there is no shortage of food at the global level, even allowing for overconsumption in the industrialized nations. Nor can overpopulation be blamed for hunger in specific countries or regions. The incidence of hunger bears no relationship to population density; at the national level, malnutrition is a problem in neither Holland nor China. The issue of population is discussed more fully in chapter 6. Two points are worth making here, however. Firstly, 'overpopulation' as a concept must be discussed in relation to the technical, political and economic context of the society in question; the amount of people a given area can support depends on the way in which production and the distribution of the benefits are organized. The Chinese case is a useful example. As a country of 500,000,000 people, it experienced famine somewhere nearly every year. In the mid-1970s it furnished over 2,300 calories a day to a population of 800,000,000 (George, 1976, p. 59). Whilst continuing to grow, its population has become better fed. Secondly, programmes of birth control, seen by many as the panacea for high population growth rates, cannot be effective if the underlying causes of high fertility rates are not addressed. Birth control policies are misguided because they attempt to tackle the problem of food consumption and not food distribution. For further discussion and critiques of the myth of overpopulation see chapter 6 of this volume, Cassen (1976), George (1976) and Lappé and Collins (1982, 1988).

Climate and Natural Hazards are to Blame

As a result of events in the Sahel and Ethiopia since the mid-1970s, droughts are possibly the phenomenon most associated with famine and hunger. The attention of the western media has undoubtedly helped to foster such an attitude. North-east Brazil has suffered recurrent droughts throughout the same period, but these have received less coverage. Excess rather than lack of water provoked the famine in Bangladesh in 1974. Nevertheless explanations have been sought in long-term climatic trends and their interaction with desertification which in turn, it is supposed, has resulted from overpopulation.

Climatic fluctuation is not new. Traditional societies coped with drought through a number of mechanisms including grain storage after good harvests or, in areas particularly prone to drought, nomadism. Few observers ask why these strategies have been disrupted, or why rural peoples overexploit the natural resources upon whose maintenance they depend. The answers lie not with overpopulation or overstocking but with the subjugation of traditional livelihoods under colonial rule, the increasing pressure from commercial agriculture on resources once held by the community for common use and the increasing inequality of access to the means of production and basic services. Natural hazards such as droughts cannot be blamed for anything but the aggravation of problems whose causes are largely economic, social and political. Whilst the drought in the American Midwest in 1988 caused extensive crop failure, nobody, at least not in the United States, was expected to starve. By far the majority of deaths from hunger and malnutrition each year, which are mostly amongst children, have nothing to do with the weather. They continue day in, day out, even during the best of harvests.

Traditional Agriculture is Incapable of Producing Enough Food

It is now well established that peasant agriculture is highly productive, much more so that on the large estates which characterized the 'White Highlands' of Kenya or the Latin American latifundia (World Bank, 1975). Traditional agriculture has the potential to produce adequate supplies of food without being 'modernized'. In terms of the energy inputs required it is a far more efficient producer of calories than modern mechanized agriculture in the core, the latter's being viable only through energy subsidies in the form of cheap oil supplies. Failing productivity in traditional agriculture, as a result of overuse of the soil and erosion, is not an inherent characteristic. It is merely a further symptom of the problem, with its causes rooted firmly in the workings of the world-economy (Blaikie, 1985).

Science is the Solution

Following on from the above, another commonly espoused argument is that a transformation to western-style modern agriculture, with its strong scientific basis, is the solution to the problem of world hunger. Agricultural researchers have achieved spectacular yield increases for the major grain crops, and it is argued that this can increase food supplies and raise farmers' incomes. As we have seen, food supply is not the problem: although a continued increase in world population may require that food production be increased, there is currently enough to go round. What then of the income issue? Crop research aims to benefit farmers, whereas many of the rural poor are landless. In theory, more intensive cultivation increases the demand for labour, but it has been shown for the so-called 'Green Revolution' that the landless do not benefit from this. Instead, the introduction of a package of seeds and inputs to increase yields has often led to increased mechanization and the displacement of labour. The reason is that many absentee landlords and middle-class urbanites have taken advantage of the higher profits to be made, rents and land prices have increased as a result, and many smallholders have been evicted or forced to give up rented land (Pearse, 1980). In other words, the Green Revolution has aggravated rather than ameliorated the problems of landlessness, and the increased income generated has accrued largely to the better off.

New agricultural technology which is introduced into unequal societies will only increase inequality; this is as true for the United States, as that country's current farm crisis demonstrates, as it is for any Third World nation (Lappé and Collins, 1988; see Gotsch, 1972, for an example from Pakistan). During the last decade or so international agricultural research has focused more strongly on smallholders' crops and on low-cost innovations in an attempt to raise the incomes of poor farmers. Nevertheless, such technology cannot be biased towards the poor; it is available to anyone who wishes to use it – more so to those with greatest access to information and the advice of extension services. The fact that this research is helping to make crops traditionally cultivated by peasants (such as tubers and pulses) commercially viable may well favour their adoption by those who have access to land and capital and are able to produce most cheaply, and thus increase the inequalities of income distribution.

Increases in food production have not benefited the urban poor significantly. In the countries which were hailed as the Green Revolution success stories in South and South-east Asia the extra food production merely helped to reduce imports: per-capita food consumption in India is no higher now than it was 20 years ago (Lappé and Collins, 1988, p. 43). There are exceptions: Colombian rice production increased dramatically in the 1970s, and rice prices to consumers fell in real terms. Against this apparent success, however, must be weighed the significant decline in labour use which accompanied the intensification and mechanization of production, and the loss of income amongst peasant upland rice producers which resulted from the introduction of new irrigated varieties (de Janvry and Dethier, 1985, pp. 72–3).

Whatever the current capacity of agricultural technology to increase food production, the continued trend towards industrial farming will increasingly involve smallholders in the market, making them more dependent upon economic factors over which they have no control. They must buy product 'X' to ensure the success of crop variety 'Y'. Whilst researchers working on peasant agriculture have sought to reduce the need for pesticides and herbicides, the aggressive marketing strategies of western agrochemical firms are having a very large impact upon farmers' decision-making. Production is becoming more inefficient with regard to the energy required to produce a calorie of food. The environmental pollution, soil compaction and erosion which result from heavy use of agrochemicals and machinery imperil the sustainability of food production (WCED, 1987; Redclift, 1987). Yet there have been few attempts to build upon traditional agricultural practices – the accumulated knowledge of many generations of farmers – to grow food in ways which are ecologically more sound. Far from being the solution to hunger, science, as currently applied to food production, is contributing to a sharpening of the crisis.

Food Distribution Systems are Inadequate

While there is little doubt that, at a technical level, the transport systems of the Third World lack the speed and efficiency of those in the core, there is nevertheless a wide range of goods that circulate through peripheral states. Even a state such as Niger, frequently diagnosed as too far from the coast to allow development, somehow manages to export uranium from deep within the Sahara Desert. It seems that raw materials inevitably find a way out, while consumer imports such as torch batteries, matches and carbonated drinks are able to penetrate to the most isolated regions. It is false therefore to claim that food cannot circulate in the same fashion. Where we see the most sweeping assertions of transport failure is in the context of emergency relief, in which massive quantities of food need to be transported at great speed. Obviously such extreme demands cannot be accommodated, but it would be wrong to argue that unusual circumstances of this nature demonstrate a more general failure. The fact remains that transport networks are able to cope with demand at a more continuous and steady level. For average conditions, then, we can argue that the lack of modern transport technology is not a significant impediment to the transfer of commodities, and is still less an acceptable explanation for patterns of food distribution.

Many putative explanations are offered for the continued growth of hunger: disasters, underinvestment and a lack of modernization, peasant conservatism and administrative inadequacy could be added to the above list. The point to be made is that they all fail to get to the root of the problem. Their focus is the visible 'face' of the food crisis, not the cause. Thus when the wretched overcrowding and evident malnutrition in the slums of Calcutta or São Paulo claim public attention, the quick answers are that there are too many people for too little food, that a 'western-style' agricultural modernization programme is needed, and so on. But these are not explanations, and such pronouncements fail to get to grips with the problem.

THE FAILURE OF FOOD PRODUCTION – CAUSES AND CASES

Purchase or Production?

Essentially, people throughout the world can obtain food in two ways. They can either grow it or buy it. It is with the factors which control their ability to do one or the other that the real causes of hunger lie. To grow food, people must have access to the means to produce it, i.e. land, water, seed and tools, and be able to retain enough of what they produce to meet calorific requirements. Access to the means of production does not necessarily imply that they will satisfy their biological needs, since they may have to pay a proportion of what they produce as tribute or sell some to pay taxes. In South-east Asia high tax levels in the 1930s contributed greatly to famines by preventing peasant rice growers from storing grain for times of scarcity (Scott, 1976).

In the industrialized countries most farmers no longer grow their own food; rather, they purchase it like everybody else. In many Third World countries food production for subsistence is still an important aspect of farming. Hyden (1980) in particular argues that a 'peasant' mode of production still pertains, at least in Africa, and that a central feature is food production for internal consumption. Few, if any, agricultural societies have been totally self-sufficient. Moreover, for many centuries these pre-capitalist formations have engaged in some form of exchange, both internally and to a limited extent with trading partners. In general, farmers in Latin America are more orientated towards cash crop production than are African farmers. What is significant is that growing food for subsistence is everywhere becoming a less important component of production. The few remaining elements of an autonomous economy are rapidly being replaced by a more complete array of capitalist relations of production and exchange.

Those who cannot or do not produce sufficient food for subsistence must purchase some or all of it. As subsistence economies give way to those founded on commerce, the pattern of income distribution becomes a critical factor in

Table 4.2   Pattern of income distribution (share of total household income) in a selection of states from Africa, Latin America and Asia

	Accruing to poorest 20% of population (%)	Accruing to richest 20% of population (%)
India (1975–6)	7.0	49.4
Indonesia (1976)	6.6	49.4
Philippines (1970–1)	5.2	54.0
Malaysia (1970)	3.3	56.6
Malawi (1967–8)	10.4	50.6
Tanzania (1969)	5.8	50.4
Peru (1972)	1.9	61.0
Mexico (1977)	2.9	57.7
Brazil (1972)	2.0	66.6
Argentina (1970)	4.4	50.3

Source: World Bank, 1981

the provision of food. Food distribution depends on the ability to buy, not on physical infrastructure (which in any case is readily accommodated to 'market' demand). Income distribution within most Third World societies is extremely unequal. Table 4.2 shows a number of examples. Large sections of the urban and rural populations have too small a share of wealth to be able to meet their needs through purchase. Put simply, those who are poor cannot buy what there is on the market.

Amongst those who produce crops primarily for cash and buy food, their income, and hence their ability to feed themselves, depends on how much they receive for what they grow. As we shall see, smallholders have no say in price determination. They are little different from wage labourers – the landless rural and urban proletariats; their wages are dependent on what the market requires, with the result that many receive insufficient returns for their labour even to provide an adequate diet. Inequitable allocation of the results of production is a necessary functional component of the present world economic system. It cannot be wished or even programmed away. Without it the world-economy would be totally transformed.

Agriculture within the World-Economy

To understand more fully why rural populations in particular neither produce enough food nor have enough money to buy it, we must examine agriculture and its changing role in the world-economy in greater detail. As a starting-point from which to consider the implications of agrarian transition for the mass of rural people, we shall first distinguish between two poles, modern and peasant agriculture. The former represents production in the core, the latter that in the periphery. As we shall see, the distinction is not entirely valid, since 'modern' agriculture is present throughout the periphery.

In the core, agriculture is essentially another form of business no different in motivation from any other industrial enterprise. The production of wheat, barley, maize and other bulk grains in the United States, Canada and the European Economic Community (EEC) has for its first and fundamental rationale that these are commodities to be sold on the open market. The agribusinesses of the core operate just as if they were producing industrial consumer goods. Indeed, if we forget for a minute that people cannot eat light bulbs, there is no real difference in motivation, and the commercial decisions that go with it, between an electrical components company and a large scale farm enterprise. The products of both are for exchange, and the profit that goes with it, rather than for immediate and essential use. In the context of the North American wheat belt, the product has value above all else because it can be sold or exchanged on the market, and not at all because it can be eaten. The fact that it will eventually be consumed as food plays no part in the commercial decisions governing its production.

Peasant agriculture, in contrast, has as its logic the survival of the family unit – the provision of adequate subsistence, clothing, shelter and so on. Peasants have a great deal of experience with the crops which they cultivate, which can be described as traditional. Traditional values and practices are adhered to in the face of a rapid integration into a capitalist economic system. Decisions taken do not reflect profit-maximizing goals, but rather long-term survival in the face of uncertainty (Scott, 1976, p. 18). Associated with a peasant mode of production, and essential to its continued viability, are institutions such as common land and rights to the exploitation of resources, usually managed collectively.

While in appearance and motivation peasant and modern agriculture are distinctly different, they nevertheless have certain functional qualities in common. In both cases agriculture takes place in a wider sphere of economic activity and cannot be considered in isolation. This is increasingly true for both the periphery and the core. In the 1950s writers spoke of a dual economy (Boeke, 1953): a modernized urban–industrial sector, counterposed with a functionally isolated archaic rural agriculture. Such a view can no longer be sustained. Just as in the core, so too in the periphery: the independence of the two systems is illusory. A functional articulation is seen to exist. Frank (1969) writes of a cascade of relations, in which surplus is extracted from the peasantry (in the form of products, cash or direct labour power) through a sequence of intermediaries and directed to the urban centres. The relations are seen most easily where the two systems are juxtaposed. Peasants' access to land is squeezed by capitalistic producers, so that the former are forced to become increasingly dependent upon wage labouring (Taussig, 1982). However, the same is true at the global level whereby the poor ensure a continued supply of cheap labour and raw materials. A chain of linkages ensures a close integration between the two erstwhile elements of the dual economy. The most dramatic of these linkages is that of migration of rural workers either to shanty towns or, as in the case of Colombia, from one specialized crop production region to the next, providing commercial farms with cheap labour throughout the year.

The periphery is a necessary component of the capitalist world-economy. The specific social formations of the periphery influence the way in which the development of capitalism is shaped therein, and the social and production systems which are the outcome cannot be entirely foreseen (Watts, 1986). However, to a greater or lesser extent peripheral states and social formations are all strongly influenced by the political and economic demands of the core. The terms of trade, wage levels, the use of the dollar as a global currency standard and the monopolistic control of transnational companies (TNCs) all bear witness to this integration. Simply because rural societies are physically remote from the centres of economic and political power, there is no reason to suppose that they are somehow structurally removed from these global influences. Third World farmers and those who have lost their ties to the land are participants in the continuing evolution of this capitalist world-economy, and to that extent the visible differences between modern agribusiness and traditional farming cloak a more fundamental unity. We must therefore search for an explanation of the global food crisis within this context. As illustrative material we shall compare the advanced agrisystems of the core, exemplified by the North American wheat-producing systems, with peasant agriculture in two countries, Kenya and Paraguay. We must also examine the way that the systems of the core are reproduced where conditions are favourable in the periphery, since this has profound implications for peasant agriculture, food production and income distribution.

Modern Agriculture in the Core

A recurrent theme of discussions dealing with the global food crisis is, of course, the 'bread-basket' of the core: the grain belt of North America. Much is made of the role of US wheat production in feeding the world's hungry, and more particularly in exerting a level of control over the global food market. Robbins and Ansari (1976), George (1976, 1979) and Lappé and Collins (1982) all point to the fact that wheat production in the United States, and now in the EEC, is determined by both profit-motivated decisions and domestic and foreign-policy political considerations. As we have observed, in the core agribusiness is, as the term suggests, business. More than by climate or other environmental variables, fluctuations in production are determined by US price-support policy, which is predisposed to ensure a high world market price.

Thus in 1972, the crisis year of the Sahelian drought in which starvation was widespread, the US government paid farmers $3 billion to take 50 million hectares out of production (Robbins and Ansari, 1976). The intent was quite clear. In order to remedy the 'glut' of previous years, which saw the grain reserve rise to 49 million tons by 1972 with a commensurate fall in the world price, the US government determined to reduce production and create an effective shortage, and thus raise the world price again. In this case the price of wheat (US No. 2 f.o.b. Gulf) on the world market increased by 236 per cent between 1971 and 1974. Corresponding increases for rice (Thai White f.o.b. Bangkok), maize (Yellow No. 2 f.o.b. Gulf) and soya bean (US c.i.f. Rotterdam) were 347 per cent, 117 per cent and 464 per cent respectively (data from Robbins and Ansari, 1976). It has been estimated that the lost production through this policy would have averted the Sahelian crisis with plenty to spare. The deliberately induced shortage was exacerbated by sales to the USSR, not, it seems, through deliberate government action but via the combined efforts of a number of individual Russian buyers and the sales pitch of the controlling corporations in the United States. The wheat that could not be made available to the starving Sahelians instead went to feed cattle so that the supply of meat to the urban centres in the USSR could be increased.

It is worth drawing attention to several key issues in this now infamous episode, for they are constantly recurring elements in this discussion. First, we can observe the extent to which agricultural production and distribution is determined (through control of marketing and transportation) by a small number of giant corporations (in this instance the 'big six' – Cargill, Continental, Bunge, Archer–Daniels–Midland, Peavey and Cook – handle 90 per cent of all the grain shipped in the world; Hightower, 1975). Monopolistic tendencies exist in agriculture just as they do in industry, again indicating their essential synonymy. Second, the level of production was determined by government policy. In essence it was a clear attempt to buy votes, through ensuring a price structure designed to satisfy Midwest farmers. The shortage had little if anything to do with the 'free-market economy'. Third, and despite an apparent shortage, adequate supplies were made available to meet an economic rather than a physical demand, in this case Russian feedstocks. Fourth, the destination of this supply remained in the core, which by global standards is already more than adequately nourished. Moreover, this extra supply was destined for luxury meat (an acknowledged inefficiency in the utilization of grain-based protein and energy). The wider context is that in part, if not *in toto*, these observations might equally apply to dairy, sugar and meat production in the EEC throughout the 1970s and 1980s, or to sugar, tea and coffee in tropical Africa, or to beef production in the Amazon. In all these cases the driving force is the capitalist nature of the world-economy. The provision of food for the hungry has little to do with such global processes.

Peasant Agriculture in the Periphery

As an initial premise, we can conceive of peasant farming as being concerned with small-scale labour-intensive food production. By extension, it would appear that the commodification of agricultural production has yet to reach the level and extent evident in the core. Traditional practices, derived from a peasant mode of production, are adhered to in the face of a rapid integration into a capitalist economic system. However, traditional foods such as maize are becoming increasingly commercialized, both by peasants and by large-scale producers. In addition, peasants are turning more and more to the production of crops which are either inedible (such as tobacco, cotton and coffee) or which are not their staple foods (vegetable oils, soya, groundnuts, fruits and vegetables). Prior to the Second World War there was a much stronger division between cash and food crops. In East Africa, for example, commercial crops were grown on either colonial estates or plantations. In other parts of the continent, peasant production of cash crops (groundnuts in West Africa for example) had begun to emerge but, within limits, peasant-based production systems were more or less intact. The same can be said for countries such as Mexico and Chile upon the break-up of the haciendas. Elsewhere in Latin America peasants subsisted alongside latifundia and provided them with unpaid labour, either in exchange for the use of land or through coercion. Plantations were the main users of wage labour. Prior to the middle of the century the independent access which peasants had to land was an important determinant of the availability of labour for both the haciendas and commercial agriculture.

Since the war peasants have been increasingly incorporated into national and global politico-economic systems, as capital has realized the potential of smallholder production. As a consequence we now witness the commercialization of peasant food crops such as maize, rice and manioc. In addition, peasants have been firmly drawn into the production of industrial crops and other inedibles. This has had strong implications for domestic food production, although it makes no difference as far as the needs of the market are concerned. It is part and parcel of the continuing expansion of industrial agriculture from the core to the periphery – modernization no less. Capitalist relations of production are penetrating even the most isolated rural communities. The current world food crisis must be interpreted thus, in terms of the way that peasant families are suborned to a global capitalist system.

It is the contention here that peasants, by becoming integrated into a global process on which they have little or no influence, have lost control of their means of production and in fact have very little freedom of action; from this perspective we can perceive the causes of the food crisis more clearly. The specific processes by which this has come about are well documented elsewhere: for example, Franke and Chasin (1980) on West Africa, Leys (1975) on Kenya, and Frank (1969) and Pearse (1975) on Latin America. At the risk of simplification we may mention the following: the enforced monetization of peasant production processes by colonial taxes; the resultant need to grow commercial crops in order to generate that money; the control of credit and other inputs, and of marketing, first by chartered companies and later by transnational corporations; the burgeoning urban populations whose political quiescence is bought through a cheap food policy at the expense of rural people; the requirement on the part of the state to earn foreign exchange through export crops; the alienation (and commercialization) of land, coerced labour, and later the development of wage labour and thus the transformation of peasants to proletarians, and so on. In fact we see the gradual commodification of all factors of production as well as the produce itself – of land, of labour and, of course, of crops. All these pressures have been experienced in one way or another throughout the periphery (the particular sequence and combination depend upon individual histories and circumstances). The net result is that, for the most part (but again dependent on the same individual circumstances), the agricultural societies of the Third World have been integrated into the capitalist world-economy. The question of whether individual farmers can themselves be considered as capitalists is not at issue. The fact remains that they operate in a social formation increasingly dominated by capitalist relations of production, and that these relations are expressed through national and international processes beyond their control.

Smallholder production in Kenya   As a first example we consider the smallholders of Kenya, where a number of factors combine to force the farmer's hand. For one thing, the state is almost totally reliant on agricultural products to pay or substitute for imports. As a proportion of Kenya's total exports, agricultural products averaged 64 per cent between 1976 and 1980. In one year, 1977, they contributed 78 per cent, with tea and coffee contributing 67.8 per cent (Barve, 1984). Accordingly, through specific programmes of assistance, farmers are encouraged to cultivate crops such as tea, coffee, sugar, tobacco and pyrethrum. All these export commodities divert resources from food production.

Further, Kenya has been a prime focus for capitalist modernization since the 1950s, with all that that entails in the way of the penetration of market forces. Land has been privatized, following the Swynnerton Plan of 1953, thus making it a capital (and thereby commercial) asset. The intention, in addition to the obvious one of furthering the commodification process, was to enable the peasant to offer security for a loan, and so invest in the enterprise. By so doing, one of the theoretical constraints of improved productivity was, in principle, removed. Of course the loan had to be repaid in cash, obliging the peasant to cultivate commercial crops.

Finally, through the economic policies followed by the state (which actively facilitate the release of market forces with few or no safeguards), modern Kenya is dominated by a rapid circulation of money which is exchanged for virtually all material needs. In order to survive, Kenyans of all occupations (perhaps with the single exception of some of the pastoralist groups) have a pressing need to generate money. One of the principal means is to release labour from 'subsistence' agriculture on to the free labour market. Hence we see a classic case of active men deserting the rural areas in search of wage employment in the cities, leaving their families behind to maintain the farms. In order to reproduce this system, and particularly with the hope of improving on these dim prospects, education is perceived as the only available route. This too demands a considerable monetary outlay, and further deepens the crisis. At the cultural level, modernization also incorporates western attitudes, and thus we see the wholesale adoption of western clothes, corrugated iron roofs and a whole panoply of consumer items, all available only through cash purchase.

All these processes of change bind the farmer to a system which requires the production of goods which can be sold on the open market, in this case involving the replacement of food crops for those with export potential. It is therefore not surprising that, when viewed on the national scale, the production of Kenya's staple food crop – maize – is failing to keep pace with population growth. Shortages of this nature have little if anything to do with technical incapacity, peasant conservatism or even environmental pressures such as declining soil fertility, soil erosion, desertification or whatever. The reality is that Kenyan peasants are simply less and less able to apply their resources to the production of food.

The food crisis amongst smallholders in Paraguay   Many of Paraguay's smallholders have increasing difficulty in meeting subsistence requirements. In 1985–6, unnoticed by the world's press, drought precipitated acute food shortages in the south-west of the country's Eastern Region. The causes lay neither with environmental problems nor with overpopulation; in 1982 Paraguay, roughly one and a half times the area of the United Kingdom, had a population of just over three million. As in most countries in Latin America, the distribution of land is grossly unequal. At the last census, taken in 1981, farms of 1000 hectares or more comprised 0.92 per cent of the total number of farms whilst accounting for 79 per cent of the censured land area. Fifty-six per cent of farms were smaller than 10 hectares and in total occupied less than 2 per cent of the land (Ministerio de Agricultura y Ganadería, 1985). Land reform has been effectively resisted by the landed ruling elite. The availability of land is not the sole determinant of food production, however. Many farmers with 3 or 4 hectares of land do not grow enough food. Peasants face an acute shortage of cash which becomes readily apparent when either food is in short supply or medical attention is required; the latter is a strong drain on cash, since few rural areas have any medical services and the costs of just travelling to an urban area for treatment, as well as the costs of the treatment itself, are high. Production of cash crops, particularly cotton, is encouraged by the government, which gives credit to landowners. Many peasants do not own the land which they work, however. Shortage of land forces a farmer to make a choice between food crops and a cash crop such as cotton. Whilst cotton offers the possibility of earning enough money to buy food, prices, which are determined in London, are notoriously unstable. They are kept locally low by middlemen, on whom peasants who do not own land are dependent for unofficial credit at the beginning of the planting season. The high rates of interest charged and the low prices paid by middlemen often prevent farmers from breaking even. Once they become indebted, farmers are trapped and must continue to sow cotton each year, forfeiting food production. Thus they can neither grow enough food nor buy it.

The situation is aggravated by shortages of labour. Labour has been increasingly attracted by temporary employment opportunities on two large dam projects on the Paraná river, and by the construction boom that these schemes have fuelled in Asunción, the capital. In rural areas, cash payment has replaced the traditional practice of exchanging labour. This has made it increasingly difficult for farmers to obtain labourers, and has reinforced the trend towards cash crop production in order to provide payment because of their shortage of cash.

Most peasants are still confined to the degraded soils of the Central Region around the capital. More fertile lands along the Brazilian border which have only recently been colonized have mostly been sold by members of the government and armed forces, to Brazilian land companies and other multinationals such as Gulf and Western (Nickson, 1981; Miranda, 1982). The Eastern Border Region has been incorporated into the Brazilian soya belt, of which more below. Ostensibly, colonization of the country's vast reserves of subtropical forest from the 1960s was to provide land for the peasantry; in practice it has gone to the highest bidder. As a result, the Paraguayan peasantry has been converted into a vast reserve of labourers in the service of local and international construction firms and agribusiness, increasingly dependent upon cheap wage foods – rice and wheat products – and increasingly hungry as a result.

The situation is not atypical of Latin America. Similar developments have occurred in north-east Brazil, Mexico, coastal Ecuador, Colombia and Guatemala, to name but a few examples. Cultural and social differences fail to mask the essential similarities in the process of incorporation throughout this continent and the rest of the periphery.

Commercial Agriculture and Agribusiness in the Periphery

Industrial agriculture is currently expanding rapidly into the periphery. Agricultural export crops have dominated the economies of most Third World nations since the early colonial days: groundnuts, oil palm, cocoa, coffee and tea in Africa; coffee, beef, sugar, cocoa and bananas in Latin America; rubber, tea and jute in Asia. Since independence, however, there has been a renewed penetration of agriculture by agribusiness, encouraged by technical developments in tropical agriculture. The markets for most agricultural produce are not in the Third World. In fact the market for foodstuffs in the Third World is extremely limited. This does not hinder capital's interest in the cheap land and labour available there; what matters is how profitable production can be. The fact that agriculture is no longer even concerned with food production is irrelevant; the trend started with the production of luxury beverages and fibres and vegetable oils for the industrial markets. If these markets now require flowers and exotic fruits, then that is what they get. Much of the renewed penetration of peripheral agriculture takes the form of contract farming, whereby the farmer grows what the company requires. Effectively, the farmer leases both land and labour; the farmer bears the risks and the company takes the profits. A classic example of this type of venture is described by Feder (1978) in his analysis of the strawberry scheme in Mexico.

The effects of this increased penetration on peasant agriculture and food production for local needs are grave. Large-scale commercial enterprises corner the benefits of credit and irrigation schemes, and can influence government policies on subsidies and imports, as Crouch and de Janvry (1980) have shown for the countries surrounding the Caribbean. 'Capitalist' crop production in the region has grown markedly, whilst 'peasant' crops have stagnated. The political power of local capitalists is a major determinant of this pattern of growth (see Hewitt de Alcantara, 1974, for an analysis of the Mexican case).

Thus, the gains in production of these capitalist [crops] are due precisely to the fact that they are produced by capitalists. And this, in turn, reinforces capitalist social relations in these crops. (Crouch and de Janvry, 1980, p. 10).

The technological change which accompanies increased capitalist penetration in agriculture contributes to the reduction of peasants' choice and control over resources. Peasant crops are pushed to more marginal lands, production declines and prices increase. More and more peasants are proletarianized and become purchasers of wage foods produced by capitalists.

Where capitalist agribusiness is concerned with food crops for domestic use, these are wage foods such as rice, whose production depends on a careful balance between government policies on food imports and subsidies on inputs such as fertilizers, pesticides and insecticides. Governments must keep wage-food prices as low as possible to maintain the political acquiescence of the urban proletariat, as well as to keep wage levels as low as possible. Cheap wage foods are neither produced nor intended by governments to alleviate the problems of rural and urban hunger and malnutrition. If prices were reduced to such a level, or even only to that prevalent in the world market, capitalists would not produce these crops. Subsidized production and imports of cheap grain merely increase the squeeze on peasant producers who swell the ranks of migrant rural labourers and increase the flow to the cities. The result is to increase rather than reduce hunger.

Agribusiness has expanded dramatically where cheap land has been available. This has been common in Latin America, from where examples are drawn here, but, as suggested above, opportunities have been seized upon in many African countries. The internationalization of agribusiness, and particularly its spread to the Third World, is one of the main features of the world food economy. TNCs have a central role in this process, which determines how food is distributed and shapes the lives of millions of people in the Third World who depend on agriculture for a livelihood. In Latin America intense capitalist development has created conditions conducive to the rapid expansion of agribusiness. Tens of millions of peasants have been pushed out of agriculture as a result (Burbach and Flynn, 1980, p. 14).

The case of soya production in Brazil is illustrative. Since the mid-1960s the number of people living solely off the land in Brazil has decreased by 50 per cent, and the rural population now constitutes only 30 per cent of a total of 135 million (Bennett, 1987, p. 144). One of the causes is the growth of commercial agriculture and the concentration of land in the hands of fewer and fewer landowners. The southern states of Paraná and Rio Grande do Sul have been the location for a massive increase in soya production, encouraged by government incentives such as cheap loans to promote the production of export crops. Brazil is now the world's second largest producer after the United States. This high-protein crop contributes nothing to the diet of Brazil's malnourished population. Smallholders who cleared the land in the 1950s have been ousted by large commercial farms run privately or by TNCs. Agribusiness is less interested in the production aspect than in the supply of inputs and in processing and sales: US grain companies invested more than $50 million in soya bean processing facilities in Brazil between 1973 and 1980 (Burbach and Flynn, 1980, p. 10).

Rising production costs and falling world prices brought about a growth in the average farm size in the soya belt. Small-scale producers were squeezed out, adding to the ranks of unemployed workers who were displaced as production became increasingly mechanized. Many of these people went either to eastern Paraguay, where soils were as fertile but the land was not yet cleared, or to Amazonia. In the former case, after purchasing and clearing land or after buying out poorer Paraguayan colonists, Brazilian farmers have again begun to be displaced as large Brazilian and foreign concerns move into the Paraguayan soya zone. In Amazonia, large-scale cattle production has taken over on the land cleared by peasant colonists, with state credit and subsidies and World Bank loans again favouring wealthy ranchers and TNCs (Fearnside, 1987; Burbach and Flynn, 1980). Livestock production, like soya, is not intended to enrich the protein-deficient diets of the poor, but rather to supply the wealthy and the core with additional and unnecessary beef. The apparently healthy food production situation in Latin America (and Asia) suggested by figure 4.2 must be reinterpreted. Most of the increase is due to the expansion of agribusiness and large-scale capitalist farming. It has contributed little or nothing to a reduction in hunger.

The above examples from Latin America illustrate a more direct influence on peasant food production and livelihood than incorporation into the world-economy – one which tends to speed up the process of depeasantization, particularly in regions where conditions are favourable for the establishment of commercial agriculture. Whilst these trends are not as advanced in Africa, we can observe their progress. The end result can be foreseen without too much difficulty: the continued penetration of areas of peasant food production, and an acceleration in the rate of its decline.

Hunger: the Irrelevance of Need in the Capitalist World-Economy

The processes which we have described, and which chart the integration of commercial agriculture and the expansion of agribusiness in the world-economy, suggest that the prospects for adequate provision of food in the periphery are poor. Three issues have emerged in the discussion. There is a continuous and persistent diversion of agricultural activity away from the production of locally needed food crops. The cash crop economies that have taken over are vulnerable to the dynamics of a wider market over which peasants have little if any control. The underlying function of the transformation of peasant agriculture is its commodification and the proletarianization of the majority of the peasantry itself.

If food production for immediate consumption continues to decline, the gap can only be filled by purchase, but in an economic or commercial sense the markets for food products do not include the needy. Thus we see the paradox of increasingly severe grain shortages in the Third World at the same time as we see an increasing proportion of total global grain production diverted to animal feeds. This grain comes not only from the core but also from the industrialized production systems of the periphery. Thus in Mexico, 'more basic grains are consumed for animal forage than are consumed by 20 million peasants' (quoted by Frank, 1981, from *International Herald Tribune*, 9 March 1978). If not the product of agricultural land, then the land itself is diverted from food production to animal pasture. In Central America, beef production increased at an annual rate of 5 per cent between 1962 and 1975 whereas beef exports accelerated at a rate of 18 per cent per annum (George, 1979). In Costa Rica per capita beef consumption declined from 49 lb in 1959 to 33 lb in 1971, yet in 1965 60 million pounds of beef were exported to the United States (Lappé and Collins, 1982).

The literature is full of such examples. The conclusion is that food is destined for markets that can afford it, and not for where the nutritional need is most evident. The locus of product is immaterial as long as costs are minimized and a profitable sale can be made. As one US rancher commented on the costs of production, 'Here's what it boils down to – $95 per cow per year in Montana, $25 in Costa Rica' (quoted by Lappé and Collins, 1982, p. 203). Thus in the midst of hunger, food is exported for profit. If consumers in the core are prepared to pay more for meat than peasants or proletarians in the periphery can afford for grain, then it should come as no surprise that the market fails to include the poor. In the context of the capitalist world-economy demand is only effective when expressed in terms of available purchasing power. As expressed in profit-based exchange relations the economic demand (i.e. that which is recognized by the global production system, and to which it can respond) is much more limited. Rural households are simply too poor: 'there is therefore nothing surprising in the combination of poverty, hunger and even starvation on the one hand and food exports on the other – a reality in many African countries. It shows once again that within the capitalist system food is grown for profit, not to feed people – especially hungry ones' (Bondestam, 1976, p. 207).

We have come full circle. As food is now a commodity, the forces which govern its production are those which relate to buying and selling at a profit. If the poor of the Third World cannot afford to purchase it, then production and distribution will fail to adjust to their real needs, even though there may be widespread malnutrition and starvation. Thus we observe a paradox whereby a global production system has the capacity to feed the world, but does not do so because people are too poor. *In extremis*, such is the logic of capitalism that the very fact that the poor of the Third World are starving is the reason why they cannot be fed.

CONCLUSION

By placing the food crisis within the context of the capitalist world-economy we have necessarily dealt with a far more complex set of relationships and processes than would have been the case had the discussion been confined to technical matters. Despite this complexity, a world-systems approach carries with it a logical structure that is denied to an idiographic technical approach. It is also more fundamental, in that it focuses on causes rather than symptoms. Furthermore, because of its holistic framework, such an approach is less easily deflected to a reductionist perspective. Therefore the discussion is best considered in the context of the remaining chapters of this book, for they are all different aspects of the same base. Accordingly, we can summarize the discussion and, at the risk of oversimplification, itemize a number of central axes or elements.

The first is that the present food crisis has a long history. Its origins can be traced back to the emergence of a capitalist world-economy in the sixteenth century. The first search for global pillage laid the foundations for the current problem. Perhaps the most telling example is that of the sugar-economy of South America and the Caribbean in the seventeenth and eighteenth centuries. Even in those early times an agricultural crop was produced solely for profit. Such a production system contained within it all the elements that we can observe today in a modified form: export and the transfer of surplus to the emergent core; transformation and, in that particular case, extermination of a pre-existing rural economy; the provision of displaced labour through the slave trade; the expropriation of resources in answer to external forces, and so on.

Second, and via a number of specific processes which nevertheless have a common eventuality, we can trace the subjugation of what has become the periphery to the core. In this case we are concerned with food. The different processes – monetization, commoditization, the manipulation of trade, control of the means of production through state apparatuses, penetration of foreign companies in allegiance with a *comprador* bourgeoisie, a global financial structure refereed by the International Monetary Fund (IMF) – all point to the same conclusion. We observe the transformation of rural societies, whose economies were based on some form of reciprocity in their exchange relationships, to a capitalist model of which the central characteristic is one of surplus value extraction and profit. The net result is that, by being more or less forcibly wedded to this capitalist suitor, peasant societies of the Third World have lost the freedom to determine their own futures. They are on an express train without the means to jump off. The power to grow food and ensure adequate nutrition has been wrested from them, while the meagre rewards they earn for accommodating to a profit-based exchange system leave them too poor to purchase the very commodities that they have been obliged to produce.

Third, we can assume with some confidence that remedial measures emanating from the core, such as aid and development schemes and trading cartels, have as their ultimate goal the preservation of the status quo. The central beneficiary of the global trading system is hardly likely to promote changes which lead to its demise. It is in this context that we can locate the North–South debate. It is merely an attempt to re-energize the dynamism that is so evidently lacking during the current global depression. By stimulating trade, it does no more than refuel a motor starved of energy. Technocratic solutions are part of the same process: watering the deserts, farming the seas, bacterial cultures and the like derive from the core. They are as remote as the sun to the peasantry of the periphery, and, like the Green Revolution, will only result in further immiseration of the hungry.

Finally we can extend these processes into the future. All reason points to a further deepening of the crisis. Short of a fundamental structural change in the world system, we can expect the penetration of capitalist relations of production to intensify. The perspectives for food production in the future are themselves threatened by the continuance of this process. There is increasing concern with the sustainability of agricultural production in the face of the global environmental pressures created by 'modern' agriculture: deforestation, pollution of soils and water supplies, salinization, soil erosion and compaction, the loss of plant genetic resources and the displacement of peasants to marginal lands unsuited to sedentary agriculture. Some of these issues are dealt with in greater detail in the following chapter. The irony is that much degradation has been a result of efforts by the state to increase food production (WCED, 1987, p. 125). There is little to suggest that modern agriculture will become any less environmentally destructive in the near future. We can be sure that any further decline in food production which results from the continued organization of agriculture along capitalist lines will be felt in the periphery, by the poor, rather than in the core.

References

Barve, A. G. 1984: *The Foreign Trade of Kenya. A Perspective.* Nairobi: Transafrica.
Bennett, J. 1987: *The Hunger Machine. The Politics of Food.* London: Polity Press.
Blaikie, P. M. 1985: *The Political Economy of Soil Erosion in Developing Countries.* Harlow: Longman.
Boeke, J. H. 1953: *Economics and Economic Policy of Dual Societies.* New York: International Secretariat, Institute of Pacific Relations.

Bondestam, L. 1976: The politics of food in the periphery with special reference to Africa. In *Political Economy of Food. Proceedings of an International Seminar*. Tampere: Peace Research Institute.

Burbach, R. and Flynn, P. 1980: *Agribusiness in the Americas*. New York: Monthly Review Press.

Cassen, R. 1976: Population and development: a survey. *World Development*, 4, 785–830.

Crouch, L. and de Janvry, A. 1980: The class basis of agricultural growth. *Food Policy*, 5 (1), 3–13.

FAO (Food and Agriculture Organization) 1975: *Production Yearbook*. Rome: FAO.

FAO 1981: *Trade Yearbook*. Rome: FAO.

FAO 1984: *Land, Food and People*. Rome: FAO.

FAO 1986: *Production Yearbook*. Rome: FAO.

Fearnside, P. M. 1987: Causes of deforestation in the Brazilian Amazon. In R. E. Dickinson (ed.): *The Geophysiology of Amazonia. Vegetation and Climate Interactions*. New York: Wiley, 37–61.

Feder, E. 1978: *Strawberry Imperialism: An Enquiry into the Mechanics of Dependency in Mexican Agriculture*. The Hague: Institute of Social Studies.

Frank, A. G. 1969: *Capitalism and Underdevelopment in Latin America*. Harmondsworth: Penguin.

Frank, A. G. 1981: *Crisis: In the Third World*. London: Heinemann.

Franke, R. W. and Chasin, B. H. 1980: *Seeds of Famine*. Montclair, NJ: Allanheld, Osmun.

George, S. 1976: *How the Other Half Dies: The Real Reasons for World Hunger*. Harmondsworth: Penguin.

George, S. 1979: *Feeding the Few: Corporate Control of Food*. Washington, DC, Amsterdam: Institute for Policy Studies.

Gotsch, C. 1972: Technical change and the distribution of income in rural areas. *American Journal of Agricultural Economics*, 54 (2), 326–341.

Hewitt de Alcantara, C. 1974: The Green Revolution as history: the Mexican experience. *Development and Change*, 5 (2), 25–44.

Hightower, P. 1975: *Eat your Heart Out: Food Profiteering in America*. New York: Crown.

Hyden, G. 1980: *Beyond Ujamaa in Tanzania: Underdevelopment and an Uncaptured Peasantry*. London: Heinemann.

ILO (International Labour Organization) 1978: *Yearbook of Labour Statistics*. Geneva: ILO.

de Janvry, A. and Dethier, J. 1985: Technological innovation in agriculture. The political economy of its rate and bias. *CGIAR Study Paper No. 1*. Washington, DC: World Bank.

Lappé, F. M. and Collins, J. 1982: *Food First*. London: Abacus.

Lappé, F. M. and Collins, J. 1988: *World Hunger: Twelve Myths*. London: Earthscan.

Leys, C. 1975: *Underdevelopment in Kenya: The Political Economy of Neo-colonialism*. London: Heinemann.

Ministerio de Agricultura y Ganaderià 1985: *Censo agropecuario 1981*. Asunción: MAG.

Miranda, A. 1982: *Desarrollo y Pobreza en Paraguay*. Rosslyn, VA: Inter-American Foundation.

Nickson, R. A. 1981: Brazilian colonisation of the Eastern Border Region of Paraguay. *Journal of Latin American Studies*, 13 (1), 111–31.

Pearse, A. 1975: *The Latin American Peasant*. London: Frank Cass.

Pearse, A. 1980: *Seeds of Plenty, Seeds of Want. Social and Economic Implications of the Green Revolution*. Oxford: Oxford University Press.

Redclift, M. 1987: *Sustainable Development. Exploring the Contradictions*. London: Methuen.

Robbins, C. and Ansari, J. 1976: *The Profits of Doom*. London: War on Want.

Scott, J. C. 1976: *The Moral Economy of the Peasant. Rebellion and Subsistence in Southeast Asia*. New Haven: Yale University Press.

Taussig, M. 1982: Peasant economics and the development of capitalist agriculture in the Cauca Valley of Colombia. In J. Harriss (ed.) *Rural Development*. London: Hutchinson, 178–205.

United Nations 1980: *Population Studies 70*. New York: Department of International Economic and Social Affairs.

Watts, M. 1986: Geographers among the peasants: power, politics and practice. *Economic Geography*, 62 (4), 373–86.

World Bank 1975: *Ad Hoc Consultation on World Food Scarcity*. Washington, DC: World Bank.

World Bank 1981: *World Development Report*. Washington, DC: World Bank.

World Bank 1986: *Poverty and Hunger: Issues and Options for Food Security in Developing Countries*. Washington, DC: World Bank.

WCED (World Commission on Environment and Development) 1987: *Our Common Future*. Oxford: Oxford University Press.

5

The Use of Natural Resources in Developing and Developed Countries

PIERS BLAIKIE

THE NATURE OF CRISIS

Titles of books and their chapters are intended to attract readers and inform them of content. As such they are important devices and should not be neglected. The title of this book is *A World in Crisis?*, and this chapter sketches out an approach to answering that question in relation to the use of natural resources. The editors of this book in its first edition suggested that this chapter might be entitled 'The Rape of the Earth', which implies some sort of answer to the book's title. That my own title has turned out more ambiguous – and less dramatic – arises from my view that much of the radical critique of natural resource use has chosen to fight crucial debates on uncertain ground. Except for the uncritically converted, it needs careful argument to distinguish 'rape' of natural resources from just plain 'use'.

It is as well to define the word 'crisis' as a starting-point. It is an emotive term and is commonly used in a number of different ways. It is used loosely in the popular media to mean that things are getting worse and something will have to be done, and as such it is a term to catch the eye and mobilize people for action. It also has a meaning from pathology and refers to the point in the progress of a disease when important development or change takes place which is decisive of recovery or death, and from this it has broadened its meaning to include a vital or decisive stage in the progress of something, or simply a turning-point (see also the discussion in chapter 1). There are also other meanings of the word that need not concern us here.

From this definition, two distinct elements in a crisis can be distinguished. The first is the nature of the disease or threatening process itself. In terms of processes which affect society or the environment, it refers to the structural determinants of this process and also to the factors which impel a society or an environment towards threat and possible irrevocable change. The second element is time based and refers to how far along the structurally determined path the society or environment has gone. The second element is, of course, always defined in terms of the first and cannot, in conceptual terms, have an independent existence.

Imagine that smoke starts to appear from the windows of two adjacent houses at the same time. One house has a sprinkler system but the other does not. It could be argued that there is no crisis in the first house, but there is in the second. The fire is at the same point since its outbreak in the two houses, but the structural determinants of whether the fire will engulf each house are different. This distinction is most important in discussing crisis in the environment, since it focuses on whether there are the technical and social means of averting irrevocable change. The same processes may be going on in different parts of the world, for example land degradation or the nitrification of drinking water, and, while it is important to measure the point to which these threats have advanced, it is equally important to assess the future capability of different societies to halt or reverse these processes. Present and future paths of social change thus become central in any discussion of natural resource use and to the assessment of whether a crisis exists or not. Some societies have sprinkler systems to stop threatening fires, but others may not.

LEVELS OF ABSTRACTION, VALUES AND VERIFICATION

Any approach to the study of the use of natural resources which includes the world scale, as this book implies, must first clarify a number of conceptual issues. Three are briefly examined here. The first concerns the level of abstraction and geographical scale of the analysis, the second concerns the values and ideology behind the assumptions of how society uses, or *should* use, natural resources now and in the future, and the third is the problem of scientific verification, particularly the difficulties involved, the value-laden scientific environment in which verification takes place and how such verification enters the political arena.

The issue of level of abstraction and geographical scale has received some treatment in chapter 1 and passing mention in other chapters. Some impacts are global because the environmental symptoms are world-wide. The present debate concerning climatic change as a result of the greenhouse effect (Jaeger, 1988) must, by the nature of its physical symptoms, be conducted on a systematic and world-wide scale. Likewise, the alleged damage to the ozone layer also has universal implications. Other symptoms are expressed regionally, e.g. downstream effects of deforestation and soil erosion, 'acid rain' and onshore marine pollution, while yet others occur locally and, although sometimes widespread, their immediate cause and effect is contained in a smaller geographical area. Examples of the latter include long-term low-level contamination from nuclear power or reprocessing plants and nitrogen leaching of the water-table.

The issue of scale and level of abstraction can also lead to problems of communication between different disciplines, particularly between natural and social scientists but also between social scientists with different approaches (Rosewaal, 1987). However, much of the variety of scale and level of abstraction in the analysis of the use of natural resources need not mean non-communication or schism between different researchers. The various theories of the world-economy of such writers as Baran (1969), Amin (1976), Wallerstein (1979), Frank (1980, 1981) and Amin et al. (1982), although pitched at the world scale, need to be used with discretion and skill when applied to a particular instance. None the less, neglect of national and regional factors is commonplace. A charge has been made by Laclau (1971, 1977) against Frank, and by implication against the other world-systems writers, that class relations at the national level and below have been neglected. This only underlines the point that the danger exists that such high-level abstraction in world-systems theories can easily lead to a low level of focus and resolution when it comes to the local level.

The second issue in the analysis of the use of natural resources is that of values and ideological underpinning. In the discussions which follow in this chapter, there are irreconcilable views on how natural resources *should* be used and on the interpretation of past and present patterns. The main lines of schism are not independent of each other but intersect, with the result that unexpected ideological bedfellows abound in a number of debates about natural resources, as will be apparent in the next section!

The classification of views on natural resource use is linked to wider issues of social change and therefore is itself a major task. A useful starting-point is that developed by Cole et al. (1983) and Edwards (1985). They identify three fundamentally different strands of economic thought – the subjective preference (neo-classical) school, the costs-of-production (neo-Ricardian) school and the abstract labour (Marxist) school. The neo-classical school takes as its prime policy objective ensuring the smooth operation of a global free market without interference in exchange relationships. The determinants of international trade should be different national factor endowments. Crises are caused merely by irresponsible governments and their misconceived policies, particularly in instituting subsidies, imposing import and export quotas and encouraging inflation (Friedman, 1977; Hayek, 1975). According to this school, there is no global crisis of capitalism, merely local crises brought on by bad luck (e.g. drought) or bad management. The cost-of-production school, in contrast, casts doubt on the ability of present world markets to achieve rapid development, particularly in the case of less developed countries (LDCs). Therefore there is a constant need to adapt international institutions to correct distributional inequalities and to promote improved technology. Crises are seen to be caused by rigidities in the economy and conflicts in distribution through (necessary) technological change and can therefore exist at the national as well as the global level. The Brandt Commission's programme *North–South: A Programme for Survival* (1980) is an example which advocates the reform of international institutions such as aid-giving agencies, trade and tariff agreements and institutional strengthening of LDCs. Lastly, for the abstract labour theory of value (or Marxist) school, the social relations of production and the specific nature of the development of capitalism expose the fundamental reality of power relationships. The crisis exists in world capitalism, which is expressed through multinational corporations, banks and the foreign policies of the major capitalist imperial powers. *The* crisis, which is the ultimate one for capitalism, arises as the result of the inevitable tendency of the organic composition of capital to rise and therefore the rate of profit to fall (Cole et al., 1983). There are countervailing influences which slow this process down, but, in terms of the first element in the definition of crisis, capitalism's internal contradictions constitute its terminal disease. In terms of the second element (how far to the turning-point), there are of course many interpretations (see chapters 1 and 2 of this book). Crises can also be seen in less abstract terms and to occur for a variety of conjunctural reasons, but always as a result of the inherent contradictions in capitalism.

While these three schools of economic thought provide a useful and intellectually rigorous starting-point for distinguishing ideological positions in the explanation of how society works or should work, they do not provide a sufficient set of criteria. Another approach is suggested here which does not contradict this schema of views but identifies which issues in natural resource use are highest on the political agenda – expressed in any of the ideological terms of the schools of thought just described. This axis is defined by the level of economic and social development. Although national comparisons of wealth run into all sorts of problems of measurement and aggregation (since they may disguise enormous disparities in living standards within a single country), it is undeniable that the issues surrounding natural resource use in developed countries (say, the 24 countries belonging to the Organization for Economic Co-operation and Development – OECD) are very different from those in the Soviet Bloc and the LDCs. In developed countries the major issues of resource use are those which affect the people living there and reflect the level of the development of productive forces. They may include safety at work and employment in primary industry and agriculture, pollution, including the management of toxic wastes and marine pollution, conflicting land uses, particularly with reference to amenity and the conservation of nature and natural beauty, and the issue of nuclear power on environmental, safety and political grounds. Other international issues which are also on the political agenda are the sustainability of current extraction rates of natural resources, global climatic changes due to, in the case of developed countries, the emission of hydrocarbons, damage to the ozone layer and trading relations with LDCs. The issues confronting LDCs in natural resource use are quite different, and revolve around how the use of natural resources contributes to development. In turn this involves control and ownership of the natural resource itself, the technology adopted and the use of surpluses from primary production, and therefore the conditions under which production and exchange are conducted world-wide. Also, as a result of the recognition of growing environmental degradation, the sustainability of current agricultural and pastoral systems has become a leading concern, and links in directly with the ownership and control issue above.

The third issue concerns the scientific verification of some so-called crises, usually referring to the impact of the misuse of natural resources (e.g. pollution, the health hazard posed by radiation, land degradation, acidification of soils due to industrial pollution and so on). It is an issue which primarily concerns itself with the second element in the definition of crisis: how serious has the problem become? Is it still reversible and at what cost? In short, has the problem arrived at the decision point and therefore is it a crisis or not?

The role of science as an objective arbiter in conflicts and problem-solving of this kind is undoubtedly crucial but is far from straightforward and far from the apolitical value-free claims made for it by scientists 20 years ago. As Stocking (1988) has written about the measurement of soil erosion and Thompson et al. (1986) about the measurement of physical 'facts' about ecological stress in the Himalayan region, there are formidable technical problems in the measurement of soil erosion (with reference to Africa) and crop yields or fuel requirements in the Himalaya. There is the celebrated case of the highest estimate of the annual per-capita fuel-wood requirement's being 56 times as great as the lowest estimate. In another study (UNDP, 1980) into the reliability of single-interview farm visits in Nepal, it was found that error on crop yields exceeded 180 per cent on more than 20 per cent of the farms (quoted by Thompson and Warburton, 1985a,b. 119). To take another example, we still do not know if and how climatic change in certain parts of Africa (particularly the western and central areas of the Sahel) is related to the global greenhouse effect (Farmer, 1988). Also, the role of human agency there on climatic change is still not clear. Possibly, overpasturing has caused an albedo effect, permanently triggering lower rainfall (Charney, 1975; Rasool, 1984). Possibly, the deforestation of the Guinea Coast has led to a reduction in the availability of recycled moisture further inland (Walker and Rowntree, 1977). Possibly, the explanation involves purely global processes and climatic change. The climatic change may even be due to a complex interactive effect of all three explanations. However, a convincing explanation is needed to direct attention to what, if anything, can be done in the way of formulating a policy. The situation in western and central Sahel looks like being a regional crisis, precipitated by regional as well as global processes, but this still remains in the realm of supposition. A final example drawn from developed countries indicates that similar problems of verification exist even when there are far more financial and technical resources. Neither the long-term impact of low-level radiation (UNEP, 1986; Loprieno, 1986; McGill, 1987) nor the impact of polychlorinated biphenyls and other pollutants on the suppression of the immune system of seals of the North Sea (McGourty, 1988) is fully known. Pollution can frequently have long lag effects before symptoms appear, and scientific evidence often takes many years to build up sufficiently for there to be few excuses for ignoring it.

A quotation about the 'most important global environmental issue of the 21st century' is apposite here:

A picture begins to emerge of a future global climate that will be radically warmer than anything hitherto experienced by mankind. This knowledge is tempered, however, by our inability to state, unequivocally, that the greenhouse effect has been detected . . . This inability frustrates attempts to formulate policies for dealing with the problem. (Jones and Warrick, 1988, p. 15)

Uncertainty in scientific evidence does not provide a firm platform from which to launch environmental campaigns nor to design development projects to reduce pollution or enhance sustainability (e.g. the UK government's scepticism over its contribution to acid rain). There are powerful structurally and politically induced forces for inertia in large organizations, and their ingestion of new and potentially embarrassing scientific data is never easy (Dahrendorf, 1988). International organizations, large government departments and private firms seldom wish to reduce profits by paying for externalities created by pollution or safety precautions for workers and the public, about which new scientific evidence may have produced some (but not unequivocal) evidence. Where there are few scientists and little research, as in LDCs, the problem is even more acute. Often problems need to become crises before anyone notices and acts.

GLOBAL RESOURCES–THE CURRENT DEBATE

The debate which followed the publication of the 'Limits to Growth' reports by the Club of Rome from 1972 is now more than a decade old, and it is not my intention to rehearse it here. However, it is a relevant starting-point since it has been the victory of the right over liberals, conservationists and advocates of no growth or redistribution and the debunking of most of the 'Limits to Growth' theses which must provide the ground for any effective critique today. Full accounts of this debate are given by Daly (1972), Maddox (1972), O'Riordan (1976), Freeman and Jahoda (1979), Smith (1979) and Sandbach (1980). Here the discussion will focus upon the aspects of the debate which deal with the impact of incorporation of economies and societies into the world-economy upon resource use, and of the growth of the world-economy itself.

Since the early computer simulations of the growth of the world-economy by Meadows et al (1972) and Mesarovic and Pestel (1975) for the Club of Rome, there have been a number of thorough debunkings by Beckerman (1974), Kahn et al. (1976), Simon (1981) and others. The grounds on which these have been based are the failure of the simulations to specify multiple relations and feedback loops in the predictions of future populations, food supply, industrial production and pollution. For example, population growth can adjust (and has adjusted) to changing economic and social conditions, and pollution control can be initiated either through the operation of the free market or through state intervention. Also, technological advances which enhance productivity per worker and per unit of natural resource (land, minerals etc.) are insufficiently accounted for: indeed, 'there is no technical reason why technical advance should not continue indefinitely' (Hagen, 1972, p. 12). Further, the future availability of natural resources cannot be based upon known reserves, since these, for most of the widely used minerals at least, have almost consistently increased as a percentage of annual use over the past 150 years or so (Alexandersson and Klevebring, 1978, p. 17; Simon, 1981; see also chapter 3 of this book). In those cases where this has not been so, substitution or completely new technologies have been able to *create* new resources and are confidently expected to do so in the future – hence, no crisis. The basic elements of this debate will be known to most readers (Du Boff, 1974, provides a good summary), and it is not these which directly concern us here. Instead, we focus upon the ideological grounding of this debate.

There is an enormous range of views on the issue of growth, equity and the use of natural resources within the world-economy, and they can be distinguished on other grounds than their fundamental economic logical assumptions. Kahn et al. (1976, pp. 9–16) define the range of views by a number of categories, each of which has an internally consistent package of positions upon such issues as technology and capital, management and decision-making, resources, income gaps, innovation, industrial development and so on. While such a classification may be helpful in identifying broad categories of optimism and pessimism (the summary criterion for distinguishing the categories), it does so within a status quo framework, for two basic reasons. The first is that the optimist–pessimist criterion does not specify some fundamental differences between the various theories of social change which the so-called 'convinced neo-Malthusian', 'guarded pessimist', 'guarded optimist' and 'technology and growth enthusiast' all hold. Thus a 'convinced neo-Malthusian' (so-called) can believe that natural resources, the heritage of poor countries, are being consumed by the rich countries, denying the poor any real hope for better living conditions (Kahn et al. 1976, p. 14), and a Marxist can also do so. However, a convinced Malthusian can believe that 'future population growth will hasten and increase the magnitude of the future tragedy', whereas (very few) Marxists would do so. Lastly, many Marxists who may be most pessimistic about many of the shorter-term implications of capitalist growth may also feel there is no short cut to progress and that capitalist technology and growth offer the best long-term strategy for the socialization of the means of production and the eventual transition to socialism (Hyden, 1983; Warren, 1973, 1980).

An alternative classification of views with much more analytical power must start with a statement of strategic social objectives in the short and longer terms and the means by which these are to be reached. These will both be based on a theory of social change and economics. To start with, let us examine some views from the neo-classical school. The theories of Beckerman (1974), Bauer (1976, 1981), Kahn et al. (1976) and Simon (1981), who have all addressed the 'Limits to Growth' debate, are built upon an affirmation of capitalism in which unfettered market forces should largely determine natural resource use and economic activity in general, without any significant intervention on the part of the state. Simon (1981, p. 154) shows, for example, how the price of many minerals has declined in real terms over the past 80 years; thus it will benefit resource-rich countries to open up their economies to exploit these resources as soon as possible so that maximum revenue can be gained from them. Conservation of mineral resources is considered wasteful and a missed opportunity; therefore incorporation into the world-economy *now* is the best path to development.

With regard to the issue of environmental degradation, for example soil erosion, alkalinization, salination and water-logging of soils, many similar 'optimistic' views can be found:

> Of course arable land in some places is going out of cultivation because of erosion and other destructive forces. But taken as a whole the amount of arable land in the world is increasing year by year. (Simon, 1981, p. 81)

> The vague counter-argument that more intensive cultivation will ruin the soil is hardly convincing in view of the fact that soil has been farmed with increasing intensity in Europe for about 2000 years and there is still no sign that it is exhausted. Most of the world's cultivable areas, by comparison, have hardly been touched or not yet touched at all. (Beckerman, 1974, pp. 239–40)

It is worth mentioning in passing that subtropical soils bear so little resemblance to those in Europe as to render Beckerman's observations invalid on technical grounds alone. Also, there are serious concerns over the sustainability of agriculture in Europe, North America and Australia (Morgan, 1986; Science Council of Canada, 1986; Messer, 1987). However, the ideological content is the same here as for their observations on other natural resources – that the free market should operate world-wide and resources are thereby never 'raped', merely used rationally; 'rape' occurs only in cases of political or economic irrationality, and specifically with tinkering with the operation of the free market.

RADICAL CRITIQUE –
EXPOSÉ WITHOUT ALTERNATIVES

Radical critiques of the 'Limits to Growth' debate have usually been content with an exposé of the evils of capitalism, although Ensenberger (1974) went further and pointed out the bourgeois origins of the reports in the sense that it was only when environmental deterioration started to foul capitalism's own backyard, and maybe even threaten the process of accumulation, that much attention was paid. However, he admits to a view which I share, when he says:

The attempt to summarize the left's arguments has shown that the main intervention in the environmental controversy has been through the critique of ideology. This kind of approach is not completely pointless, and there is no position other than Marxism from which such a critical examination of the material would be possible. But, an ideological critique is only useful when it remains conscious of its own limitations. It is in no position to handle the object of its research by itself. As such it remains merely the interpretation of an interpretation of real conditions, and is therefore unable to reach the heart of the problem. . . . [Thus] Marxism [can become] a defensive mechanism, as a talisman against the demand of reality, a collection of exorcisms – these are tendencies we all have reason to take note of and combat. The issue of ecology is but one example.

Almost all critics of capitalism until the 1970s, with the possible exception of William Morris, paid little attention to the use of natural resources and environmental issues. Neither Marx nor Engels paid much attention to the environment, more interested as they were in the development of capitalism in the nineteenth century when the environment was seen to perform merely an 'enabling function' for rapid capitalist development (Redclift, 1983, p. 7). Marx did mention that 'all progress in capitalistic agriculture is a progress in the art, not only of robbing the labourer but of robbing the soil; all progress in increasing the fertility of the soil for a given time is a progress towards ruining the lasting resources of that fertility' (Marx, quoted by Bagchi, 1982, p. 213). However, it was not until later that major catastrophes in the developed world, such as the dust bowl in the United States in the 1930s, and writers on resource use at the periphery, particularly about the Sahel (discussed below), drew attention to the very serious problems of environmental use under capitalist agriculture – Malcolm (1938), Jacks and Whyte (1939), Glover (1946), Rounce (1949) and Hyams (1952) are some earlier examples of empirical accounts of environmental degradation in Africa and Asia.

There are two problems, however, which radical critiques of environmental degradation under world capitalism have to face. The first is that degradation also occurs at the present time in China and the USSR too – quite disastrously so – and there is little evidence that any battle is being won to conserve soil, water and forest resources in Mozambique, Angola, Guinea or Vietnam. In the USSR there is widespread evidence of oil spills, extinction of fishing grounds, high levels of air pollution and widespread loss of topsoils from the dry steppe (for favourable accounts see Gerasimov and Armand, 1971, and Pryde, 1972, and for a swingeing critique see Komarov, 1981). Pryde observes that Marxian economics assumes that only labour produces value, and therefore all natural resources are considered 'free' inputs of production. However, the explanation of wasteful and harmful use of natural resources in the USSR probably derives more from the nature of the state and the context in which it evolved. The forced march to industrialization, insensitive and bungled centralized planning and also perhaps a careless optimism (MacEwan, 1984) all contributed. The ecological implications of the following extract from 'The Great Plan' (1929) – quoted by Burke (1956) – can be imagined: 'It is a tremendous country [referring to the USSR] but not yet entirely ours. Our steppe will truly become ours only when we come with columns of tractors and ploughs to break the thousand year old virgin soil. We must plough the earth, break rocks, dig mines, construct houses, we must take from the earth'. At the same time, set against a probable 'destruction of nature', as Komarov (1981) calls it, as a result of ignorance, incompetence and the suppression of the free flow of important information about pollution and other failures, must be the nation's tremendous successes in industrial production, raising the standard of living of its people and making itself secure against its enemies. The tremendous changes which Pres. Gorbachev hopes to achieve in terms of *glasnost* and *perestroika* can only but help the acknowledgement of these serious problems. Komarov (1981) gives many examples of the suppression of information on pollution and waste on the grounds that its publication was anti-Soviet. Presumably *glasnost* has helped to some extent the international sharing of the impact of the Chernobyl disaster, and has facilitated the airing of the very serious problems of salinization and desertification in the southern republics which have recently been reported in the British press.

The record of the use of soil, water and forest resources in China has also left a mixed commentary, from the very favourable (e.g. Sandbach, 1980, chapter 6) to the unfavourable (Dequi et al., 1981; Howard, 1981; Delfs, 1982; Smil, 1983, 1987). Perhaps there was a tendency for Chinese achievements to stand for the sole instance of proof that socialist reconstruction was not a figment of the imagination and that it was going on right now – an onerous and overworked role that did not come through unscathed upon detailed examination. Smil's account of environmental decline in China is marked by almost unrelieved pessimism:

> the magnitude of China's accumulated environmental problems owing to the legacy of ancient neglect and recent destruction is depressing. The dimensions of the future tasks in population control, food and energy supply and overall social modernisation are overwhelming, and the potential for further accelerated environmental degradation is quite considerable. (Smil, 1983, p. 198)

Smil's unrelenting criticism of and cynicism towards *all* China's efforts in development, its institutions and its western admirers ('embarrassingly misinformed and naive') makes me hesitant in believing all I am being told. However, there are other, less ideologically aggressive, accounts of many of the unfortunate processes of environmental decline, and so the general conclusion that the revolution in China has not dramatically improved the situation is probably a safe one.

The problem which radical critiques must face up to, then, is that there will always be struggles and contradictions over natural resources even in a 'socialist state', and that problem-solving is not the technocratic preserve of status quo neo-classical economics. The directions that lead away from generalist calls to the barricades towards practical solutions and the art of the possible may seem less than revolutionary and sometimes make tame and reformist reading.

The second problem is that it is not true in all cases that capitalist natural resource use is wasteful, polluting and socially irresponsible, although at the most abstract level capitalism undoubtedly is all these things. As Baran and Sweezy (1966) have pointed out, competition between producers leads to chronic overproduction which itself is wasteful. Crises of overproduction are constantly resolved and reproduced by the creation of needs through advertising, and by the manufacture of arms whose built-in obsolescence and destruction in warfare require constant replacement of both the armaments themselves and industrial equipment and also require the capture of new markets (new consumers). At this abstract level, then, capitalist resource use is inherently wasteful. However, action to alter resource use is always 'contexted', and taken not just with abstract and global theories and objectives in mind. The questions which arise in any specific case concern alternative and possible ways of using natural resources which are less wasteful and more socially useful than at present. There are many capitalist farms and plantations, mines and hydroelectric stations which are not particularly wasteful and may in the longer term be socially progressive. Therefore rather less abstract criteria must also be established which can be used in individual cases, and these should be related to the type of capitalist development involved in each case and also to a longer-term strategy of socialist transition. An examination of the problematics of underdevelopment theories and their major critics (Warren, 1973, 1980; Hyden, 1983; Kitching, 1983) is particularly relevant here.

One of the major problems with critics of underdevelopment theory such as Warren is that the role of struggle against *present* iniquities and inequalities, to better the quality of life for workers and peasants the world over, tends to be obscured. In the long term the prodigious ability of capitalism to increase production can promote socialist transition by systematically socializing the means of production, by encouraging workers to organize into unions and to move towards directing and controlling the means of production, and by using capitalist technology to provide an ample sufficiency of material goods for all. In such an analysis there is often, but not always, a conflict over the long-term objective of winning better working conditions, and more control over the way and rate of use of natural resources in any particular case. Therefore a socialist strategy for the use of natural resources in LDCs has to be based upon a theory of socialist transition. There are three crucial elements which have been debated, all of which depend on whether the socialist transition is possible without a fully developed capitalist stage.

The first element is the need to increase production in LDCs so that the population achieves its basic needs. A failure of a population to get enough to eat and to clothe and house itself may lead to an irruptive gesture – a riot or uprising – but has seldom led to either successful *coups d'état* or socialist government thereafter. Socialist construction has been extraordinarily difficult to secure when the population is preoccupied with finding enough to eat. Certainly the falterings of socialist initiatives in Tanzania and Mozambique (Coulson, 1982) were partly due to the failure of agriculture to feed their populations satisfactorily, although the successful efforts of destabilization by Pretoria and its puppets were also involved in the latter case. Therefore, as long as capitalist enterprises do not undermine the capability for future production (erode soils, deplete forests, pollute life-support systems) but lead to improvements in living standards, there seems little ideological and economic alternative but to accept them.

Second, capitalist enterprises can (under circumstances to be described) socialize the means of production, and help to organize the workforce into unions and to involve workers in the experience of working in, and eventually taking over, the capitalist enterprises themselves. As long as the enterprise technically trains its workers and builds up the confidence that workers can run it with the required skill and discipline in the longer term, then again it should be accepted.

Third, capitalist enterprises can provide funds for governments through royalties or concessions to create physical infrastructure and encourage other enterprises through industrial linkages, which accelerate the two processes above.

The problem is that few capitalist enterprises in LDCs fulfil even one of these criteria which may be conducive to socialist transition. Many mining, timber and agricultural enterprises in LDCs financed by both transnational and indigenous capital have been responsible for widespread depletion of soil fertility, deforestation, pollution and the rapid working out of minerals with little in the way of royalties or taxes and the construction of port facilities, roads, railways etc. They have often casualized the workforce, frequently awarding workers unused to factory conditions very short contracts which are terminated before unionization and a struggle for better wages and working conditions can begin. Furthermore, transnational companies (TNCs) can reduce royalties and taxes paid to government by a variety of means. That these cases form the majority, and struggle (or adept negotiation) cannot substantially reduce the majority, is of course the argument of underdevelopment theorists such as Frank (1980) and Amin (1974). In this chapter I take the view that the experience of capitalist resource use is mixed, but that a variety of means, including armed struggle, nationalization and bargaining over prices, training, participation and the like, *can* be employed to raise living standards and the experience and consciousness of the workforce. However, every case must be taken in context and evaluated for its advantages or disadvantages in furthering socialism. Such a cautious, longer-term and ambivalent view does not lend itself to rhetorical anti-capitalism, nor to the universal means of achieving socialist transition. As the next section indicates, there are many cases where world capitalism certainly has *not* provided the three conditions for progressive social change, and here the analogy of 'rape' of the earth (and exhaustion of human labour) is most appropriate.

Thus much of the left critique of natural resource use has been characterized by Utopianism and neo-populism. There is an unwritten assumption in much radical writing about the world economy that under socialism class contradictions would somehow disappear, allowing a more far-sighted use of the natural environment. While critiques of the impact of world capitalism are useful, sins of omission leave the reader wondering whether pre-capitalist forms of resource use *can* manage to provide for a socialist future, and whether or not autarky, collectivization or co-operatives are on the agenda. There will always be conflicts of interest and techno-administrative problems of natural resource management in every future social formation, and these have to be addressed alongside the political struggles of the day.

My own view is that monolithic theories of development with inflexible blueprints should be avoided. This does not mean an abandonment of theory altogether, but a recognition of the opportunities to further the cause of socialism when they arise and also of constraints on achieving this end. Strategic and military considerations will frequently be important (e.g. as in Nicaragua, Mozambique and Angola today), but not in all instances. The size of the country itself and its natural endowment are other relevant factors. Where small size and military vulnerability endanger socialist objectives, the extent to which support from the USSR is forthcoming and principled may be decisive (as in Cuba, for example). No one would propose that opportunities for a leftist government to achieve power in Africa, Asia or Latin America should wait for industrialization and/or modernization according to our uncritical 'application' of classical Marxism. It is just that, in the absence of capitalist development and a degree of development of state functions and infrastructure, further progress towards socialism will be very difficult, although my own view stops short of Kitching's (1983), which is that backward societies produce backward socialism. I disagree not so much because this may not be true, but because the 'backward' socialism *may* lead to something more progressive more quickly than backward capitalism.

Moving now to developed countries, the distinction between the two sets of values about natural resource use and the environment is immediately apparent. While those values of commentators from or about LDCs revolve around the various paths to development, those from or about developed countries revolve around a different set of priorities and principles altogether, which can be labelled environmentalism. This set has been discussed by a number of writers (O'Riordan, 1981, chapter 1; Redclift, 1984, chapter 3), and their reviews show that little of environmentalism in the West is derived from Marxist thinking, radical though parts of this movement may be. Bahro (1982) is one of the very few exceptions. He is a citizen of the German Democratic Republic, and offers a critique of capitalism that maps the way forward in one of the few socialist blueprints for the use of natural resources in his *Socialism and Survival*. His critique centres on the production ethos of capitalism, and particularly on the compulsive consumption of capitalist countries, stimulated by advertising and the needs of capitalism to find markets. This in turn leads to degradation of the resources of LDCs, and brings about environmental crisis and exploitation of the working classes in the South. The threat of nuclear war is also seen by Bahro to be a central issue. The way forward, however, seems curiously Utopian and owes little to classic notions of class struggle. Instead, no-growth ethics and individual volition to act in accordance with a conversion to ecological politics are offered as solutions which find common ground with many other pressure groups and Green political parties in western Europe (see Redclift, 1984, for a further discussion).

DESTRUCTIVE NATURAL RESOURCE USE IN AGRICULTURE AND FORESTRY

Starting with LDCs, one of the earliest and best critiques of the impact of capitalism upon natural resource use focused upon the Sahel. Here the radicalization of the 'drought issue' exposed the unfavourable effect of commercial expansion of beef, cotton and peanut production during the French colonial period and up to the present time upon pre-capitalist strategies of adequate storage, sophisticated cropping and pastoral techniques, risk-sharing and risk-aversion. All were undermined by commercialization. This reduced the size of stored food reserves, increased reliance on cash surpluses from the sale of commercial crops to buy in essential foods, substituted monocropping and deep ploughing for intercropping and other moisture- and soil-retaining strategies, and broke up communal risk-sharing groups to replace them with mutually competitive entrepreneurs (Meillassoux, 1974; Copans, 1975; Franke and Chasin, 1980). Similar studies in Nigeria (Watts, 1983, 1984) and Kenya (O'Keefe et al. 1977) show the destructive aspects of both commercial crops and commercial farmers who displaced and spatially marginalized indigenous crops and sometimes the cultivators themselves. Some of the most extreme cases have been carefully documented by Dinham and Hines (1983), who have shown how TNCs have pillaged the soil, pushed farmers off their land and pressed the most outrageously one-sided arguments upon national governments in Africa; an example is summarized below. In many of these cases, local organizations for communal production, and mutual aid between households, villages and larger groups in times of food shortage, were weakened or eliminated. It is a most important issue in socialist construction as to whether these organizations can transcend their frequently inegalitarian and feudal origins as well as provide a basis for increased production. The issue is similar to that faced by the mid-nineteenth-century Russian populists, such as Herzen and Chernyshevsky, who saw the *obshchina* (the institution which organized peasant labour and rent payments in favour of the landlord) as collectivist or proto-socialist, capable of forming the basis for a progressive transition to socialism (Kitching, 1982, p. 35).

The most recent developments of this debate have been centred on the occurrence of the appalling series of famines in Ethiopia and the Sudan in 1983–4, and at the time of writing in Darfur, Kordofan and the Upper Nile Provinces of the Sudan. Somehow the tenor of the present debate is less radical and less abstract, and concerns itself with more practical matters. The action (and non-action) of the government and of international agencies involved, the impact of war, possible climatic change and the role of human agency in overstocking and desertification are the main issues. Somehow, to use these events as a stick to beat world capitalism does not seem to be good copy these days – although the structural conditions which brought famine about have remained unchanged.

Another widely publicized case of destruction of natural resources by international capital is the logging of tropical and subtropical forests. Here the direct and indirect effects of capital upon deforestation must be spelt out: they are broadly similar to the effects upon agriculture, which can also lead to environmental degradation. There has been much debate about whether it has been the logging companies or the shifting cultivators who have been responsible for very rapid deforestation throughout the world (reported to be 10–12 hectares per minute by the Food and Agriculture Organization (FAO), or up to 40 hectares per minute in the view of Myers, 1983). The problem is probably much more serious than was once thought. Satellite pictures and improved photo-interpretation have revised upwards the rate of deforestation and have often cut national estimates of remaining forests of various grades by as much as a factor of 30 per cent. For example, in the Philippines the actual percentage of forest of total land surface is only 38 per cent, as opposed to the official figure of 58 per cent (Grainger, 1980). Smil (1983, p. 12) estimates that the actual forested area in China is only 35 per cent of the official figure, and even existing forests are of lower productivity than claimed.

What is clear is that forests are disappearing fast (for global and regional surveys see Grainger, 1980, Routley and Routley, 1980, Myers, 1983, and Plumwood and Routley, 1982). While local circumstances are bound to vary greatly, it is noticeable that the 'blame' for tropical and subtropical deforestation is placed either on indigenous cultivators or logging companies as a matter of ideological temperament rather than detailed local analysis. A comparison of the papers by Lanly (1982, for FAO) and Plumwood and Routley (1982) shows that whenever both refer to the same area they come to very different conclusions as to who is to blame. However, the connection between forest contractors and indigenous cultivators and pastoralists is strongly made by the loss of cultivators' land, and the closure of common property resources by private ownership or private contracts and forest concessions, which *force* the latter to cut down the forest (for a fuller account see Blaikie, 1985). Undoubtedly, rapid population growth of rural populations is also a contributory factor, given that other political economic determinants of production and reproduction are fixed, given and unexamined. Hence the implications of a large logging concession (or purchase of forest for clearing to start up cattle ranches) in a populated area spread beyond the boundaries of the concession itself and can set up complex and damaging 'knock-on' effects in other areas and parts of the rural economy.

Logging companies often seek to reduce the flow of economic benefits to the host government from logging by neglecting or even blocking forward linkages such as sawmills, plywood or chipboard manufacturing. Second, they often pay very low prices for timber by a number of honest and dishonest strategies, among the most important being the use of superior knowledge of the details of timber resources. Third, they attempt to maximize the repatriation of profits, as do most TNCs, and avoid contributing to the cost of either reafforestation or training local personnel in forest management. Fourth, they 'cream' forests and take little care over the conservation of other trees not valuable enough to fell. Lastly, they frequently fail to honour clauses in agreements once logging has been started, since local monitoring is difficult for many LDCs (Leslie, 1980).

The damage from indiscriminate logging under the conditions implied above can be enormous. The loss of livelihood of local inhabitants, siltation of reservoirs, damage to hydroelectric plant and increase in flash floods with consequent damage to property, crops and livestock are obvious results, and are documented in depth all over the world.

TOWARDS A PROGRESSIVE USE OF NATURAL RESOURCES

For a natural resource to be used at all, four conditions have to apply. First, there has to be positive identification of the natural resource as defined by either existing or known but undeveloped technology. Second, there has to be the technical expertise to use the resource. Third, political control by the would-be user appropriate to the resource itself and to the technology and social relations of production is necessary. Some resources do not require much political control: where a timber concession is signed, the timber is cut and the whole operation can be over in less than 18 months. Others such as minerals or estates and plantations require a longer-term political security of agreements, operation and profit-taking by the user. Fourth, the resource has to have utility relative to other occurrences of the same resource (or a substitutable one) elsewhere. This usually refers to the quality and location of the resource, but also to the political conditions which may help to determine the other three conditions (for a general discussion, see Blaikie, 1985).

Some of the most important areas of struggle to win some of the preconditions for future socialism and for a better standard of living for workers today in developing countries are outlined in the following paragraphs. However, two qualifying remarks are necessary. First, it is through struggles across a wide area that a transition to socialism is made, and many of these will lie in different arenas altogether from struggles over natural resource use. However, it is also true that natural resources use *is* a very important element in realizing the general social objectives discussed in this chapter, as well as other objectives which are desirable in themselves, whether or not they serve a longer-term strategy, e.g. legislation for working conditions to prevent occupational diseases.

A second qualifying remark is necessary because little mention has been made of the nature of the state and therefore of internal struggles between contending classes. There can be no assumptions that a national government represents a benign and unified front fighting socialist or nationalist battles over, say, mineral rights against a TNC. There are also continuous struggles within all states on union representation, a free press, civil rights, women's rights and so on, which only indirectly concern international capitalist enterprises, institutions and governments. What is also clear is that there are immense variations in the form and strength of the state, and the relative importance of TNCs in the struggle over the use of natural resources (i.e. in primary industries). In sub-Saharan Africa, for example, the state as a rule is perhaps weaker than in, say, Morocco or Algeria, and 'has left representatives of international capital to exercise more influence than they have been able to do in other societies where an indigenous class structure has been in place' (Hyden, 1983, p. 195).

To return to the preconditions of resource use, the first is positive identification. The political economic circumstances of resource identification have largely to do with technical expertise (particularly geological surveying, often with test drilling), aerial surveys of forests and sample botanical field surveys, and other natural resource surveys and assessments. The issue here is control over that information and sometimes over its ideological content also. From early imperial times, 'trickery, deceit, astute lobbying and brute force' have been the means by which mining companies have secured concessions in developing countries (referring here to Rhodes's concessions from King Lobengula in Southern Africa; Lanning and Mueller, 1979, p. 59). A part of the trickery and deceit was then, as it is now, a knowledge of the mineral resources of a country which is withheld by one party in a negotiation from another. Collection of data about minerals is notoriously expensive and is also risky (Howe, 1979). Exploration for some minerals, particularly in ocean mining, is so expensive that even large TNCs have to form consortia (United Nations, 1980c). Therefore information about natural resources generated by exploration has an extremely high value. In the same way, natural resource surveys often carried out in remote areas of developing countries are held only by the party wishing to negotiate the logging concession with government bureaucrats. On the one side of the negotiation there is a TNC with good information which it withholds completely, and on the other side an imperfect but open administration where the inconsistencies and imperfect knowledge of the bureaucrat are already known by the TNC (Leslie, 1980). Furthermore, the level of technical knowledge about mining engineering, metal refining and possible environmental impacts (particularly pollution) has to be of a high standard, not necessarily possessed by negotiating teams of LDCs.

Clearly technical training as well as training in management, plus knowledge at the highest level, are part (and only part) of the solution. The opening up of all information about the natural resources under negotiation or the carrying out of alternative surveys are other small but not insignificant ways with which information can be put more squarely at the service of national governments. This helps those governments to obtain more reliable criteria on which national or local control and use of natural resources could be planned, but by no means assures this. This course of action also assumes that, in many cases, TNCs are a necessary evil.

The point which is being made here is not a defence of multi-national corporations. They are, in the view of the present author, a necessary evil that Africa has to accept in the absence of any advanced economic structures of its own. (Hyden, 1983, p. 20)

This probably does not hold true for some Latin American countries or for India where there is much more evidence of these more advanced local economic structures, and therefore the need for a more complete knowledge of natural resources is less acute. But the reliance of many African countries on foreign expertise is considerable, as in the catastrophic decline in Angola's ability to produce diamonds after white technicians withdrew in 1975.

Turning now to the question of the expertise which is required to use natural resources, it is clearly in mining and ore refinement that it is most crucial, because it is scarce. It is here that the inequality in bargaining power between the national governments of the developing countries and the TNCs is most acute, and therefore attempts to institute international legislation or codes of practice tend to fail in the bilateral details of negotiations: 'The need for an international code has been recognised by the international community and the Commission [on TNCs] has decided that, whatever its nature, the code should be effective' (United Nations, 1976, p. 39). Here is the response of a TNC to the Chilean government's attempts at negotiating a joint interest in the production of plastics with the TNC: 'Our response was that we were interested in a joint venture if we could have a controlling interest so that we could protect our exceptional technology and ensure the proper efficient management of the business with co-operation between TNCs and governments' (Gunnemann, 1975, p. 10). It is logical that technical expertise is shared only when it is obsolete and of low value. This problem is the more intractable because TNCs are usually in a powerful position, and only training to an extremely high degree of expertise and retention of those trained in the developing countries will start to solve the problem. Also, the experience of gaining expertise through lengthy foreign training has shown that those fortunate few naturally want to work for TNCs themselves and not to be relatively poorly paid government negotiators.

The question of political control is perhaps one over which LDC governments have the most leverage. In the past, extremely large revenues have accrued to many companies who are able to secure long agreements from puppet and

other *comprador* leaders of national governments who found themselves in contractually inferior positions during negotiations. In some cases strong political
domination of a national government has ensured extraordinary large benefits
to flow to many companies up to the present day, and for TNCs to dominate
the politics of the country and mould the labour market to their requirements.
For example, Namibia has very large deposits of lead, cadmium, zinc, diamonds
and uranium. Various transnational mining companies and the South African
government have combined to plan an area of black homelands to provide
labour for these mines (Lanning and Mueller, 1979, p. 455). However, political
independence has given a very important extra leverage for national governments
to secure a greater share of both revenues and control over production of mineral
resources as well as agriculture (when farmed by white settlers in Central and
East Africa). Many mineral concerns require large capital outlays and high
entry costs, which make it difficult for local firms to compete. It has also
been difficult to persuade TNCs to commit themselves to backward- and
forward-linking manufacturing undertakings, so much so that they have
'conspicuously failed' to accelerate local development (Girvan, 1976, p. 30).

The ability of a government to play one TNC off against another in these
negotiations is lessened because there is colossal concentration of the share of
production in many minerals (e.g. nickel and copper; United Nations, 1981b).
In addition, there undoubtedly exist unofficial 'fidelity' contracts and informal
understandings between TNCs which enable them to deal more effectively
with intransigent national governments. There is also considerable vertical
integration of TNCs in all stages of mineral extraction. This is so with bauxite
production in the Caribbean, enabling TNCs to control all stages of mining
and refining of aluminium to such an extent that they virtually control the
political economy of the entire Caribbean region. Sometimes the United States
itself becomes involved to assert control for its TNCs, as in the case of Chile
(Girvan, 1976, chapter 2; United Nations, 1981b).

However, despite these difficulties the situation of the political control by
developing nations over minerals has gradually improved. Before the last war,
royalties, which were calculated on the level of production of refined metal,
were the most common means of retaining revenue. This gave way in the 1950s
to income tax, which many TNCs preferred because production costs and profit
can be withheld or manipulated. Recently, there has been a tendency for many
more joint ventures, often financed from sources other than the TNC itself,
with considerable local participation up to a high level. Further developments
have recently taken place in which TNCs withdraw to minority holdings and
leave national governments to take the risks in production and in keeping down
labour costs. In this way TNCs organize service contracts to provide technical
services and marketing, while the national government is left with direct control
of the workforce. The same problems of contractually unequal relations between
the two parties still exist, and it is not clear that there is any obvious advantage
for workers of industrial concerns or peasant smallholders (growing crops for
TNCs) in these new arrangements.

Turning now to the issue of political control over land and water resources in developing countries, some of the problems discussed in relation to the mining industry are similar, particularly with regard to highly mechanized plantation growing of commercial crops on a large scale. Displacement of peasants and pastoralists from what is usually the most fertile land from which they derived their livelihood, their partial or complete proletarianization (so that they have to work part or full time on the commercial farm or plantation) and the frequent depletion of natural fertility are all symptoms of many but by no means all capitalist agricultural enterprises. Dinham and Hines (1983) give one example, among many, of a TNC called Bud Antle Inc., a large Californian-based food conglomerate which negotiated a large-scale highly capital intensive scheme to grow quality vegetables in Senegal, which were flown to markets in Europe. By 1976 the project had virtually failed and the Senegalese government (which had formed a joint enterprise with Bud Antle Inc.) was left with eroded soils, imported machinery which could not be maintained and dispossessed pastoralists/peasants. Surely one lesson to learn from this is that the Senegalese government was ill advised, in a legal sense, not to insist on clauses in the agreement which would have prevented this débâcle. Joint ventures with TNCs must be buttressed with hard-headed agreements, based upon reliable knowledge of the issues and resources involved.

This is one extreme case. Taking together all the circumstances of TNC involvement in land and water resources, there is a continuum of commitment to conserve these resources over a long period, through short-lease contracts of 10 years or so, to very short-term logging contracts. Clearly, for reasons of resource conservation as well as for more general objectives of social development, longer leases with penalty clauses for running down natural resources (depletion of soil fertility, destruction of forest areas etc.) are necessary, but require a well-developed bureaucracy with a considerable degree of technical training and sense of public duty. It is this latter development which is a necessary, although not sufficient, condition to implement a natural resource policy which avoids the depredations of TNCs and the expansion of commercial agriculture into unsuitable areas.

CONCLUSION

The relevance of emotive, but none the less useful, words like 'crisis' and 'rape of the earth' depends upon the values and ideology of the user, the level of abstraction of the analysis and, linked to this, the degree of its radicalness from the left or right. The problems of scientific verification are crucial and cannot be relied upon to provide timely, value-free and independent validation or refutation of a state of crisis. It is likely that 'silent emergencies' (to use a term originally applied by Fraser, 1988, to famines) stalk the future of many people and are, in a sense, unrealized crises. This is occurring because of a failure of the free flow of scientific information on the condition of the biosphere.

It is tempting to distinguish between *problems* of the developed countries and *crises* of LDCs. In the former the highly developed nature of civil society makes it possible for workers' and consumers' interests to be politically represented. Information on the state of the biosphere, hazards at work and so on reach governments in quantities unimaginable in most LDCs. Certainly there is no room for complacency and there is a constant struggle on various issues between trade unions, environmentalist groups, consumer organizations and Green parties on the one hand, and state corporations, private businesses and bureaucratic structures on the other. However, to refer back to the starting-point of this chapter, the structural determinants of crisis are far less strongly present in developed countries. These determinants are the lack of political power of workers and consumers, the lack of awareness and knowledge of the silent emergencies which are not obvious to all, gross profiteering, and corruption and cynicism in public life, the failure of the rule of law and weak government. This list of evils which constitute the structural determinants of crisis can occur under all modes of production, of course, but they are less marked in developed bourgeois democracies. It is also possible that a restructured and more open USSR will begin to reduce these structural determinants of crisis which were undoubtedly present under the old regime.

For most LDCs, the path of transition to capitalism is crucial to establishing the means for averting crises. The debate between Frank and Warren, outlined in this paper, is important here. Empirical evidence since Frank's publications gives considerable weight to Warren's counter-arguments, as also, in a narrower context, do those of Hyden in Africa. In contrast, many countries in Southeast Asia are experiencing exceptional capitalist growth. Along with that come, of course, many threatening developments of non-sustainable agriculture, deforestation through commercialization of logging and pollution. However, along with these should also come the benefits of political power to workers and consumers, better monitoring of the environment and a more open and informed society. For those countries which have not experienced such growth (and the majority of countries in Africa come into this category), all the worst results of poverty for the majority and evidence of irresponsible resource use can be found. For those countries who can make no transition to capitalism or some sort of socialism, crisis – for once – is not an overworked word.

Acknowledgement

I wish to acknowledge the help of Sheila Chatting, who acted as my research assistant and suggested many useful ideas.

References and Bibliography

Alexandersson, G. and Klevebring, B. I. 1978: *World Resources*. Berlin: Walter de Gruyterm.

Amin, S. 1974: *Accumulation on a World Scale: a Critique of the Theory of Underdevelopment*, vol. 1. New York: Monthly Review Press.

Amin, S. 1976: *Unequal Development: An Essay on the Social Formations of Peripheral Capitalism*. Brighton: Harvester Press.

Amin, S., Arrighi, G., Frank, A. G. and Wallerstein, I. 1982: *Dynamics of Global Crisis*. London: Macmillan.

Bagchi, A. K. 1982: *The Political Economy of Underdevelopment*. Cambridge: Cambridge University Press.

Bahro, R. 1982: *Socialism and Survival*. London: Heretic Books.

Baran, P. A. 1969: *The Longer View: Essays Towards a Critique of Political Economy*. New York: Monthly Review Press.

Baran, P. A. and Sweezy, P. M. 1966: *Monopoly Capital*. New York: Monthly Review Press.

Bauer, P. T. 1976: *Dissent on Development*. London: Weidenfeld & Nicolson.

Bauer, P. T. 1981: *Equality, the Third World and Economic Delusion*. London: Weidenfeld and Nicolson.

Bauer, P. T. and Yamey, B. S. 1957: *The Economics of Underdeveloped Countries*. Cambridge: Cambridge University Press.

Beckerman, W. 1974: *In Defence of Economic Growth*. London: Jonathan Cape.

Belshaw, D. 1980: Taking indigenous knowledge seriously. The case of inter-cropping techniques in East Africa. *IDS Bulletin*, 10, 24–7.

Bertleman, T. et al. 1980: *Resources, Society and the Future*. Oxford: Pergamon Press.

Blaikie, P. M. 1981: Class, land-use and soil erosion. *ODI Review*, 2, 57–77.

Blaikie, P. M. 1984: *Natural Resources and Social Change. Social Sciences: Changing Britain, Changing World*. Milton Keynes: The Open University.

Blaikie, P. M. 1985: *The Political Economy of Soil Erosion*. London: Longman.

Blaikie, P. M. and Brookfield, H. C. 1987: *Land Degradation and Society*. London: Methuen.

Bojö, J. 1988: Sustainable use of land in developing countries. In *Perspectives of Sustainable Development*. Stockholm: Stockholm Studies in Natural Resources Management No. 1, 39–54.

Brandt Commission 1980: *North–South: a Programme for Survival*. London: Pan Books.

Brokensha, D. W., Warren, D. and Werner, O. (eds) 1980: *Indigenous Knowledge Systems and Development*. Washington, DC: University Press of America.

Burke, A. E. 1956: Influence of man upon nature: the Russian view. In W. L. Thomas et al. (eds), *Man's Role in Changing the Face of the Earth*. Chicago, IL: University of Chicago Press.

Charney, J. G. 1975: Dynamics of deserts and droughts in the Sahel. *Quarterly Journal of the Royal Meteorological Society*, 101, 193–202.

Cole, K., Cameron, J., and Edwards, C. B. 1983: *Why Economists Disagree: The Political Economy of Economics*. London: Longman.

Copans, J. 1975: *Qui le nourrit de la famine en Afrique* and *Sécheresses et famines du Sahel* (2 volumes). Paris: Maspero.

Coulson, A. 1982: *Tanzania: A Political Economy*. Oxford: Clarendon Press.

Dahrendorf, R. 1988: Science and politics: expectations, errors and clarifications. *Interdisciplinary Science News*, 13 (1), 12–17.

Daly, G. H. E. (ed.) 1972: *Towards a Steady-state Economy*. San Francisco, CA: Freeman.

Delfs, R. 1982: The price of neglect. *Far Eastern Economic Review*, 20 August, 20.

Dequi, J., Leidi, Q. and Jusheng, T. 1981: Soil erosion and causation in the Winding River Valley, China. In R. P. C. Morgan (ed.), *Soil Conservation: Problems and Prospects*. Chichester: Wiley.

Dinham, B. and Hines, C. 1983: *Agribusiness in Africa.* London: Earth Resources Research.

Du Boff, R. B. 1974: Economic ideology and the environment. In H. G. T. Van Raay and A. E. Lugo (eds), *Man and the Environment Ltd.* The Hague: Rotterdam University Press.

Edwards, C. B. 1985: *The Fragmented World: Competing Perspectives on Trade, Money and Crisis.* London: Methuen.

Ensenberger, H. M. 1974: A critique of political ecology. *New Left Review,* 8, 3–32.

Farmer, G. 1988: *The Rainfall Climatology of the Sahel from Senegal to Somalia. Report to IUCN.* Climatic Research Unit, University of East Anglia, Mimeo.

Frank, A. G. 1978: *Dependent Accumulation and Underdevelopment.* London: Macmillan.

Frank, A. G., 1980: *Crisis in the World Economy.* London: Heinemann.

Frank, A. G. 1981: *Crisis in the Third World.* London: Heinemann.

Franke, R. and Chasin, B. 1980: *Seeds of Famine.* New York: Universe Books.

Fraser, C. 1988: *Lifelines for Africa Still in Peril and Distress.* London: Hutchinson.

Freeman, C. and Jahoda, M. (eds) 1979: *World Futures: The Great Debate.* Oxford: Martin Robertson.

Friedman, M. 1977: *Inflation and Unemployment.* London: Institute of Economic Affairs.

Gerasimov, I. P. and Armand, D. L. 1971: *Natural Resources of the Soviet Union: Their Use and Renewal.* San Francisco, CA: Freeman (first published Moscow, 1963).

Girvan, N. 1976: *Corporate Imperialism: Conflict and Expropriation.* New York: Monthly Review Press.

Glantz, M. H. (ed.) 1977: *Desertification.* Boulder, CO: Westview Press.

Glover, Sir H. 1946: *Erosion of the Punjab: Its Causes and Cure.* Lahore: Feroz Printing Works.

Goudie, A. 1981: *The Human Impact: Man's Role in Environmental Change.* Oxford: Blackwell.

Graham, R. 1981: *The Aluminium Industry and the Third World.* London: Zed Press.

Grainger, A. 1980: The state of the world's tropical rain forests. *Ecologist,* 10, 6–52.

Gunnemann, J. P. (ed.) 1975: *The National State and Transnational Corporations in Conflict.* New York: Praeger.

Hagen, E. E. 1972: Limits to growth reconsidered. *International Development Review,* June, 10–12.

Hayek, F. A. 1975: *Full Employment at Any Price?* London: Institute of Economic Affairs, Occasional Paper No. 45.

Howard, P. H. 1981: Impressions of soil and water conservation in China. *Journal of Soil and Water Conservation,* 36, 122–4.

Howe, C. W. 1979: *Natural Resource Economics Issues, Analysis and Policy.* New York: Wiley.

Hyams, E. 1952: *Soil and Civilization,* 2nd edn. London: John Murray (1976).

Hyden, G. 1983: *No Shortcuts to Progress: African Development Management in Perspective.* London: Heinemann.

Jacks, G. C. and Whyte, R. O. 1939: *Vanishing Lands: A World Survey of Soil Erosion.* New York: Doubleday Doran.

Jaeger, J. 1988: *Developing Policies for Responding to Climatic Change.* Summary of discussions held at Villach, 28 September–2 October 1987, and Bellagio, 9–13 November 1987). World Meteorological Organization/United Nations Environmental Programme, WMO/TD No. 25, Mimeo.

Jones, P. D. and Warrick, R. A. 1988: Greenhouse effect and the climate. *Atom,* 381, 13–15.

Kahn, H., Brown, W. and Martel, L. (eds) 1976: *The Next 2000 Years.* New York: William Morrow.

Kitching, G. 1982: *Development and Underdevelopment in Perspective: Populism, Nationalism and Industrialisation.* London: Methuen.

Kitching, G. 1983: *Rethinking Socialism.* London: Methuen.

Komarov, B. 1981: *The Destruction of Nature in the Soviet Union.* London: Pluto Press.

Laclau, E. 1971: Feudalism and capitalism in Latin America. *New Left Review,* 67, 19–38.

Laclau, E. 1977: *Politics and Ideology in Marxist Theory.* London: New Left Books.

Lanly, H. P. 1982: *Tropical Forest Resources.* FAO Forestry Paper 30. Rome: FAO.

Lanning, G. and Mueller, M. 1979: *Africa Undermined.* Harmondsworth: Penguin.

Leslie, A. J. 1980: Logging concessions: how to stop losing money. *Unasylva,* 32 (129), 2–7.

Loprieno, N. 1986: Radiation knows no frontiers. *European Environment Review* 1 (1), 2–9.

MacEwan, M. 1984: The greening of Britain. *Marxism Today,* July, 23–7.

Maddox, J. 1972: *The Doomsday Syndrome.* London: Macmillan.

Malcolm, E. W. 1938: *Sukumuland: An African People and Their Country.* London: Oxford University Press for International African Institute.

McGill, S. M. 1987: *The Politics of Anxiety: Sellafield's Cancer-Link Controversy.* London: Pion.

McGourty, C. 1988: Seal epidemic still spreading. *Nature, London* 333, 553.

Meadows, D. H. Meadows, D. L. and Anders, J. 1972: *The Limits to Growth.* London: Earth Island.

Meillassoux, C. 1974: Development or exploitation: is the Sahel famine good business? *Review of African Political Economy,* 1, 27–33.

Mesarovic, M. and Pestel, E. 1975: *Mankind at the Turning Point.* London: Hutchinson.

Messer, J. 1987: The sociology of land degradation in Australia. In Blaikie and Brookfield (1987), 232–8.

Morgan, R. P. C. 1986: Soil erosion in Britain: loss of a resource. *Ecologist,* 16 (1), 40–2.

Myers, N. 1983: The tropical forest issue. In T. O'Riordan and K. Turner (eds), *Progress in Resource Management and Environmental Planning,* vol. 4, pp. 1–28.

OECD (Organization for Economic Co-operation and Development) 1972: *Problems of Environmental Economics.* Paris: OECD.

O'Keefe, P., Wisner, B. and Baird, A. (eds) 1977: *Kenyan Underdevelopment: A Case Study of Proletarianization.* London: International African Institute.

O'Riordan, T. 1976: *Environmentalism.* (2nd edn, 1981) London: Pion.

Orr, D. W. and Soroos, M. S. 1979: *The Global Predicament.* Chapel Hill, NC: University of North Carolina Press.

Plumwood, V. and Routley, R. 1982: World rainforest destruction – the social factors. *Ecologist,* 12, 4–22.

Pryde, P. R. 1972: *Conservation in the Soviet Union.* Cambridge: Cambridge University Press.

Qu Geping 1980: Deserts in China and their prevention and control. *Mazingira,* 4, 74–9.

Rasool, S. I. 1984: On dynamics of deserts and climates. In J. Houghton (ed.), *The Global Climate.* Cambridge: Cambridge University Press.

Redclift, M. 1983: *Development and the Environmental Crisis: Red or Green Alternatives?* London: Methuen.

Redclift, M. 1984: *Sustainable Development: Exploring the Contradictions.* London: Methuen.

Richards, P. (ed.) 1975: *African Environment: Problems and Perspectives.* London: International African Institute.

Richards, P. 1985: *People's Science: Ecology and Food Production in West Africa.* London: Methuen.

Rosewaal, T. 1987: *Scale and Global Change*. SCOPE Report No. 35.

Rounce, N. V. 1949: *The Agriculture of the Cultivation Steppe of the Lake, Western and Central Provinces*. Capetown: Longmans.

Routley, R. and Routley, V. 1980: Destructive forestry in Melanesia and Australia. *Ecologist*, 10, 56–7.

Sandbach, F. 1980: *Environment, Ideology and Policy*. Oxford: Blackwell.

Science Council of Canada. 1986: *A Growing Concern – Soil Degradation in Canada*. Ottawa: Science Council of Canada.

Simon, J. L. 1981: *The Ultimate Resource*. Princeton, NJ: Princeton University Press.

Smil, V. 1979: Controlling the Yellow River. *Geographical Review*, 69, 251–72.

Smil, V. 1983: *The Bad Earth: Environmental Degradation in China*. London: Zed Press.

Smil, V. 1987: Land degradation in China. In P. M. Blaikie and H. C. Brookfield (eds), *Land Degradation and Society*. London: Methuen.

Smith, V. (ed.) 1979: *Scarcity and Growth Reconsidered*. Baltimore, MD: The Johns Hopkins University.

Stewart, F. 1977: *Technology and Underdevelopment*. London: Macmillan.

Stocking, M. 1988: Recognising environmental crises in Africa. In P. M. Blaikie and T. Unwin (eds), *Environmental Crises in Developing Countries: How Much, For Whom, By Whom?* DARG Monograph Series No. 5. London: Institute of British Geographers.

Svedin, U. 1988: The concept of sustainability. In *Perspectives of Sustainable Development*. Stockholm: Stockholm Studies in Natural Resources Management No. 1, 5–18.

Thompson, E. P. 1968: *The Making of the English Working Class*. Harmondsworth: Penguin.

Thompson, M. and Warburton, M. 1985a: Uncertainty on a Himalayan scale. *Mountain Research and Development*, 5 (62), 115–35.

Thompson, M. and Warburton, M. 1985b: Knowing where to hit it: a conceptual framework for the sustainable development of the Himalaya. *Mountain Research and Development*, 5 (3), 203–20.

Thompson, M., Warburton, M. and Hatley, T. 1986: *Uncertainty on a Himalayan Scale*. London: Milton Ash Editions, Ethnographica.

United Nations 1976: *Transnational Corporations: Issues Involved in the Formulation of a Code of Conduct*. New York: UN Centre on Transnational Corporations.

United Nations 1980a: *Sea-Bed Mineral Resource Development: Recent Activities of the International Consortia*. New York: United Nations Department of Economic and Social Affairs.

United Nations 1980b: *The Activities of the Industrial Mining and Military Sectors of Southern Africa*. New York: United Nations Centre on Transnational Corporations.

United Nations 1980c: *The Nickel Industry and the Developing Countries*. New York: United Nations Department of Technical Cooperation for Development.

United Nations 1981a: *Transnational Corporations in the Bauxite/Aluminium Industry*. New York: United Nations Centre on Transnational Corporations.

United Nations 1981b: *Transnational Corporations in the Copper Industry*. New York: United Nations Centre on Transnational Corporations.

UNDP (United Nations Development Program) 1980: *Project Document, National Farm Management Study, No. NEP/80/035/OP/A*.

UNEP (United Nations Environmental Program) 1986: *Radiation: Doses, Effects, Risks*. Geneva: UNEP.

Walker, J. and Rowntree, P. R. 1977: The effect of soil moisture on circulation and rainfall in a tropical model. *Quarterly Journal of the Royal Meteorological Society*, 103, 29–46.

Wallerstein, I. 1979: *The Capitalist World-Economy*. Cambridge: Cambridge University Press.

Warren, B. 1973: Imperialism and capitalist industrialization. *New Left Review*, 81, 3-44.

Warren, B. 1980: *Imperialism: Pioneer of Capitalism*. London: New Left Books.

Watts, M. 1983: On the poverty of theory: national hazards research in context. In K. Hewitt (ed.), *Interpretations of Calamity*. London: Allen & Unwin, 231-62.

Watts, M. 1984: The demise of the rural economy: food and famine in a Sudano-Sahelian region in historical perspective. In E. Scott (ed.), *Before the Drought*. Boston: Allen & Unwin.

Williams, G. 1976: Taking the part of peasants: rural development in Nigeria and Tanzania. In P. Gutkind and I. Wallerstein (eds), *The Political Economy of Contemporary Africa*. London: Sage Publications.

Young, O. R. 1982: *Resource Regimes*. Berkeley, CA: University of California Press.

6

Malthus, Marx and Population Crises

ROBERT WOODS

Malthus's principle implies, without explicitly stating, the existence of an optimum population, i.e. a population the size of which maximizes the level of real wages or income per head. It also suggests the possibility of sub-optimum and supra-optimum populations which, if excessively low or high, will lead to population crises. Such crises may therefore result from either underpopulation or overpopulation. Malthus himself was concerned almost entirely with overpopulation, the geometrical rate of population growth, the arithmetical rate at which food supply could be increased and the discrepancy between the two rates. Crises would be characterized by 'misery and vice' but also by substantial increases in mortality which would act to redress the imbalance between supply and demand by reducing the latter, thereby easing population towards the optimum size. However, these Malthusian subsistence crises do not represent the only forms of population crises that are possible, nor were they in any sense inevitable for, according to Malthus, the tension between the rate of population growth and that of food supply would lead to other preventive checks on population increase.

Malthus's first *Essay* of 1798 and his second *Essay* of 1803 were particularly influential amongst his contemporaries and have proved so again in the late twentieth century when the problem of world overpopulation seems acute.* In this chapter I shall first review the current state of the world demographic system using this neo-Malthusian perspective. Second, the concept of crises of numbers will be questioned in the context of both other potential causes of population-related crises and the long-term recurrence of similar events. Third, I will examine the Malthusian principle at greater length by contrasting

* The literature on Malthus's principle of population and his political economy (Malthus, 1970, 1973) is voluminous, but see especially Spengler (1972), Petersen (1979), Dupâquier et al. (1983), Coleman and Schofield (1986) and Winch (1987).

Table 6.1 World population growth since 1950

	Population (millions)								1980–5 Growth rate (% p.a.)	1980–5 CBR	1980–5 CDR	Projections (millions) 2000	Projections (millions) 2050
	1950	1955	1960	1965	1970	1975	1980	1985					
World	2525	2757	3037	3354	3696	4066	4432	4837	1.7	27	11	6206	9973
Africa	220	245	275	312	355	407	470	555	2.9	46	17	903	2297
North America	166	182	199	214	226	236	248	264	0.9	16	9	286	320
Latin America	164	188	216	248	283	322	364	405	2.3	32	8	543	868
East Asia	673	738	816	899	994	1096	1175	1250	1.2	19	7	1443	1709
South Asia	716	786	877	988	1117	1256	1404	1568	2.2	34	12	2164	3810
Europe	392	408	425	445	459	474	484	492	0.3	14	11	527	570
Oceania	13	14	16	18	19	21	23	25	1.5	21	8	28	37
USSR	180	196	214	231	242	253	265	279	1.0	19	9	312	362

Sources:: United Nations *Demographic Yearbook*, 1981, 1985; for projections Demeny, 1983, p. 110

Table 6.2   Percentage share of world population

	1950	1980	2000	2050	Increase 1980–2000
Africa	8.71	10.60	14.55	23.03	24.42
North America	6.57	5.60	4.61	3.21	2.14
Latin America	6.50	8.21	8.75	8.70	10.10
East Asia	26.65	26.51	23.25	17.14	15.12
South Asia	28.36	31.68	34.87	38.20	42.87
Europe	15.52	10.92	8.49	5.72	2.43
Oceania	0.51	0.52	0.45	0.37	0.28
USSR	7.13	5.98	5.03	3.63	2.65

Source: based on table 6.1

it with the concept of surplus population advanced in Marxian political economy.

THE CONTEMPORARY WORLD DEMOGRAPHIC SYSTEM

The population of the world in the late 1980s is approaching 5 billion. In 1950 it was 2.5 billion, and in 2050 it may be 10 billion. In these circumstances it is obvious why Malthusianism is again influential and why population is believed to pose a 'global problem' yet, as table 6.1 reveals, there are extreme regional variations in the rate and character of population growth. During the early 1980s the rate of world population increase was about 1.7 per cent per annum, but whereas in Africa, Latin America and South Asia it was 2 per cent or more, in Europe, the USSR and North America it was 1 per cent or less. One of the implications of these differential rates is illustrated in table 6.2 which shows the changing percentage of the global population in each region.

   The world can be divided into three. In Africa and South Asia growth is rapid and will remain so well into the twenty-first century, by which time this area will contain over 50 per cent of the global total. In Latin America and East Asia growth has been rapid, but its pace has slackened, especially in East Asia, and in consequence this area is likely to maintain its 30 per cent share. The remaining 20 per cent of the world's population will live in Europe, the USSR, North America and Oceania, but this area will contain only 8 per cent of world population growth between 1980 and 2000 compared with two-thirds and a quarter in the other two areas.

   The immediate cause of these different growth rates is also obvious from table 6.1. The crude death rate (CDR) is now relatively low, but the crude birth rate (CBR) remains particularly high in many regions. Figure 6.1 illustrates international variations in CDR and CBR and thus the level of natural increase

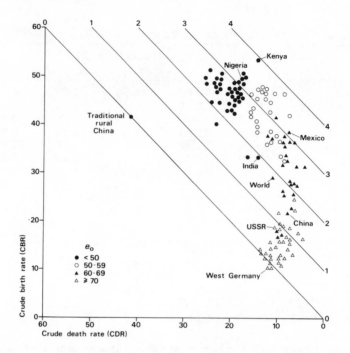

Figure 6.1   International variations in estimated crude birth and death rates, 1975–1980
Source: table 6.1

which is shown on the diagonal lines as the annual growth rate per cent. Figure 6.1 also classifies countries by their estimated or known life expectations at birth, e_0. The four classes of e_0 used in figure 6.1 are also illustrated in figure 6.2. Clearly there are substantial differences in the mortality levels experienced, yet these differences are not as important as those in the pattern of fertility since it is the latter that largely determines the population growth rate.*

In the late 1970s the population of Kenya was growing by nearly 4 per cent per annum and the CBR was higher and the CDR lower than even African standards, of which Nigeria was more typical (Dow and Werner, 1983; Page and Lesthaeghe, 1981). In Mexico growth rates exceeded 3 per cent with mortality nearly as low as in Europe (in terms of CDR at least) (Hicks, 1974; Seiver, 1975). The case of India, which dominates the South Asia region, raises a number of interesting issues for, whilst the growth rate was at about 2 per cent, e_0 had probably only just reached 50 by 1980 and although the CBR had

---

* The following provide important reviews of contemporary international demographic patterns: United Nations (1979); Tabah (1980, 1982); Coale (1982, 1983a,b); Demeny (1982, 1983). In this chapter we will not be particularly concerned with the general relationship between population and development, for which see Coale and Hoover (1958), Cassen (1976), Simon (1977) and Birdsall et al. (1979).

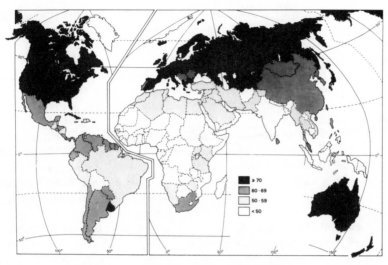

Figure 6.2   International variations in estimated life expectation at birth e_0, 1975–1980
Source: table 6.1

declined the pace of the fertility transition had not proved to be as rapid as in parts of East and South-east Asia (Cassen, 1978; Chaudhry, 1982).

These three countries – Kenya, Mexico and India – provide contrasting examples of areas which might be thought to be experiencing a population crisis, because of its rate of increase, or to be overpopulated, because of its density. However, there are other areas with rather different demographic conditions where the terms population crisis and overpopulated are just as appropriate. In the late 1970s China appears to have had only a moderate growth rate, yet in the two preceding decades the disparity between a rapidly falling death rate and a persistently high birth rate led the Chinese government to adopt an extreme neo-Malthusian policy. In Western Europe, in contrast, there are examples of national populations with death rates in excess of birth rates where depopulation is a legitimate fear (Federal Republic of Germany in figure 6.1, for example; Bourgeois-Pichat, 1981; Kirk, 1981; van de Kaa, 1987).

Apart from these countries which, in terms of the implied Malthusian definition of population crises, may be appropriate candidates for the use of the term, there are other instances in which the phrase might be justified. Within the USSR there are extreme differences between the contemporary demographic systems of European Russia (especially the Baltic States) and Soviet Central Asia. Whilst the latter is similar to South-west Asia, the former is far closer to Eastern and Western Europe. One consequence of this demographic division will be the declining proportion of Russians in the Soviet population and the exacerbation of a nascent ethnic problem in Soviet society (Heer, 1965; Coale et al., 1979; Jones and Grupp, 1983). In parts of Europe the problems of demographic aging are particularly acute, whilst the age structures of most Latin American, African and Asian countries are such that the phrase 'youth revolution' has been widely used. In the former the escalating cost of welfare payments to the elderly imposes strains on the economically active, whilst in the latter the increased costs associated with child health care and education are immediate, and the costs linking underemployment and unemployment in a rapidly expanding labour force are likely to prove critical to the prospects for political stability.

The term over-urbanization has been used in instances where it seems that there is a maldistribution of population towards a rapidly expanding urban sector whose employment base is not increasing in step (Gugler, 1982). Rural depopulation is not an automatic consequence, although even where the absolute size of the rural population also increases some relief from pressure may be experienced (Clarke and Kosinski, 1982; Grigg, 1980). Although these three additional examples – nationally divisive demographic subsystems, extreme age structures, over-urbanization and redistribution – all in their own ways create conditions that if extreme might be regarded as population crises, they are in a sense peripheral to the central issues in the Malthusian and even the popular conception of a population crisis. The central issues are rate of growth, size and thus density of population. Figure 6.1 shows that on at least one of these traditional criteria conditions appear ripe for a crisis of numbers in several countries, but for others in similar or only slightly better economic circumstances neither criterion is met or, if either is, then in not such an obvious manner. Has the population crisis been averted in China, for example? If so, by what means? Can these means also be applied in Nigeria, Kenya, Mexico or even India? Our ability to answer these questions depends upon an understanding of the processes which produce fertility decline. Given that a return to the former high levels of mortality is unacceptable, a fall in fertility is the only possibility for the reduction in the rate of population growth. (Bulatao and Lee, 1983, survey the most recent work on the determinants of fertility.)

China and India

Figure 6.3 provides a summary of the recent changes in Chinese mortality and fertility in terms of the CBR, the CDR and the total fertility rate (TFR) (Coale, 1981a; Pressat, 1982, p. 33). All three rates should be subjected to close scrutiny; they are estimates based on Chinese sources which may always be checked and are likely to be revised in the light of both the population census and the Chinese State Family Planning Commission's fertility survey of 1982 (Aird, 1982; Pressat, 1982, 1983; Caldwell and Srinivason, 1984). The figure appears to show both the substantial reductions in mortality and fertility that have occurred since 1949, and the effects of disruption and famine during the Great Leap Forward of the late 1950s and early 1960s. The demography of traditional rural China has been reconstructed by Barclay et al. (1976) using survey data collected in 1929–31. Their estimates suggest a CBR of 41.2, a CDR of 41.5, an e_0 of

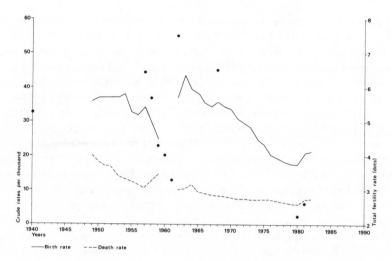

Figure 6.3   Changing estimated crude birth and death rates in the People's Republic of China
Source: based on Pressat, 1982

24.2 and a TFR of 5.50. The mean age at marriage was 21.32 for men and 17.52 for women, with virtually all women marrying before age 50 (I_m was 0.874). Marital fertility was of the natural fertility form with no evidence for parity-specific control, yet it was particularly low by most historical European standards (I_g and I_f was 0.446). Some of these measures are now available for the early 1980s, again in estimated form. Life expectation at birth e_0 is probably in the low sixties and infant mortality is about 50 per thousand (Banister and Preston, 1981; Banister, 1987). Figure 6.3 shows that the TFR was 2.24 in 1980 and 2.63 in 1981, but in the urban areas it was only 1.15 and 1.39 respectively in those years. It may well be that the fertility transition began in the mid-1950s in these urban areas. Its progress can be linked with the growing involvement of women in the labour force and perhaps the economic and social upheavals that followed the founding of the People's Republic.

   The first birth control campaign was probably largely ineffective and certainly ended when the Great Leap Forward was initiated in 1958. The second campaign (1962–6) does not appear to have been very effective either, but the third, which began in 1968 and was re-emphasized in 1972–3 and 1979, has proved a most important vehicle for persuasion or coercion (Aird, 1978a; Caldwell and Srinivasan, 1984). The aims of the present campaign as stated in 1979 were to reduce the growth rate of the Chinese population to 0.5 per cent by 1985 and to zero by 2000, to increase the age at marriage to 25 for men and 23 for women in rural areas, and to 27 and 25 respectively in urban areas, and to reduce the completed family size to one at best and two under certain circumstances, but certainly not three or more (Coale, 1981a,b). Several attempts have been made to examine the long-term consequences if these goals were to be achieved. Chen and Tyler (1982), for example, have made the necessary projections which show that immediate adoption of the one-child family norm would stabilize the population by 2000, and that zero growth would be achieved by 2020 with the two-child family. A three-child family norm would ensure the doubling of population to about 2 billion by 2030. The implication of these projections is clear: unless the TFR is reduced to, and maintained at, approximately 2, the accelerated growth of the Chinese population which has occurred in the period since the 1960s will not be curtailed.

Establishing the one- or two-child family norm has posed particularly difficult problems for the Family Planning Commission. First, it has been necessary to employ all the resources of the state's propaganda machinery to persuade couples that one child is sufficient when Chinese tradition lays particular emphasis on the family, children – especially boys – and the family economy (Baker, 1979, Watson, 1982). However, the possession of a 'one-child certificate' (acquired after sterilization or the elapse of four years from the birth of a first child and the couple's promise to have no more) entitles the child and parents to special educational and financial benefits. The penalties for breaking the terms of the certificate are severe, as are those imposed on couples who have two or more children. The balance of carrot and stick has been such that nationally some 56 per cent of eligible couples were reported to have certificates in 1980, an increase from 34 per cent in 1979, but in Beijing the figure was 79 per cent (Goodstadt, 1982). Second, couples have had to adapt to the use of new methods of birth control, and to submit to sterilization or abortion in a third or subsequent pregnancy. The results of the survey, reported by Caldwell and Srinivasan (1984, p. 77), suggest that amongst married couples where the wife was under 50 up to 25 per cent had been sterilized, 35 per cent used intra-uterine devices (IUDs), 6 per cent used oral contraceptives, 1 per cent used condoms, and 31 per cent were not using birth control methods. Many of the women in the last-mentioned group would be in their twenties. Third, there are striking differences between the urban and rural sectors of Chinese society. Family size targets are lower in the urban areas, the proscribed marriage ages are higher and the family planning programme is more effective. However, only about 20 per cent of the Chinese population lives in towns. The problem is illustrated quite effectively in the following way: in 1981, 86 per cent of all births to urban couples were first births, whereas among rural couples the figure was 40 per cent, with only 44 per cent in China as a whole (Caldwell and Srinivasan, 1984, p. 79). In the rural sector in particular there is a lasting desire for early marriage, often arranged by parents, and for large families, so that the kin-related labour supply can be maintained, the wishes of grandparents fulfilled and some measure of security in old age guaranteed. Whilst these traditional and entirely rational objectives remain they represent the most important restriction on the success of the family planning programme and the general demographic strategy of which it is the major part (Qian Xinzhong, 1983).

Some of these remarks could also be applied to rural India. However, as figure 6.1 suggests, the pace of demographic change in India has been relatively slow during the last two decades with population growth at about 2.20 per cent per annum in the 1960s and 2.23 per cent in the 1970s. Although a family planning programme has existed in India since 1952, and has been much expanded since 1966, in the late 1970s only 24 per cent of couples in the reproductive age group were protected by some effective method of birth control (Cassen, 1978). Of these 24 per cent, sterilization represents by far the most important category. The impact of the programme has not been spectacular nationally since between the 1960s and 1970s the CBR fell by only 13–16 per cent from about 42 to 38–39, thereby only keeping pace with the decline in the CDR (Jain and Adlakha, 1982, p. 595; see also Dyson and Crook, 1984). On reflection, some of the reasons for the programme's limited influence are obvious. For example, Gwatkin (1979) provides an account of the government's difficulties in 'enforcing' a rigorous sterilization scheme during the mid-1970s and the state of national emergency to which it contributed. However, the underlying causes are less obvious. There is considerable demographic diversity amongst the Indian states in terms of both mortality and, particularly, fertility variations (Woods, 1982, pp. 72–85). This diversity is clearly not merely a function of economic development, levels of living or urbanization, but seems to reflect cultural divisions between north and south (Dyson and Moore, 1983). In the north – Rajasthan and Uttar Pradesh, for example – mortality and fertility rates are well above the national average and the take-up rate for family planning is low, but in the south – especially Kerala – mortality and fertility are lower and the family planning programme has been quite successful. These differences can be associated with, amongst other variables, regional differences in female social status. In the south marriage and kinship arrangements favour a higher level of female autonomy – female literacy is far higher for instance – and the interests of child and mother are of greater importance than appears to be the case in the male-oriented northern societies.

The simple comparison of China and India is not particularly helpful, but it is reasonable to observe certain apparently significant differences (Coale, 1983b). The ability and will of the state to enforce a rapid change in family planning practice are certainly important. Both governments have acted in a neo-Malthusian fashion, but in China the political machine has been effective and the public will supportive or at least compliant, whereas in India the democratic institutions have been used to thwart excessive haste. However, it is possible that the fertility levels that pertain in rural south India are equivalent to those in many areas of contemporary rural China. In India the attainment of a stationary population must remain a long-term goal which will be achieved with economic development, social modernization and the encouragement of government planners. In China the same goal can be realized more quickly, but probably not as soon as is hoped for in government circles.

Latin America

The rate of population growth has been particularly rapid in recent decades
in Latin America. Mortality has declined abruptly since the 1940s and 1950s
(Palloni, 1981; see also United Nations, 1982), but fertility has remained
persistently high in many of the larger countries and is only now beginning
to decline. The reasons for the international and regional variations in fertility
levels together with the leads and lags in decline have proved particularly
difficult to explain. They are not simply and universally a reflection of socio-
economic differences, literacy, education, urbanization or the availability of
an effective family planning programme (Beaver, 1975), but each of these
influences can be important in certain circumstances. To illustrate the complex
nature of the problem let us examine Stycos's summary of his results for
Costa Rica.

> The case of Costa Rica is of special interest because the fertility fell so
> fast despite the fact that in other aspects it is more typical of Latin
> America. While its literacy levels have been unusually high for some time,
> the substantial improvements in post-primary schooling in the 1950s
> probably effected an educational breakthrough critical for fertility.
> The initiators of the decline were young couples, living in the more
> economically developed cantons, especially those undergoing the greatest
> reduction in agricultural employment. Such couples cut the number of
> their children by such means as *coitus interruptus* and the condom, methods
> unencumbered by legal restrictions. Information about contraception was
> diffused by freer communication than elsewhere between spouses – a
> product of higher educational levels in Costa Rica, and possibly of cultural
> norms more permissive of sex-related discussions. This set a favourable
> scene for the family-planning programme, which both stepped up
> communications and made new contraceptives available – IUDs and pills.
> Once the programme had been launched, older or less educated women
> were attracted to contraception, and the fertility of the more backward
> cantons began to decline. (Stycos, 1978, p. 424)

Despite these developments Costa Rica still had a CBR of 31.1 and a CDR
of 4.1 in the late 1970s (e_0 was 68 and the annual growth rate was 2.70 per
cent), but in Mexico the CBR was 38.3 and the CDR was 7.8 (e_0 was 66 and
the growth rate was 3.05 per cent).

From the Malthusian perspective Mexico appears to provide a classic example
of a society experiencing a population crisis where the rate of growth is out
of balance with economic growth or development (Coale, 1978). Until 1973
there was little or no effort on the part of the government to encourage family
limitation, although the provision of medical facilities was extended and, like
other Latin American countries, mortality declined. Those analyses based on
data for 1960 and 1970 have pointed to the perpetuation of high fertility (the

CBR was 45 and the CDR was 9 in the early 1970s) despite rapid economic development, increases in literacy, education, urbanization and an increasing female labour force participation rate (Hicks, 1974; Seiver, 1975). Although many of these classical demographic transition theory variables did help to account for cross-sectional variations in fertility – especially the share of the labour force employed in agriculture (positively related to fertility) and the percentage of the population speaking an indigenous language (negatively related to fertility) – fertility decline was limited and it appeared to be related to reduction in the agricultural labour force and increase in life expectation, if any of the variables. More up-to-date accounts await the availability of the 1980 census, but even in Mexico there are some indications of the first stages of a fertility transition which may have elements in common with the experience of Costa Rica.

A CRISIS OF NUMBERS?

The governments of China, India and, rather more recently, Mexico have each identified a population problem in their respective countries. By this they affirm their belief in the need to keep population under control, the beneficial influence that a slow rate of growth may have for economic development and their fear of the excesses which a crisis of numbers may bring for economic, social and political stability. Having reviewed the state of the contemporary world demographic system and the reaction of certain governments to their current position, we can now reconsider the concept of a crisis of numbers and population crises in general. This will be done in two ways: by examining the structure of demographic systems, and by reviewing the influence of crises in historical perspective (Woods, 1987).

Figure 6.4 places most of the important elements which must operate by definition in any demographic system in a form of order. It should be read from the top downwards and outwards. A population may appear to have reached a crisis point if either the rate of its growth or its size are out of balance with the prevailing economic and social conditions, yet a number of the elements in figure 6.4 are by the same token able to generate forms of crisis in their own right. Therefore the use of the word 'crisis' is appropriate at two levels. First, we can distinguish a general crisis of numbers which will affect the nation or regions thereof. Second, more particular tensions which are linked with subelements of the demographic system may arise. It is also obvious that some of the general crises stem from imbalances between rates of change in the three major demographic components: mortality, fertility and migration.

A small number of examples must suffice to illustrate these distinctions. Contemporary international labour migration serves to redress the imbalance between levels of regional economic development since it is attuned to the most obvious spatio-temporal differentials, but it is also associated with political and social crises. In the United States the political reaction to illegal Mexican immigration has been substantial (Grebler, 1966; Dagodag, 1975), and the governments of France, the Federal Republic of Germany and Switzerland have all been faced with a new migrant ethnic minority problem to which they have reacted with a variety of restrictions designed to meet perceived national self-interest and political expediency (Castles and Kosack, 1973; White and Woods, 1983). In the Republic of South Africa the need for a pliant workforce together with the threat posed by extreme differences in rates of growth between white and non-white populations have encouraged the development of a strictly controlled system of internal circulatory migration (Smith, 1983). The crisis associated with the Sahel drought apparently provides a classic example of the consequences of overpopulation coupled with environmental degeneration. In this example we have a modern demographic crisis (see figure 6.4) of the subsistence variety having its origins in food shortages and ultimately overpopulation. However, such a simple interpretation, which equates famine with natural disasters, conceals complicated arrangements for the allocation of food resources, termed entitlements by Sen (1981). Sen contrasts the 'food availability decline' interpretation, which emphasizes the production and supply of food, with the 'exchange entitlements' approach, which emphasizes the distribution of food to consumers and the relative position of those consumers in the market for food. Famines can thus occur when the exchange entitlements of a population decline and need not merely be a consequence of a reduction in the availability of food. Sen (1981) provides detailed examples of this phenomenon drawn from Bengal (1943), Ethiopia (1973), the Sahel (1968–73) and Bangladesh (1974).

These examples suggest, amongst other things, the variety of forms which a population crisis can take and, by implication, the imprecision of the phrase

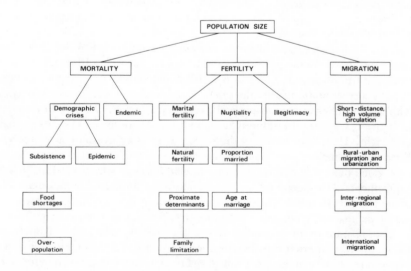

Figure 6.4   Elements of the general demographic system

as a technical term in modern usage. In contrast, the use of the equivalent phrase 'demographic crisis' in the historical literature is normally restricted to the direct consequences of a sharp, usually short-term, increase in mortality. As figure 6.4 shows, there are two major sources for these demographic crises. The most straightforward of these relates to outbreaks of epidemic disease, such as bubonic plague, typhus, smallpox and influenza. The second source – often referred to as subsistence – is apparently coincidental with Malthus's concept of the positive check. Here food shortages are a direct effect of overpopulation: famine, or at least widespread malnourishment, and increased mortality are the inevitable consequences. However, absolute food shortages also arise when inclement weather affects the harvest or when livestock suffer from disease, whereas relative food shortages amongst certain social categories relate to the system of allocation and entitlement mentioned above.

It has proved particularly difficult to establish an empirical definition of demographic crises, to infer the most likely causes of particular events and to assess the importance of crisis-related mortality for historical demographic systems. Most of these difficulties are related to the vagaries of interpreting historical data. The key definition variable for such demographic crises has been the extent to which crisis mortality exceeds the normal level of mortality. Crisis periods are therefore those in which mortality exceeds the normal by an arbitrarily determined amount (Schofield, 1972). The application of this definition is dependent upon the availability of reliable mortality statistics but breaks down where and when there is only a general impression of high mortality or merely reference to famine, death or disease. For example, the return of bubonic plague to Western Europe in 1348–9 represents a clear instance of a substantial demographic crisis yet the extent of mortality above the average is beyond accurate measurement, although it is generally presumed that from 25 to 50 per cent of the population died in the first outbreak (Hatcher, 1977; McNeill, 1977). Where parish registers provide a more accurate means of identifying mortality crises – their intensity, frequency and cause – there is obviously scope for a more detailed assessment. Wrigley and Schofield's (1981, pp. 645–93) analysis of data from 404 English parishes in the period from the 1540s to the 1830s is a particularly important example in this respect. In only four years did more than a quarter of the parishes experience a mortality crisis. There was no national crisis of numbers in England during these three centuries, although there were many localized crises – often associated with epidemic disease – and several regional crises. On the basis of this analysis there seems little reason to suppose that demographic crises played an important part in the general population system, and even if they did locally or for short periods regionally then they were of the essentially random epidemic disease variety rather than the subsistence kind. However, England may not have been typical even of Europe. France suffered regional famines in the seventeenth century and severe food shortages in the eighteenth, Scandinavia was racked by severe mortality crises at the end of the eighteenth century and it has been argued that much of Western Europe experienced a subsistence crisis in the early decades of the nineteenth century. (On France see, for example, Meuvret, 1977; on Scandinavia see Jutikkala, 1955, Jutikkala and Kauppinen, 1971, and Sogner, 1976, and for Napoleonic Europe see Post, 1977, and Flinn, 1981, pp. 47–64.)

The impact of demographic crises certainly varies with period and place. Low rates of population growth in fifteenth-century England have been associated with attacks of bubonic plague (Gottfried, 1978). The stationary population of Tokugawa, Japan is said by some economic historians to have been related to chronic famine conditions (Hanley and Yamamura, 1977). In general, however, the various pre-industrial demographic regimes may have been influenced more by the background level of mortality – especially high infant mortality – than by the dramatic but irregular additional element of crisis mortality. Cassen and Dyson (1976) provide an interesting illustration of the negligible long-term impact of demographic crises in their study of Indian population dynamics. Bubonic plague is possibly the most significant exception to this general point. However, the scheme outlined in figure 6.4 also requires us to examine the significance of the balance between mortality and fertility for the changing rate of population growth. The north-west European marriage pattern, as defined by Hajnal (1965, 1982), developed from a number of interrelated social institutions amongst which the nuclear family system, the formation of new households at marriage and the necessity to ensure the independent household's economic viability before marriage can be contracted were the most important. Given the effective operation of these social 'rules', the consequent marriage pattern is likely to reflect prevailing economic conditions in the longer term. In the good years couples will marry relatively young, in their early twenties perhaps, but in the bad years marriage will be delayed to the late twenties or early thirties and celibacy will increase. By these means the demographic and economic systems will be linked via nuptiality since, although natural fertility predominates, the level of overall fertility will mainly be controlled by the age at first marriage and the proportion of the population marrying. After a short-run demographic crisis opportunities for household formation would increase, marriage would be contracted at a younger age and the birth rate would rise, thus replacing the lost population within a generation or two. The effective operation of this mechanism within an essentially agrarian society like Tudor England would tend to nullify the long-term impact of a single mortality crisis, but not the repeated recurrence of such events (Wrigley and Schofield, 1981).

Taking this historical perspective on pre-industrial western societies helps us to appreciate the complex nature of the crisis concept and to see that, although mortality crises were by no means rare, they were not necessarily of great importance for the working of the population system. This is not to say that at certain times villages, towns, and even whole regions were not devastated by epidemics or famines, but rather that the long-term influence of the dramatic and acute on population and economic growth must be brought into question.

OVERPOPULATION AND SURPLUS POPULATION

So far we have considered the incidence, nature and influence of population crises – variously defined – using an essentially Malthusian framework and largely within the terms of the demographic rather than the related economic, social or political systems. In this section we turn again to reconsider Malthus's *Principle of Population* and to compare his term *overpopulation* with Marx's concept of a *surplus population*. This comparison will enable us to examine wider aspects of crises beyond those that stress rates of population growth, size and acute events in the relatively short term. The possibility of a 'permanent crisis' affecting subpopulations also needs to be considered.

Malthus's theories of population and political economy often appear unsophisticated in the 1980s, yet the ideas contained therein are just as important and as debatable as they were over 150 years ago. Given a zero or negligible rate of technological progress in a predominantly agrarian society the functional relationship between income per head (or real wages) and the size of a population will tend to be positive when densities are low; population growth stimulates effective demand for commodity production, but a turning-point will be reached above which diminishing returns will transform the positive to a negative relationship. An extra increase in population size will depress even further the level of income per head to a point at which the material necessities of life are in such short supply that life itself is endangered. Once that point is reached then overpopulation is apparent (Grigg, 1980, p. 62; Schultz, 1981, pp. 9–61; Wrigley and Schofield, 1981, p. 460). Malthus himself compared the possible geometric rate of increase of population (using the United States as his example) and the known arithmetical rate of increase of agricultural production (in the absence of counter-examples) in order to illustrate his point that the former would inevitably outstrip the latter if there were no checks on the rate of population growth. Since it seemed that there was generally some balance between the rate of growth of subsistence and the level of real wages, effective checks must be in operation to make the balance possible and to avoid excessive overpopulation. These checks have come to be known as positive, mortality related (misery and vice), and preventive, nuptiality related (moral restraint). In the first *Essay* of 1798 the positive check was especially emphasized, but the second *Essay* of 1803 altered the balance of checks towards the preventive (see footnote to p. 151). As we have already seen, the preventive check appears to have worked in early modern England; life was not in general endangered by the positive check, but elsewhere and in theory the positive check, via an increase in mortality, could effect a reduction in population size and a return to some form of resources–population equilibrium. In short, Malthus argues that overpopulation (simply an excess of population compared with food supply or a population of such a size that an increase will adversely affect the level of real wages or income per head) is an ever-present possibility in human society which, when it occurs, will be rectified by an increase in mortality (a mortality

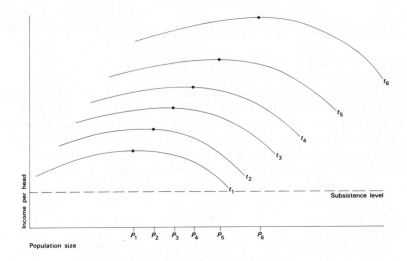

Figure 6.5   The influence of technological advance on the relationship between income per head and population size

crisis), but which can be averted by delaying marriage and thus effectively reducing the growth rate of a population by lowering its fertility.

The logic of the 'ever-present possibility' seems inescapable, but there are a number of other ways in which overpopulation can be avoided. Figure 6.5 illustrates one possibility. The t_1 function shows the classic curvilinear relationship between income per head and population size where the former is low and relatively close to the level of subsistence. The optimum population under these conditions is P_1. If we now assume, in contrast with Malthus, that technological advance is possible, that it even accelerates over time, then the functions $t_2, \ldots, t_6$ become appropriate. The related optima $P_1, \ldots, P_6$ show an increase, as of course does income per head. Although the principle of diminishing returns and overpopulation is retained in the functions $t_1, \ldots,$ t_6, by t_6 population increase beyond P_6 must be enormous before the subsistence level is reached and the material necessities of life are endangered. In these circumstances, although the logic of the ever-present possibility remains unbroken, technological advance makes its direct consequences unlikely. In oversimplified form figure 6.5 portrays the experience of the western economies over the last two centuries, except that marginally suboptimum or supraoptimum population sizes were actually involved (Schultz, 1981, pp. 47–51).

An additional obvious method of avoiding overpopulation would be to employ alternative more effective forms of the preventive check. Natural fertility began to disappear in France in the late eighteenth century and in the rest of Europe from the last decades of the nineteenth. The use of birth control methods is now universal in the West, and even in underdeveloped countries, as we have seen, substantial numbers limit their marital fertility to the extent that completed family size varies from two to four children. This neo-Malthusian strategy has been favoured by western governments and international agencies in their dealings with underdeveloped countries, but it has also been freely adopted and made most effective in China.

These two ways of avoiding overpopulation can still be discussed within Malthus's original framework; they merely require the relaxation of one of his most important assumptions concerning the limited ability of the rate of subsistence to increase and the extension of moral restraint to cover what for Malthus represented an immoral act – the deliberate use of methods of birth control. Marx's 'law of population' was developed from a radically different perspective.*

First, Marx was intent on countering the idea that the regulation of human population is controlled by a natural law. Rather, we should seek to understand the economic, political and thus social system which provides the context for changes in demographic patterns and structures. Since 'Capitalist production can by no means content itself with the quantity of disposable labour-power which the natural increase of population yields. It requires for its unrestricted activity an industrial reserve army which is independent of these natural limits' (Marx, 1976, p. 788), this emphasis on population as a dependent variable, albeit an important one, is particularly significant in the Marxian framework. Second, in *Capital* at least Marx aims to unravel the laws that govern the development of capitalism and is thus especially interested in the problem of capital accumulation and its relation to the supply of labour. Whilst the accumulation of capital feeds on appropriated surplus value the working population 'produces both the accumulation of capital and the means by which it is itself made relatively superfluous' (Marx, 1976, p. 783). The creation of a surplus population of unemployed

> is a necessary product of accumulation or of the development of wealth on a capitalist basis, this surplus population also becomes, conversely, the lever of capitalist accumulation, indeed it becomes a condition for the existence of the capitalist mode of production. It forms a disposable industrial reserve army, which belongs to capital just as absolutely as if the latter had bred it at its own cost. (Marx, 1976, p. 784)

The inherent phases of production in the capitalist industrial system create, but also require, a pool of unemployed or semi-employed 'hands', the size of which relates to the current stage of the cycle. Within the working class – that section of the population that does not own or control means of production, but gains its subsistence by selling its labour power – relations between members of the reserve and active armies, as well as the ratio of the former to the latter,

---

* Once again the literature of Marx's economic and social thought is enormous, but relatively little has been written concerning his approach to the population question which is to be found mainly in chapter 25 of *Capital* (Marx, 1976). Some interesting comparisons of Malthus and Marx are to be found in Hayes (1954), Petersen (1964), Daly (1971), Harvey (1974) and Woods (1983), but see also Meek (1953) for Marx and Engels on Malthus.

have important consequences. 'The over-work of the employed part of the working class swells the ranks of its reserve while, conversely, the greater pressure that the reserve by its competition exerts on the employed workers forces them to submit to over-work and subjects them to the dictates of capital' (Marx, 1976, p. 789). Marx also argues that the movement of wages will be determined by the ratio of members of the reserve army to those in the active army and not by variations in the absolute numbers of the working population. The higher the proportion of the working class in the reserve army, the lower will be the level of wages for those in the active army (Junankar, 1982, p. 85). The extent of pauperism will also be influenced by this reserve-to-active ratio, and those in the reserve army will obviously be most prone to poverty. There is a further claim: both fertility and mortality will be inversely related to the level of wages.

Marx concludes his discussion of the general law of capitalist accumulation by distinguishing four separate elements in the surplus population. The *latent* surplus population is to be found mainly in rural areas from whence it is drawn into the towns, when need arises, to become part of the manufacturing proletariat. The *floating* surplus population emigrates as capital emigrates and its size can be increased by the laying-off of apprentices or female workers. The *stagnant* element suffers from extremely irregular employment often associated with the decaying branches of industry. This partially active population is liable to suffer from a maximum of working time but a minimum of wages. The fourth element, that is 'the lowest sediment of the relative surplus population, dwells in the sphere of pauperism'. Again, there are four divisions: the lumpenproletariat (vagabonds, criminals, prostitutes), those who are able to work, orphans and pauper children, and 'the demoralized, the ragged and those unable to work'. Accordingly, for Marx, 'Pauperism is the hospital of the active labour-army and the dead weight of the industrial reserve army (Marx, 1976, p. 797).

Two additional points need to be examined a little more carefully before we turn to a more general evaluation of the implications of the Marxian model and its relation to the Malthusian principle. One of the most important features of Marx's economic thinking is the argument that short-run trade-cycle-like crises are an inherent feature of the capitalist system. During a crisis the size of the reserve army increases as workers are laid off, wage rates are depressed, and eventually profits begin to rise again and the 'beautiful trinity of capitalist production', as Marx (1976, p. 787) calls overproduction, overpopulation and overconsumption, can repeat itself. Economic crises, fluctuations in the absolute and relative size of the surplus population, and changes in the rate of capital accumulation are all closely linked in this scheme. The size and structure of the labour force responds to and effects in turn the system of economic change. Economic crises and population structures are thus closely linked in this scheme. Marx was further concerned to establish that 'in fact every particular historical mode of production has its own special laws of population, which are historically valid within that particular sphere' (Marx, 1976, p. 784). Of these laws he only outlined the one peculiar to the capitalist mode. We should be warned not to expect this 'capitalist law of population' to operate under the ancient slavery, feudal, Asiatic or socialist modes (Marx, 1964).

The contrasting concepts of overpopulation and surplus population can bear scrutiny from a variety of perspectives. Only two will be explored here and related to the contemporary world demographic system. It is obvious that overpopulation may not actually be present (depending on the way we wish to define the list of symptoms), but that it is *always* an 'ever-present possibility'. The principle from which it is derived is meant to be general and universally applicable, but even in its original Malthusian form over-population is by no means inevitable. Within the capitalist system the creation of a surplus population is inevitable. The limits to its existence are determined by the creation and destruction of an entire economic framework. Further, Marx's scheme emphasizes the importance of divisions within society which are to be defined in essentially economic terms. Owners of the means of production are distinguished from sellers of labour power, but even within the working class the active and reserve armies are distinctive, as are subdivisions within the surplus population. Access and allocation are of crucial importance here, rather than the size of the labour pool or the total population. A fall in income per head (one form of pauperization) is not a simple function of the rate of population increase or its size, but stems from the tensions between employers, employees and unemployed. Certain individuals may remain permanent, or at best long-term, members of the surplus population whereas larger numbers will be temporary members, part of the stagnant reserve army perhaps. Whether an individual family lives in poverty will be determined by the length of time spent by its potential wage earners as part of the surplus population.

It appears, therefore, that in Malthus's model population crises stem from excessive rates of population growth, but that in Marx's model crisis conditions exist for that alienated group whose economic position renders them socially and politically weak. The former would regard a population crisis as a severe, acute yet avoidable event, but the latter model envisages a permanent stage of crisis for a subpopulation of varying size.

Do either of these models, laws, schemes bear any relation to contemporary or historical reality? We could justifiably claim that both models provide interesting interpretive frameworks, but that their emphasis, structure and scope differ radically. Neither is able to capture the complexities of any one particular situation. Malthus's model works best in pre-industrial western Europe. Marx's model has specific value for early industrial capitalism. In approaching the population crisis in contemporary Africa, for instance, we would do well to advance via political economy (entitlement theory perhaps; Sen, 1981), whilst remembering the logic of the 'ever-present possibility' and its apparent realization in large regions. It is also clear that many governments frame their population policies in an explicitly neo-Malthusian fashion and that in the case of China this is not regarded as counter-revolutionary.

CONCLUDING REMARKS

Five points are worth making by way of conclusion. First, the concept of a population crisis is particularly difficult to define. There is a variety of forms, some of which are particularly conditioned by ideological stance. Second, it appears that the world is now passing through one form of population crisis, but, as Coale (1982, p. 15) has remarked, 'if 2000 years from now a graph were made of the time sequence of the rate of increase of world population, the era of rapid growth that began a couple of centuries ago would look like a unique and narrow spike'. Third, there are clear signs that the rate of world population growth is now slackening, but also that there are many large countries with particularly high growth rates in which fertility is only just starting to be reduced. There are thus in certain places 'crises of numbers', but even in the West some structural characteristics are giving rise to economic and social problems. Fourth, the neo-Malthusian approach to population issues predominates in the late twentieth century, yet it owes only loose affiliation to Malthus's original principles. Fifth, Marx's political economy, together with notions of allocation and entitlement, offers an additional perspective on contemporary demographic issues which emphasizes the differential characteristics of classes together with the economic origins of what for some groups is a permanent population crisis.

References

Aird, J. S. 1978a: Population growth in the People's Republic of China. In *Joint Economic Committee (Congress of the United States), Chinese Economy Post-Mao*. Washington, DC: US Government Printing Office, 439–75.
Aird, J. S. 1978b: Fertility decline and birth control in the People's Republic of China. *Population and Development Review*, 4, 225–53.
Aird, J. S. 1982: The preparation for China's 1982 census. *China Quarterly*, 91, 369–85.
Baker, H. D. R. 1979: *Chinese Family and Kinship*. New York: Columbia University Press.
Banister, J. 1987: *China's Changing Population*. Stanford, CA: University Press.
Banister, J. and Preston, S. H. 1981: Mortality in China. *Population and Development Review*, 7, 98–110.
Barclay, G. W. Coale, A. J. Stoto, M. A. and Trussell, T. J. 1976: A reassessment of the demography of traditional rural China. *Population Index*, 42, 606–35.
Beaver, S. E. 1975: *Demographic Transition Theory Reinterpreted: An Application to Recent Natality Trends in Latin America*. Lexington, MA: Lexington Books.
Birdsall, N., Fei, J., Kuznets, S., Ranis, G. and Schultz, T. P. 1979: Demography and development in the 1980s. In P. Mauser (ed.), *World Population and Development: Challenges and Prospects*. Syracuse, NY: Syracuse University Press, 211–95.
Bourgeois-Pichat, J. 1981: Recent demographic change in Western Europe: an assessment. *Population and Development Review*, 7, 19–42.

Bulatao, R. A. and Lee, R. D. (eds) 1983: *Determinants of Fertility in Developing Countries*, vol. 1, *Supply and Demand for Children*. New York: Academic Press.

Caldwell, J. C. and Srinivasan, K. 1984: New data on nuptiality and fertility in China. *Population and Development Review*, 10, 71–9.

Cassen, R. H. 1976: Population and development: a survey. *World Development*, 4, 785–830.

Cassen, R. H. 1978: *India: Population, Economy and Society*. London: Macmillan.

Cassen, R. H. and Dyson, T. 1976: New population projections for India. *Population and Development Review*, 2, 101–36.

Castles, S. and Kosack, G. 1973: *Immigrant Workers and Class Structures in Western Europe*. Oxford: Oxford University Press.

Chaudhry, M. 1982: Demographic transition theory and developing countries: a case study of India. *Demography India*, 11, 73–114.

Chen, C. H. and Tyler, C. W. 1982: Demographic implications of family size alternatives in the People's Republic of China. *China Quarterly*, 89, 65–73.

Clarke, J. I and Kosinski, L. A. (eds) 1982: *Redistribution of Population in Africa*. London: Heinemann.

Coale, A. J. 1978: Population growth and economic development: the case of Mexico. *Foreign Affairs*, 56, 415–29.

Coale, A. J. 1981a: Population trends, population policy, and population studies in China. *Population and Development Review*, 7, 85–97.

Coale, A. J. 1981b: A further note on Chinese population statistics. *Population and Development Review*, 7, 512–18.

Coale, A. J. 1982: A reassessment of world population trends. *Population Bulletin of the United Nations*, 14, 1–16.

Coale, A. J. 1983a: Recent trends in fertility in less developed countries. *Science*, 221 (4613), 828–32.

Coale, A. J. 1983b: Population trends in China and India (a review). *Proceedings of the National Academy of Science of the USA*, 80 (6), 1757–63.

Coale, A. J. and Hoover, E. M. 1958: *Population Growth and Economic Development in Low-Income Countries: A Case Study of India's Prospects*. Princeton NJ: Princeton University Press.

Coale, A. J., Anderson, B. and Harm, E. 1979: *Human Fertility in Russia Since the Nineteenth Century*. Princeton, NJ: Princeton University Press.

Coleman, D. and Schofield, R. S. (eds) 1986: *The State of Population Theory: Forward from Malthus*. Oxford: Blackwell.

Dagodag, W. T. 1975: Source regions and composition of illegal Mexican immigrants to California. *International Migration Review*, 9, 499–511.

Daly, M. E. 1971: A Marxian–Malthusian view of poverty and development. *Population Studies*, 25, 25–37.

Demeny, P. 1982: Population policies. In J. Faarland (ed.), *Population and the World Economy in the 21st Century*. Oxford: Blackwell, 206–28.

Demeny, P. 1983: A perspective on long-term population growth. *Population and Development Review*, 10, 103–26.

Dow, T. E. and Werner, L. H. 1983: Prospects for fertility decline in rural Kenya. *Population and Development Review*, 9, 77–97.

Dupâquier, J., Faure-Chamoux, A. and Grebenik, E. (eds) 1983: *Malthus Past and Present*. London: Academic Press.

Dyson, T. and Crook, N. (eds) 1984: *India's Demography: Essays on the Contemporary Population*. New Delhi: South Asia Publishers.

Dyson, T. and Moore, M. 1983: Kinship structure, female autonomy, and demographic behaviour in India. *Population and Development Review*, 9, 36–60.

Flinn, M. W. 1981: *The European Demographic System, 1500–1820*. Brighton: Harvester Press.

Goodstadt, L. F. 1982: China's one-child family: policy and public response. *Population and Development Review*, 8, 37–58.

Gottfried, R. S. 1978: *Epidemic Disease in Fifteenth-Century England: The Medical Response and the Demographic Consequences*. Leicester: Leicester University Press.

Grebler, L. 1966: *Mexican Immigration to the United States: The Record and its Implications*. Los Angeles, CA: University of California Press.

Grigg, D. B. 1980: Migration and overpopulation. In P. E. White and R. I. Woods (eds), *The Geographical Impact of Migration*. Harlow: Longman, 60–83.

Gugler, J. 1982: Overurbanization reconsidered. *Economic Development and Cultural Change*, 31, 173–89.

Gwatkin, D. R. 1979: Political will and family planning: the implications of India's emergency experience. *Population and Development Review*, 5, 29–59.

Hajnal, J. 1965: European marriage patterns in perspective. In D.V. Glass and D. E. C. Eversley (eds), *Population in History*. London: Arnold, 101–43.

Hajnal, J. 1982: Two kinds of preindustrial household formation system. *Population and Development Review*, 8, 449–94.

Hanley, S. B. and Yamamura, K. 1977: *Economic and Demographic Change in Preindustrial Japan, 1600–1868*. Princeton, NJ: Princeton University Press.

Harvey, D. W. 1974: Population, resources and the ideology of science. *Economic Geography*, 50, 256–77.

Hatcher, J. 1977: *Plague, Population and the English Economy, 1348–1530*. London: Macmillan.

Hayes, D. R. 1954: Neo-Malthusianism and Marxism. *Political Affairs*, 33, 41–57.

Heer, D. M. 1965: Abortion, contraception and population policy in the Soviet Union. *Demography*, 2, 531–9.

Hicks, W. W. 1974: Economic development and fertility change in Mexico, 1950–1970. *Demography*, 11, 407–21.

Jain, A. K. and Adlakha, A. L. 1982: Preliminary estimates of fertility decline in India during the 1970s. *Population and Development Review*, 8, 589–606.

Jones, E. and Grupp, F. W. 1983: Infant mortality trends in the Soviet Union. *Population and Development Review*, 9, 213–46.

Junankar, P. N. 1982: *Marx's Economics*. Oxford: Philip Allan.

Jutikkala, E. 1955: The great Finnish famine in 1696–7. *Scandinavian Economic History Review*, 3, 48–63.

Jutikkala, E. and Kauppinen, M. 1971: The structure of mortality during catastrophic years in a pre-industrial society. *Population Studies*, 25, 283–5.

van de Kaa, D. 1987: Europe's second demographic transition. *Population Bulletin*, 42 (1), 1–59.

Kirk, M. 1981: *Demographic and Social Change in Europe, 1975–2000*. Liverpool: Liverpool University Press.

Malthus, T. R. 1970: *An Essay on the Principle of Population* (reprint of 1798 essay edited by A. Flew). London: Penguin.

Malthus, T. R. 1973: *An Essay on the Principle of Population* (reprint of 1803 essay with amendments edited by T. H. Hollingsworth). London: Dent.

Marx, K. 1964: *Pre-Capitalist Economic Formations*. London: Lawrence and Wishart.

Marx, K. 1976: *Capital*, vol. 1. London: Penguin.

McNeill, W. H. 1977: *Plagues and Peoples.* Oxford: Blackwell.

Meek, R. L. (ed.) 1953: *Marx and Engels on Malthus.* London: Lawrence and Wishart.

Meuvret, J. 1977: *Le Problème des subsistances à l'époque Louis XIV.* Paris: Mouton.

Page, H. and Lesthaeghe, R. J. (eds) 1981: *Birth-Spacing in Tropical Africa.* London: Academic Press.

Palloni, A. 1981: Mortality in Latin America: emerging patterns. *Population and Development Review*, 7, 623–49.

Petersen, W. 1964: Marx versus Malthus: the symbols and the men. In W. Petersen (ed.), *The Politics of Population.* London: Gollancz, 72–89.

Petersen, W. 1979: *Malthus.* London: Heinemann.

Post, H. D. 1977: *The Last Great Subsistence Crisis in the Western World.* Baltimore, MD: Johns Hopkins University Press.

Pressat, R. 1982: La population de la Chine. Bilan des trente dernières années. *Population*, 37, 299–316.

Pressat, R. 1983: Premiers résultats du recensement de la Chine. *Population*, 38, 403–9.

Qian Zinzhong 1983: China's population policy: theory and methods. *Studies in Family Planning*, 14 (12), 295–301.

Schofield, R. S. 1972: 'Crisis' mortality. *Local Population Studies*, 9, 10–21.

Schultz, T. P. 1981: *Economics of Population.* Reading, MA: Addison-Wesley.

Seiver, D. A. 1975: Recent fertility in Mexico: measurement and interpretation. *Population Studies*, 29, 341–54.

Sen, A. 1981: *Poverty and Famines: An Essay on Entitlement and Deprivation.* Oxford: Clarendon Press.

Simon, J. C. 1977: *The Economics of Population Growth.* Princeton, NJ: Princeton University Press.

Smith, D. M. (ed.) 1983: *Living Under Apartheid: Aspects of Urbanization and Social Change in South Africa.* London: Allen & Unwin.

Sogner, S. 1976: A demographic crisis averted? *Scandinavian Economic History Review*, 24, 114–28.

Spengler, J. J. 1972: Malthus's total population theory: a restatement. In J. J. Spengler (ed.), *Population Economics.* Duke, NC: Duke University Press.

Stycos, J. M. 1978: Recent trends in Latin American fertility. *Population Studies*, 32, 407–25.

Tabah, L. 1980: World population trends: a stocktaking. *Population and Development Review*, 6, 355–89.

Tabah, L. 1982: Population growth. In J. Faarland (ed.), *Population and the World Economy in the 21st Century.* Oxford: Blackwell, 175–205.

United Nations 1979: *World Population Trends and Policies, Monitoring Report*, vol. 1, *Population Trends*; vol. 2, *Population Policies.* New York: United Nations Department of Economic and Social Affairs.

United Nations 1982: *Levels and Trends of Mortality since 1950.* New York: United Nations Department of International Economic and Social Affairs.

Watson, J. L. 1982: Chinese kinship reconsidered: anthropological perspectives on historical research. *China Quarterly*, 92, 589–622.

White, P. E. and Woods, R. I. 1983: Migration and the formation of ethnic minorities. *Journal of Biosocial Science, Supplement*, 8, 7–24.

Winch, D. 1987: *Malthus.* Oxford: Oxford University Press.

Woods, R. I. 1982: *Theoretical Population Geography.* Harlow: Longman.

Woods, R. I. 1983: On the long-term relationship between fertility and the standard of living. *Genus*, 39, 21–35.

Woods, R. K. 1987: Contemporary world population growth. *Geography Review*, 1 (2), 6–9.

Wrigley, E. A. and Schofield, R. S. 1981: *The Population History of England, 1541–1871: A Reconstruction*. London: Arnold.

7

World Capitalism and the Destruction of Regional Cultures

RICHARD PEET

After centuries of unreality, after having wallowed in the most outlandish phantoms, at long last the native, gun in hand, stands face to face with the only forces which contend for his life – the forces of colonialism. And the youth of a colonized country, growing up in an atmosphere of shot and fire, may well make a mock of, and does not hesitate to pour scorn upon, the zombies of his ancestors, the horses with two heads, the dead who rise again, and the djinns who rush into your body while you yawn. The native discovers reality and transforms it into the pattern of his customs, into the practice of violence and into his plan for freedom.

Frantz Fanon, *The Wretched of the Earth* (1963), p. 58

The development of capitalism as the dominant world economic system spreads its culture into all regions of the globe. A thousand interactions pit local and regional cultures, related to local environments and forms of livelihood, against international culture founded on a dynamic capitalism which first appeared in Euro-America. There are several dimensions to the resulting cultural interaction. Capitalist culture absorbed elements from the regional cultures it encountered – its conception of paradise on earth is strongly flavoured by the encounter with the Polynesians on 'unspoilt Pacific Islands'. Capitalism and regional cultures have merged into synthetic cultures – Japanese culture contains strong elements from the islands' particular version of its Asiatic feudal past. But a continuing theme of the encounter between world capitalist and regional non-capitalist cultures of the Third World is the pervasive power of the first and the transformation of the second.

Multinationalization of industry is linked with the global spread of international advertising agencies, the development of consumerism and other effects on the 'whole fabric of society' in the Third World (Janus and Roncaglio, 1979). The effect of tourism on West Indian culture is described in terms of art, music, dance and literature's becoming 'the patrimony of an expanding tourist economy. In the process, artistic expression in the West Indies loses its integrity and, indeed, its relevance to the West Indian experience' (Perez, 1975, p. 140). International publishing, dominated by multinational corporations, leads to 'cultural dependency as a complement to economic dominance' in the Third World (Golding, 1978). The international flow of films, television and radio programmes, records, cassettes and printed materials from centre to periphery carries cultural messages in persuasive formats (Guback, 1974; Varis, 1974, 1976; de Cardona, 1975; Gould and Johnson, 1980). The US 'communications complex' is shown to be a 'powerful mechanism [which] now directly impinges on people's lives everywhere' carrying 'a vision of a way of life. The image is of a mountain of material artifacts privately furnished and individually acquired and consumed' (Schiller, 1969, pp. 3, 17; Schiller, 1978). People in the periphery are shown to receive their news about themselves via the centre (Harris, 1974, 1976; Masmondi, 1979). In general, present relations between centre and periphery have been characterized in terms of 'cultural imperialism' by the capitalism centre.

Beyond the kind of empirical work reviewed above lies the need to constuct a structural-geographic theory of culture which addresses two topics: What is culture, and how is it formed? What gives capitalist culture its extraordinary powers of diffusion and destruction? In this chapter we attempt to answer these questions, mainly via an analysis of certain ideas developed in Marx's *Capital* and *Grundrisse*. However, we can only provide an outline of certain aspects of such a theory and these aspects must be bluntly and briefly stated, without the usual academic qualifications and without extensive exemplifications. A complete Marxian-geographic theory of culture, adequately supported by empirical research, is still years away.

One aspect of culture is emphasized – what Marx called the 'entire super-structure of distinct and peculiarly formed sentiments, illusions, modes of thought and views of life' which arises as the basis of the 'social conditions of existence' (Marx, 1969); that is, emphasis is placed on the consciousness dimension of culture. By consciousness we mean, quite simply, 'the way people think' – not about each specific thing that crosses their minds, but about 'things in general'. Consciousness refers to patterns and modes of thought as conditioned by the whole way a people live. In turn consciousness guides all aspects of cultural production – thought is materialized as art and literature – as well as all other aspects of material reproduction. Human experiences in different evironments yield regional consciousnesses and cultures. World capitalism, a new order and scale of experience, is represented as a powerful culture which overwhelms local and regional experience. As a result, the mass of the world's people become adherents of one global culture system. This new world cultural system, however, shares certain attributes with previous regional cultures, especially in those aspects of consciousness most obviously prescribed by domination by nature and social forces beyond human control. Domination is traced to two main sources: domination by natural forces in the early stages of human development, and domination by spontaneous uncontrollable social production in the later period. We concentrate here on one main effect of domination – the assumption by consciousness of a religious mystical form. The idea is to follow one connecting strand from natural environment through mode of production to culture and religious consciousness. Other aspects of the geography of culture and consciousness have been considered by the author elsewhere (Peet, 1980, 1982).

THE SOCIOLOGY OF CULTURE

We can approach an understanding of the geography of culture through the more developed concepts of its sociology as outlined by the late Raymond Williams (1980, 1981). Williams traces the problem in defining the 'exceptionally complex term' culture to two different kinds of understanding. An earlier idealist emphasis, dating from the eighteenth century, lay on the 'informing spirit' of a whole way of life, most evidently its language, styles of art and kinds of intellectual work. A later materialist emphasis was on culture as the direct or indirect product of the social order. Williams accepts the second, with the proviso that 'cultural practice' and 'cultural production' are not simply derived from, but are also constitutive of, the social order. Phrasing this a little differently, he sees culture as the '*signifying system* through which necessarily (though among other means) a social order is communicated, reproduced, experienced and explored' (Williams, 1981, p. 13). Any social system has a distinguishable signifying system – as a language, as a system of thought or consciousness, as a body of specifically signifying works of art. Although this signifying system is essentially involved in all forms of social activity, Williams also retains the more specialized use of the term 'culture' to refer particularly to artistic and intellectual activities, interpreting these broadly to include all 'signifying practices . . . from language, through the arts and philosophy to journalism, fashion and advertising' (p. 13).

Sociology focuses on the forms of culture resulting from the social process of cultural production. Hence Williams is interested in social institutions (feudal households, the market, the corporation), cultural institutions (craft guilds, academies, professional societies) and means of culture production (speech, dance, writing, amplificatory systems) seen in relation to art forms and products. While accepting the general determination of form by social process, Williams argues that art forms can never be reduced to mere anticipations or reflections of social processes: cultural processes are always relatively autonomous. Some cultural practices, for example those occurring within conditions of wage labour, are effectively inseparable from their determining social relations; others are only indirectly determined, and perhaps not determined at all, for example poetry and sculpture, where there seems to be considerable autonomy.

For Williams, there are a number of tendencies in cultural development of significant importance which are also pertinent to the theme of this chapter. The first tendency is the transformation of 'popular culture'. In capitalist society this is increasingly mass produced, privately under market conditions and by state educational and political systems. Cultural production engenders a major expansion of the cultural and educational bureaucracies, interlocked with more general political, economic and administrative bureaucracies. Williams's conclusion is that, while there is innovatory work in many forms of art and thought, the genuinely emergent in art has to be defined primarily in terms of its contribution to alternatives to this dominant general system.

The second tendency is the institution of an international, even a world, market. Except in some closed or subsistence societies, the processes of cultural import and export have always been important. But changes in the means of cultural production and distribution, especially cinema and television, have led beyond such simple processes to more general processes of cultural dominance and dependence. These have radical effects on national signifying systems, like languages, and lead to new forms of multinational cultural combine.

The third tendency is fundamental changes in the labour process, which have radically affected the definition of cultural production. In advanced industrial societies, direct cultural production now often involves a small and declining proportion of the working population, while the number involved in information processing has increased, in the case of the United States to as much as 50 per cent of the working population. While many of the older types of determination – state power or economic property and command – are still decisive, there are quite new complexities in the whole system of cultural production and reproduction (Williams, 1981, p. 233).

The sociology of culture thus investigates the aesthetic forms, institutions and relations of society's artistic and intellectual life. A materialist explanation understands culture as the signifying system of a social order, which structural Marxism interprets in terms of a mode of production, i.e. a determinate combination of the physical and social forces and relations of production. The main contribution of Williams is his articulation of the social order and the system of cultural production. Williams's articulation does not withdraw from the essential Marxist position – that there is a process of determination in any social totality – but rather attempts to 're-value "determination" towards the setting of limits and the exertion of pressure, and away from a predicted prefigured and controlled content' (Williams, 1980, p. 34). Williams's work is a successful account of the multiple interactions between determination and autonomy in the broad tradition of Marxian cultural analysis.

THE GEOGRAPHY OF CULTURE

Having briefly explored the sociology of culture, we are now in a position to investigate its geography. This will be carried out from a materialist perspective similar enough to that of Williams to employ his useful analytical categories.

What is the role of geography in the academic division of labour? Geography investigates two of the relations which all humans must enter: the relation with the natural environment – the ultimate source of material existence – and the relation with other humans across space. Here the structure of determination, inherent in a Marxian understanding, is deepened to include determination by the natural world. Thus for the Italian Marxist Timpanaro:

> By materialism we understand above all acknowledgement of the priority of nature over 'mind', or if you like, of the physical level over the biological level, and of the biological level over the socio-economic and cultural level; both in the sense of chronological priority (the very long time which supervened before life appeared on earth, and between the origin of life and the origin of man), and in the sense of the conditioning which nature still exercises on man and will continue to exercise at least for the foreseeable future. (Timpanaro, 1975, p. 34)

Relations between humans and nature give rise to scientific and philosophical reflection and to artistic expression. Relations with nature are long lasting, providing a continuity of cultural theme through history. Thus religion is always an illusory compensation for the fear of death and the oppression in general which nature exercises on humans (Timpanaro, 1975, pp. 48–52).

Nature, however, does not exercise the same kind, or level, of determination in all places and at all times. Place-specific natural determinations are fertile ground for geographical investigation. There are several dimensions to place specificity. First, the character of nature varies between different regional and local environments – if reflection on nature is a primary source of intellectual life, then reflection on a particular nature can be seen as a primary source of regional consciousnesses and art forms. Second, humans encounter not only local nature, but a wider natural environment, through spatial interactions from migration to the diffusion of cultural traits – hence the aspect of relative location. Third, relations with nature are always mediated by socioeconomic forces and institutions, of which the forces of production applied to natural resources are most important. As the level of the productive forces available in any regional–social group increases, determination by nature decreases to be replaced by other determinations. Space thus becomes a mosaic of different types and natural levels of determination, similar to the different levels of economic development. This takes a centre (low natural determination) and periphery (high natural determination) geographic shape under capitalism. New forms of cultural practice, developed in the hegemonic centre, diffuse over space, even when they are inappropriate in terms of the (low) level of indigenous productive development. This last (diffusion) aspect of the geography of culture will be discussed later.

Cultural geography thus extends material determination back into the relations between society and nature. It investigates the particular local forms taken by culture in response to particular environments, locations and levels of development, it sees these local cultures as preserved by the frictions of distance and it looks at the spatial interactions between localized cultural forms. Geography and sociology integrate within a holistic understanding of the totality of cultural practice.

NATURE AND CULTURE IN PRE-CAPITALIST SOCIETIES

Even when investigating such relatively autonomous dimensions of life as human cultures, Marxists insist on founding their analysis in the material production process. This is particularly the case for geographically oriented Marxists, whose special scientific function involves deliberately stressing the geomaterial aspects of the reproduction of life. Marx's work on labour and production, outlined in *Capital*, provides an introduction to this analysis.

For Marx, labour is a process which appropriates natural materials in forms adapted to human needs. Through acting on nature humans transform the external environment, but also find and develop their own inner (human) nature – the difference from other animals is that humans increasingly regulate and control their metabolism with nature. 'Man not only effects a change of form in the materials of nature; he also realizes his own purpose in those materials. And this is a purpose he is conscious of' (Marx, 1976, p. 284). The land is the universal material for human labour, both objects of labour spontaneously provided – fish, timber and ores – and objects of labour filtered through previous labour – raw materials. Similarly, an instrument of labour is an originally natural thing interposed between the worker and the object of labour. With only the slightest development of the labour process, specially prepared instruments are required. For Marx, the use and construction of tools and instruments is characteristic of the specifically *human* labour process – instruments are the means by which the relation with nature is regulated. 'It is not what is made but how, and by what instruments of labour, that distinguishes different economic epochs. Instruments of labour not only supply a standard of the degree of development which human labour has attained, but they also indicate the social relations within which men work' (p. 286). Thus automated and mechanical instruments of labour offer the most decisive evidence of the economic character of the present social epoch of production.

For Marx, therefore, the productive forces – instruments and objects of labour (means of production) and living labour – are the underlying determinants of the socio-cultural character of a historical epoch. Different social relations between producers, and between producers and the environment, are generally appropriate to different levels in the development of the forces of production. When humans rely mainly on spontaneous natural products and instruments of labour, as in hunting and gathering societies, the natural environment is regarded as the extended body of the individual, and the private ownership of nature is absent. With the development of specially prepared instruments of labour and the extensive use of natural raw materials, private ownership appears. This eventually extends, from the ownership of pieces of nature by individuals, to the ownership of the labour power of other human beings from outside the community (slavery) and then inside the community (serfdom) (Peet, 1981). The achievement of minority class control over the productive forces of a society enables nature to be conquered. Nature is increasingly transformed into useful objects (use-values) and a 'second nature' (i.e. a humanly altered environment).

For Marx, 'The mode of production of material life conditions the general process of social, political and intellectual life' (1970, p. 21). More generally, social existence determines consciousness. When social existence is characterized by dominance and servitude, including an immediate reliance on nature, human consciousness is restricted, the forces which control existence are deified and thought is clouded by religious supposition. As Marx says, the 'ancient social organisms of production'

> are founded either on the immaturity of man as an individual, when he has not yet torn himself loose from the umbilical cord of his natural-species connection with other men, or in direct relations of dominance and servitude. They are conditioned by a low stage of development of the productive powers of labour and correspondingly limited relations between men within the process of creating and reproducing their material life, hence also limited relations between man and nature. These real limitations are reflected in the ancient worship of nature, and in other elements of tribal religions. (Marx, 1976, p. 173)

The pre-scientific understanding of the natural and social conditions of existence pervades consciousness and culture at low levels of productive development. It is characterized particularly by the transposition of will, from humans just beginning to realize that they have it, to a nature which continually contradicts its freedom of action.

The different regional social formations are characterized by different levels of productive development and thus different developments of consciousness. Regional social formations are also located in entirely different environmental settings, transform different kinds of natural resources into varying useful objects ('what is made') and are subject to the varying control of dissimilar natural forces. Thus Wittfogel argues that, because different social organisms find different *means* of production in their local environments, their *modes* of production are different and economic development is channelled in different directions, resulting in various kinds of states and politics. Hence the need for control over water resources in arid and semi-arid Asiatic regions led to hydraulic civilizations dominated by strong centralized state bureaucracies and 'oriental despotism' (Wittfogel, 1929, 1957, 1985). Different natural environments provide varying physical opportunities for different kinds of specialized production in terms of climate, natural soil fertility, raw materials and energy sources. Culture is thereby influenced in two main ways: productive complexes structure cultural development – in general, industrial districts have different cultural emphases from agricultural districts – and cultural production directly finds different work materials in local environments. Thus the different economic and cultural productive forces, in combination with different natural materials, yield an array of regional socio-cultural formations across geographic space.

In the discipline of geography, the most prominent proponent of this idea was the environmental determinist Ellen Churchill Semple. For Semple, the 'physical effects of geographic environment . . . are reflected in man's religion and his literature, in his modes of thought and figures of speech' (1911, p. 40). Thus habitat influences the legal structure of a social group. The effect of environment, acting through the type of economic activity, enriches the language in one direction but restricts it in others. The mythology of a people echoes the surrounding natural environment. The cosmography of primitive people – what she calls 'their first crude effort in the science of the universe' – also bears the impress of the local habitat (pp. 40–1). The problem with Semple's natural determinism, however, is the directness of the relation between environment and culture. Thus in discussing world climatic zones as determinants of the 'girdles of culture around the earth', she passes straight from heat and moisture to the natural qualities of 'human temperament' (pp. 633–5). Missing from her explanation is an elaborated theory of the social mediation between the natural environment and a people's 'spirit', which would result from a structural-materialist analysis. We can, however, admire her willingness to take bold steps where others fear to tread.

Nature and Religious Consciousness in Pre-capitalist Societies

Let us isolate one aspect of this complex issue – the influence of environment on religious consciousness – for detailed consideration. First we must ask: What is the origin of religion, and how has religious understanding changed over time?

Religion is an attempt to understand and influence the great forces on which life depends: religion is a 'set of symbolic forms and acts which relate man to the ultimate conditions of his existence' (Bellah, 1969, p. 67). These 'ultimate conditions' include the origin of human existence, in terms of descent from a line of ancestors, and the continuation of existence, in terms of the social relation with nature. However, as the eighteenth-century philosopher David Hume, one of the first to subject the concept of religion to scientific scepticism, argued:

We are placed in this world, as a great theatre, where the true springs and causes of every event are entirely concealed from us; nor have we either sufficient wisdom to foresee, or power to prevent, those ills with which we are continually threatened . . . These *unknown causes*, then, become the constant object of our hope and fear; and while the passions are kept in perpetual alarm by an anxious expectation of the events, the imagination is equally employed informing ideas of those powers on which we have so entire a dependence. (Hume, 1968, p. 21)

Religion, therefore, results from contemplation of the conditions of existence under circumstances of a threat to existence. Hence primitive religion has two themes recurring around two conditions of that existence: the worship of the dead, under the assumption that they retain consciousness and can influence the fortunes of the living, and the worship of natural phenomena, conceived as animated, conscious and endowed with the power and will to benefit, or injure, humankind. The most universal phenomena of nature, the sky and the earth, have been most universally worshipped. Other natural phenomena enjoy more restricted regional deification – the sun, the moon, the stars, fire, water, plants and animals (Frazer, 1926). At a low level of productive development, and a low level of control over nature, the religious practices of ethnic (tribal) systems are quite directly related to local physical environments and the way that these are used:

> In the simple ethnic systems, religion often seems to be almost entirely a ritualization of ecology. Religion is the medium whereby nature and natural processes are placated, cajoled, entreated, or manipulated in order to secure the best results for man. Even at a very primitive technological level, however, every culture operates selectively in taking its sacred 'resources' from its ecological milieu. The religious behavior of such societies becomes an extended commentary on selected, usually dominant, features of their economies. (Sopher, 1967, pp. 17–18)

Archaic myths and rituals, however, persist long after the ecological elements around which they have crystallized have passed out of the focus of particular interest. Hence broad regional religions appear, based on the sanctification of different economies through ritual and the differential persistence of archaic ecologies in ritual forms (p. 19).

With the development of the productive forces and the spread of intensive product exchange, natural determination recedes and religion takes a human-centred form. Universalizing religious systems (Buddhism, Christianity, Islam) break through the restriction of a special relationship to particular places and human groups. 'As society and economy become more complex, symbolization and abstraction of ecological matter increase, the process becoming intensified in the transition from ethnic to universalizing systems' (p. 19). Thus the religious calendar of Judaism reflects the ecological characteristics of the Mediterranean, Christianity preserves some of these elements but also incorporates aspects of earlier pagan religions, such as celebration of the winter solstice (Christmas), and the Islamic calendar is more completely liberated from ecological ties – perhaps because Islam was, from the beginning, a religion of the mercantile town (p. 22).

In the pre-capitalist world, therefore, people thought about the origins and purpose of life in ways broadly similar (roughly the same things were being thought about) but particularly different (there were different natural circumstances and levels of control over circumstances). The resulting map of religious consciousness was characterized by both broad spatial order (the universalizing human-centred religions) and spatial specificity (ethnic and tribal religious systems). It was the product of multilinear historical development, based originally on greatly different physical environments, with these different historical pasts differentially preserved into the present. Human interaction with environment produced a mosaic of regional forms of consciousness in the pre-capitalist world.

EXCHANGE AND THE DISSOLUTION OF PRE-CAPITALIST SOCIETY

The pre-capitalist cultural world was thus made up of a series of regional socio-cultural formations, each with its own kind of consciousness as in the case of different religions. The capitalist mode of production, founded on the production and exchange of commodities and wage-labour relations, originates a new kind of socio-cultural order, which is universal in scope. The growth of this new order breaks down the old geography of society and culture through various economic means. One such means – the growth of commodity exchange – is examined in detail here.

Economic contact between capitalist and non-capitalist societies usually begins with the occasional trading of commodities. Exchange between the two kinds of society – the one resting on exchange, the other on direct use – has effects which transcend the mere commodity, however, acting to dissolve all the social relations of non-capitalist societies. In pre-capitalist societies, each productive individual had a presupposed relation to land, the main means of production, guaranteed by the occupation of a certain territory by a communal group and protected by warfare from surrounding communities. Pre-capitalist societies were differentiated on the basis of their mode of control of land and soil, whether this was communally owned ('primitive communism'), privately owned by dispersed individual members of a community ('Germanic mode of production'), in locally centralized private ownership ('feudal mode of production'), or in centralized private ownership ('Asiatic mode of production') (Peet, 1981). In all pre-capitalist modes, production was primarily for the immediate satisfaction of the needs of the family and community. Surpluses were transferred within the pre-capitalist class structure to the controllers of the communal territory (the warrior class and/or the state) and the guarantors of favourable natural conditions (the priests). Relations of immediate dependence on nature were paralleled by relations of personal dependence in production, with individuals being 'imprisoned within a certain definition, as feudal lord and vassal, landlord and serf, etc., or as members of a caste etc. or as members of an estate etc.' (Marx, 1973, p. 163). Generalizing, Marx regards all societies characterized by these kinds of dependence as the 'first social forms, in which human productive capacity develops only to a slight extent and at isolated points' (p. 158).

The second great historical form, personal independence founded on real dependence, begins with the exchange of surplus products at boundaries between communities. As products became commodities in the external relations of a community they also, by reaction, became products in its internal relations. Now, for Marx, 'the characters who appear on the economic stage are merely personifications of economic relations; it is as bearers of these economic relations that they come into contact with each other' (1976, p. 179). The exchange of commodities means that the 'guardians' of products must recognize each other as owners of private alienable property, as persons who are independent of each other. This 'relationship of reciprocal isolation and foreignness does not exist for members of a primitive community of natural origin, whether it takes the form of a patriarchal family, an ancient Indian commune or an Inca state' (p. 182). Exchange thus produces a new kind of human personality and new social relations. As more and more products are exchanged, the traditional relations between people in pre-capitalist societies disintegrate: 'in the developed system of exchange . . . the ties of personal dependence, of distinctions of blood, education, etc. are in fact exploded, ripped up . . . and individuals *seem* independent (this is an independence which is at bottom merely an illusion, and it is more correctly called indifference), free to collide with one another and to engage in exchange within this freedom' (Marx, 1973, pp. 163–4).

Whereas in the natural community the reproduction of the individual as a member of the community is the objective of production, with exchange the pursuit of individual wealth becomes the main objective. In pre-capitalist society, Marx argues, natural wealth supposed an essential relation between the individual and the objects which formed that wealth, but with money a world of things becomes accessible and greed limitless. 'Monetary greed, or mania for wealth, necessarily brings with it the decline and fall of the ancient communities . . . it is the antithesis to them. It is itself the community, and can tolerate no other standing above it' (p. 223). This process is completed when, as a result of the decay of pre-capitalist communal relations, land and labour are separated and become purchasable on the market. With this, money directly becomes the 'real community' since it is the general substance of survival, and social product, of all. Furthermore, Marx concludes:

in money the community is at the same time a mere abstraction, a mere external, accidental thing for the individual, and at the same time merely a means for his satisfaction as an isolated individual. The community of antiquity presupposes a quite different relation to, and on the part of, the individual. The development of money in [this] role therefore smashes this community. All production is an objectification of the individual. In money (exchange value), however, the individual is not objectified in his natural quality, but a social quality (relation) which is, at the same time, external to him. (Marx, 1973, p. 226)

Thus for Marx one form of compulsion, the domination of the individual by nature and other individuals in pre-capitalist societies, is replaced by 'an objective restriction of the individual by relations independent of him and sufficient unto themselves', conditions which 'although created by society, appear as if they were *natural conditions*, not controllable by individuals' (p. 164). This is far from the abolition of relations of dependence. It is, instead, 'the dissolution of these relations into a general form' (p. 164).

Marx does not mourn the passing of early society. He regards it as just as 'ridiculous to yearn for [the] original fullness' of the historic individual, as it is to 'believe that with this complete emptiness [capitalism], history has come to a stand still' (p. 162). Instead he anticipates a third epoch in the history of humanity, for which capitalism, with its formation of 'a system of general social metabollism, of universal relations, of all-round needs and universal capacities' creates the preconditions: 'Free individuality, based on the universal development of individuals and on their subordination of their communal, social productivity as their social wealth, is the third stage' (p. 158). In this discussion, there are strong implications of historical inevitability, with capitalism being the necessary creator of universal relations between people: 'Universally developed individuals, whose social relations, as their own communal relations, are hence also subordinated to their own communal control, are no product of nature, but of history' (pp. 161–2).

NEW FORMS FOR OLD IN THE DEVELOPMENT OF CONSCIOUSNESS

Although there are broad similarities, the original dissolution of pre-capitalist societies into the first capitalist societies differs from the destruction of contemporary non-capitalist societies through exchange and other external relations with an already advanced capitalism. In particular, sophisticated and powerful institutions, forms of culture and modes of consciousness have now developed which make the process of dissolution more rapid and effective. The periphery of the world capitalist system, where non-capitalist relations of production still exist, is therefore subjected to attack at a number of levels simultaneously: the economic level, through exchange and the spread of capitalist relations of production, the political level, through colonial control or hegemonic power, and the cultural level, through the penetration of capitalist cultural products and ideas. But the ideas that capitalism exports to the periphery lead merely to the replacement of one kind of mysticism by another. How does this come about?

The ways in which humans produce their life determine the ways in which they conceptualize it. Under capitalism, the productive forces are privately owned, with antagonistic relations between wage labour and capital, and competitive relations within the class of capitalist owners. Wage-labour relations channel economic surplus to private owners of the means of production, while competition between owners forces the reinvestment of surplus in improved efficient means of production. The social relations of production are thus exactly the source of the rapid development of the productive forces. In turn, productive development reduces the domination of nature, eventually reversing the main direction of determination in the capitalist centre, in the sense that some natural forces can be altered, controlled or even destroyed by advanced societies armed with powerful productive means. Increasingly, nature is rationally understood via a science which develops as part of the general development of the productive forces. Nature need no longer be feared as a destructive god, although at times of natural catastrophe and death the ancient ritual of prayer immediately surfaces. As continued existence comes to depend on human productive effort, rather than on the whim of nature, religious mysticism tends to be replaced by scientific rationalism. This is what Marx calls the 'great civilizing influence of capital', that is:

> its production of a stage of society in comparison to which all earlier ones appear as mere *local* developments of humanity, and as *nature idolatry*. For the first time, nature becomes purely an object for humankind, purely a matter of utility; ceases to be recognized as a power for itself; and the theoretical discovery of its autonomous laws appears merely as a ruse so as to subjugate it under human needs, whether as an object of consumption or as a means of production. In accord with this tendency, capital drives beyond national barriers and prejudices as much as beyond nature worship, as well as all traditional, confined, complacent, encrusted satisfactions of present needs and reproduction of old ways of life. It is destructive towards all of this, and constantly revolutionizes it, tearing down all barriers which hem in the development of the forces of production, the expansion of needs, the all-sided development of production, and the exploitation and exchange of natural and mental forces. (Marx, 1973, pp. 409–10)

Civilization as a whole is thus transformed by the development of the productive forces. But consciousness does not break through, from religion to science, in all aspects of this more civilized life. Instead new mysticisms develop, or old ones are updated, particularly in the realm of social explanation. The means for understanding and controlling nature (natural science and technology) outstrip the means for understanding and controlling society (social science and politics). To gain comprehension of this we have to draw out more fully the implications for consciousness of the development of capitalist relations of production. Again we shall rely on Marx for the direction the analysis should take.

Under capitalism, each person's survival is made to depend on the pursuit of self-interest by all others. Common interest is not the direct motive of production and exchange, but proceeds 'behind the back of these self-reflected particular interests, behind the back of the individual's interest in opposition to that of the other' (Marx, 1973, p. 244). Production and consumption are thus organized through a network of spontaneous relations between reciprocally indifferent individuals. These relations confront the individuals as their common subordination: 'The general exchange of activities and products, which has become a vital condition for each individual – their mutual interconnection – here appears as something alien to them, autonomous, as a thing' (p. 157). More concretely, the allocation of the society's labour power is not achieved directly, through social planning, but via the collision on the market of products made by quantities of labour:

> In other words the labour of the private individual manifests itself as an element of the total labour of society only through the relations which the act of exchange establishes between the products, and, through their mediation, between the producers. To the producers, therefore, the social relations between their private labours appear as what they are, i.e., they do not appear as direct social relations between persons in their work, but rather as material relations between persons and social relations between things. (Marx, 1976, pp. 165–6)

The commodity, in reality merely a sensuous physical thing, is at the same time supra-sensible, or social, and what are, in fact, definite social relations between humans assume the fantastic form of a relation between things. To find an analogy:

> We must take flight into the misty realm of religion. There the products of the human brain appear as autonomous figures endowed with a life of their own, which enter into relations both with each other and with the human race. So it is in the world of commodities with the products of men's hands. I call this the fetishism which attaches itself to the products of labour as soon as they are produced as commodities, and is therefore inseparable from the production of commodities. (Marx, 1965, p. 165)

Commodity fetishism thus arises from the social character of the labour which produces commodities, i.e. labour organized indirectly through the exchange of products on the market. It results in a mode of capitalist thought analogous to the natural religion of pre-capitalist times.

Reflection on the social forms of human life begins after they have already assumed the fixed quality of natural forms. Thus the economic formulae of bourgeois economics 'which bear the unmistakable stamp of belonging to a social formation in which the process of production has mastery over man, instead of the opposite, appear to the political economists' bourgeois consciousness to be as much a self-evident and nature-imposed necessity as productive labour itself' (p. 175). Hence the idea arises that it is 'natural' for society to be organized for the benefit of a minority class of individuals who pursue their own self-interest (capitalism corresponds to an eternally and exclusively selfish human nature), or that it is natural that economic development proceeds autonomously under the direction of an invisible hand ('invisible' because it does not exist!). Taking this social form for granted, science is misled – its categories incapable of explaining the structure and movement of society. A society which does not scientifically understand itself – whose economists, for example, cannot predict the autonomous course of economic change – requires some other general theory, some other account of the essential causes of things. Religion served this function in the past, when humans were dominated by nature. Religion and other mysticisms continue to serve this function in the present, when humans are dominated by their own society:

> For a society of commodity producers, whose general social relation of production consists in the fact that they treat their products as commodities, hence as values, and in this material form bring their individual, private labours into relation with each other as homogeneous human labour, Christianity with its religious cult of man in the abstract, more particularly in its bourgeois development, i.e., in Protestantism, Deism, etc., is the most fitting form of religion. (p. 172)

Instead of understanding themselves as individuals connected to each other in the production of their existence (which would occur in a co-operative democratically planned society), capitalist individuals understand their necessary connecting social relationship to occur indirectly – 'man in the abstract' or a humanized god, rather than humans directly, to control the course of history. Instead of society's being understood scientifically, as would be possible if humanly created laws guided its development, society is understood mystically as though a power outside human control directs its course. (Hence the president of the world's most powerful nation prays for economic guidance.) Instead of morality's being a set of agreed-upon principles derived from social practice, it is a set of laws written by God and revealed to prophets, who then tell others how they should live, sometimes profiting financially from doing so and not infrequently disobeying the laws themselves. Social consciousness thus remains religiously mystical in capitalist society. As society becomes increasingly unstable and uncontrollable, as during crisis, social consciousness is increasingly mystified. Religious revival occurs in situations demanding the finest in social *scientific* understanding.

CONTRADICTION, CRISIS AND CONSCIOUSNESS

On the basis of the general statements made above about the relation between economic structure and consciousness, we can outline some features of contemporary systemic development at the centre of world capitalism. The centre is the region where capitalist social relations prevail. This means not only that all economic functions are dominated by wage-labour relations and all products are made for, and exchanged via, the market, but also that commodification has penetrated deeply into culture, so that ideas are thoughts for sale and beauty is aesthetically defined by market criteria.

Under capitalism the relations which tie together the main elements of production are inherently contradictory, making society prone to repeated crisis. Competitive market relations oppose one capitalist to another, one segment of production to another, one region's people to another. Antagonistic relations within each segment of production oppose capitalist to workers. Thus, in the case of unemployment crises, competition forces employers to relocate production from unionized areas where wages are high to non-unionized areas where wages are low. Such movements cause regional unemployment directly, and also general unemployment indirectly via a reduction in worker incomes, a lack of mass markets, a lower overall rate of economic growth and thus a further lack of demand for workers (Peet, 1983, 1987). Widespread and persistently high unemployment rates call for a co-operative response at the level of the society in general. But what is 'society' under capitalism? The captains of industry 'related' through competition? Share-owners interested only in higher dividends no matter where these come from? Workers and the bosses they hate? Consumers interested in price to the exclusion of product origins? The most realistic answer is the national state. Yet under capitalism, the state does not control the investment of capital, the fundamental process through which employment is generated. The capitalist state can ameliorate the effects of unemployment, but it cannot eradicate its causes, which lie in the very way capitalism operates. As it cannot be the function of the capitalist state to destroy the basis of the capitalist system, there are limits to state policy. These limits preclude social action aimed at the *systemic roots* of crises. The result is the escalation of the level of crises of all kinds at the centre of the world capitalist system. Capitalist consciousness responds to an environment of perpetual crisis.

Past social crises were precipitated by natural causes which were worshipped as gods. Present social crises are precipitated by social causes which are equally uncontrollable, and equally worshipped as gods. Hence prayers are offered for salvation from the effects of crises. 'God' is entreated to save humankind from social and political disasters. If scientifically understood by a people in collective control of their destiny, the causes of crises could be eradicated by concerted human action. As it is, however, the social causes of crises are not understood, nor do they change ('God refuses to act'). Culture and consciousness are forced into a number of forms.

All such forms have this in common: culture must deal with the events of social life, but mainstream culture cannot explain these events scientifically for to do so would reveal society's uncontrollability. Hence graphic depictions of events are superimposed on mystical theories of their causes. To accomplish this, various natural mysticisms are preserved or even resurrected from the past. Astrology, for example, presents the human personality and life events as functions of the arrangement of the planets and stars – a scientifically ridiculous notion. Yet astrological tables and forecasts are printed in the most modern media and are read by peasant, proletarian and president alike. The popular concept of economic change, repeated nightly in the news, is that the Stock Exchange somehow controls it – when the Dow-Jones is up, that's good news. In fact stock exchanges are merely collections of capitalist gamblers responding to signs from an economy moving and changing as the result of thousands of blind and partially sighted decisions. Related to this is the fascination that culture has with implausible solutions to problems, like high technology or outer space, and the vast range of modern fetishisms these produce – for example, films and television programmes in which machines totally dominate humans, or humans become machines. Again similar is a fascination with the *effects* of social crises. Hence we have the production of art forms which depend on an escalating level of violence and destruction, even reaching the point where people are really killed rather than 'merely' seen to be in the 'video nasties' rented nightly to pre-teenagers by kindly village storekeepers who 'only supply what people want!'. Then if ordinary mortals cannot control human destiny, perhaps superpeople can – hence the propagation of superhero images, and the worship of presidents and prime ministers who eventually (inevitably!) turn out to be mortal or, better still, European royalty and nobility who have had long practice in disguising their utter mundaneness. These and many other forms of culture and thought need a mode of expression also suited to the social–structural origins of crisis – a mode which allows problems to be aired, but not taken seriously. This need is met by 'entertainment'. Hence even the news media, the main form by which social problems are communicated, are now embedded in the entertainment industry. Entertainment is the opium of the people. Culture becomes the aesthetics of anaesthesia.

Cultural production involves the production of entertaining commodities, using the same technique as material commodity production – indeed pioneering these techniques. But the links between commodity production and culture production transcend mere technique. The development of mass commodity-production techniques requires a people 'habituated to respond to the demands of the productive machinery' (Ewan, 1976, p. 25). The function of the 'culture industry' becomes the mass production of 'ready-made clichés' which overpower the consumers, stunting their powers of imagination and turning participants into listeners and observers amenable to control (Horkheimer and Adorno, 1972). The 'consciousness industry' (radio, television, cinema, recording, advertising etc.), one part of the culture industry, does more than this. Enzensberger (1974, p. 11) argues that, beyond selling products, it sells the existing social order: 'The few cannot go on accumulating wealth unless they accumulate the power to manipulate the minds of the many'. More generally, the consciousness industry operates to proscribe information and explanation, limiting the range of thought, while making its consumers dependent on the industry for those thoughts that they do have.

The theoretical difficulty here lies in the question of 'deliberate intention', i.e. explaining 'consciousness manipulation' without resorting to the naive idea that the heads of corporations meet to decide what to allow people to think. It must be remembered that under capitalism the 'general interest is precisely the generality of self-seeking interests' (Marx, 1973, p. 245). Corporations pursue their own special interest first, and the interests of all corporations second. Thus while propaganda deliberately manufactured in the general interest of capital is not unknown (see, for example, most British newspapers), this is not the main way that minds are reproduced in advanced capitalist countries.

The 'consciousness industry' holds a central position in advanced capitalism. It can use the attractions of high monetary reward and mass adulation to employ the most creative minds, the most beautiful bodies, the most convincing voices, the most appealing personalities and the latest technology to mass-produce technicolour stereophonic edited pieces of exaggerated 'experience', which the modern mind has come to prefer over real experience, just as the addict prefers drug-induced fantasy over mundane reality. It is by structuring consciousness in general, rather than spreading propaganda deliberately, that popular adherence is gained.

Concentration of economic power in Euro-America and Japan forms the material base and provides the need for a similar concentration of cultural power. The finest global resources are poured, with abandon in the case of national spectacles, into cultural production. But the centre of culture does not correspond exactly in space with the centre of economic power: the hierarchies of dominance are not exactly parallel. In the 1950s Los Angeles took over the central role in global cultural production from New York as the dominant media changed from stage and radio to the mass-visual instruments of film and television. The natural environment of Southern California became the world's favourite place, just as the 'perfect personalities' of a small group of Americans became everyone's favourite uncle, aunt, friend and lover. There are also several secondary centres of cultural production, deriving from the regional historical roles of the various national powers, the differing languages spoken and the need still to translate what is ordinary for the people of California into forms which can readily be assimilated by other people. The regional subcentres operate within their 'own' cultural regions, as London does in former British colonies, and compete for certain parts of the hegemonic global culture, as London does for stage or music. Even sub-subcentres shine for a while, as Liverpool did for example. No one doubts, however, that *Los Angeles rules* in the production of entertaining commodities.

The ideas produced at the capitalist centre, together with the cultural forms taken by these ideas, are infused with a convincing power. They take their strength from the levels of contradiction and crisis which produce intense experiences in the First World. These experiences are sifted through, the images are intensified and these images are strung together to form 'entertainment'. Entertaining programmes are then projected with all the might of high technology into the minds of a world of people. Yet such entertainment, dealing only with effects and not causes, is not satisfying. Failure to find satisfaction in consumer culture, however, only leads to more powerfully technologized forms of that culture, indeed forms so powerful that they feed back as exaggerating influences on to their originating experience. The end result of this process might be called 'superculture', had the description 'super' not already been overtaken by the very normalization of the extraordinary we seek to describe. We shall call the resulting product by its own favourite description – hence we are now in the phase of 'ultraculture'.

Ultraculture is not made as a whole planned entity, but as individual components launched on the market of minds in the pursuit of individual profit.

It is relatively autonomous in that it limits the minds of its *producers* as well as its consumers, creating further internal conditions for its own more exaggerated continuation.

The resulting hegemonic consciousness is driven by social crisis but remains fundamentally unconnected with the *origins* of crisis. In this way, capitalist consciousness can become increasingly false, yet increasingly persuasive, both at the centre of crisis (the First World) and in the periphery (the Third World).

INTERNATIONALIZATION OF CAPITALIST SOCIAL RELATIONS: THE RESULT IN ULTRACULTURE

Commodity exchange was the first force breaking apart pre-capitalist and non-capitalist societies, with commodity fetishism taking the place previously occupied in consciousness by natural religion, i.e. as the main form of mystification. As the ancient social organisms broke apart under the dissolving influence of intensive commodity exchange, new relations of production were imposed from the top of the social hierarchy and from the centre of the global capitalist system. Capitalism is a mode of production in which labour, separated from its natural conditions of existence, is forced to work for a capitalist class which monopolizes the humanly produced means of production. Private ownership of the social productive forces is then legitimated in ideology and protected by laws in turn backed by the armed might of the state. Capitalist culture is the way of life which develops within and around this framework of relations and institutions. There are hundreds of ways in which the economic and social framework structures the development of culture, and just as many instances of return influence. Let us briefly analyse one of these – the case of mass consumption.

Over and over again the early history of capitalist production has involved deprivation for industrial workers – from the women and children paid starvation wages in the early industrial revolution to modern industrial workers stacked on top of each other in crowded cities in South-east Asia. The harshness of the early industrial experience made legitimation of the capitalist social order a difficult, but not unsurmountable, task. Workers were organized into urban masses performing similar tasks on much the same set of natural resources to yield a narrow range of mass-consumption products. This meant inculcating the same kinds of industrial skills into people spread over the broad region of the industrial revolution. Legitimation devices ('God and country') which proved successful in one place could be repeated, with different nuances, elsewhere. Yet the working class was able to develop a radical consciousness and to struggle for an increased share of the social product. It was permitted to gain a higher standard of living because capitalist relations of production proved far more productive than previous social relations. In addition, the Euro-American centre could exploit weaker peripheral societies, where capitalist relations were only just beginning to form. The result was high mass consumption for an elite of, at most, a few hundred millions of central workers and a few tens of millions in the periphery. Indeed, the consumption of this working elite has gone far beyond physical need: this means that sophisticated persuasive advertising devices had to be invented to make consumers 'need' products in one year that were unknown to them a few years earlier. It meant refocusing culture on the consumption of commodities – a new meaning adhered to commodity fetishism, the more conventional sense of a perverse fascination with buying and owning objects. It meant also that the conditions of mass production had to become the determinants of aesthetic taste, so that people would like owning a product virtually identical with that of their neighbours and friends (as opposed to the earlier craft items which bore the special imprint of their makers). In the realms of consciousness, mass production and consumption have meant standardized programmed ways of thinking with little room for regional variation. Global capitalism evolves as a single way of life gradually incorporating into its social relations, ways of producing and styles of consumption, the majority of the world's population.

CENTRE AND PERIPHERY IN THE GEOGRAPHY OF CONSCIOUSNESS

Around the centre of global cultural power lies a series of peripheral regions in which articulation processes link the differentiated cultural past to an increasingly homogenized cultural future. The origin of regional differentiation is changing from the natural environment and sequence of modes of production characterizing a region's history to its present level of economic development and thus capacity to absorb ultraculture diffused from the centre. The present geography of culture is a function of both mechanisms.

The economic grounds for the penetration of capitalist ultraculture into the minds and lives of the people of the peripheries have long been prepared. There are a few areas of the world which have not long been involved in the production of commodities for the world capitalist market and which, therefore, have been subject to the dissolving effects of exchange. Involvement in exchange opens the non-capitalist community to outside influences, first by its effect on the traditional relations and purpose of production and thus on the economic basis of culture, and second by the implantation of foreign elements directly into local culture via individuals, institutions and communications media oriented to the outside world. Recently the extension of capitalist relations and forms of production to the periphery has intensified the influence of capitalist culture. In the resulting culture invasion the 'communications media' play a central role. The medium itself (radio, television etc.) is destructive of local culture, apart from the messages it carries. Advanced communications media overwhelm those of a technically less developed society – thus in India, the cinema is presently obliterating village and street theatre – even though the values carried by the local media may be vastly superior.

Beyond technical superiority, however, lie the kinds of messages carried from centre to periphery. Centre culture signifies the contradictions of advanced capitalism, but only by dwelling on effects, 'solutions' etc. A culture which endlessly revolves around effects cannot be fundamentally satisfying, yet one which plunges into causes will be threatening not only to authority and the established institutions but also to the normal lives of a comfort-seeking populace. The result is an ultraculture with intense powers of visual and aural stimulation, but little real content, which provides mesmerizing diversion at the flick of a switch. Under the desperate economic conditions of the Third World, where 'solutions' appear to flow from the centre countries in the form of aid, loans etc. rather than from the efforts of Third World people themselves, the need for diversion is equally, if not more, pervasive. The dreams presented by the Hollywood factories, reflecting the intense contradictions of the centre, invested with all the technical and economic power of the centre, and combining intense stimulation with tranquilization, are more effective diversions than those locally produced – thus 'Hollywood' keeps half the cinemas in the non-socialist world supplied with films (Varis, 1976). However, peripheral regions have to be able to import culture and afford at least simple cheap replicas of the instruments of communication – radios, television, projection equipment, video cassette recorders etc. Economic function in the global economy determines the ability to pay for what is demanded in advance by 'ultraculture hungry' people – or is it 'junk-culture addicts always mentally in advance of the physical ability to be addicted'? On the global scale, the newly industrializing countries and the oil exporters have the economic means, but not always the political–cultural will, to import ultraculture extensively, while on the regional scale the urban hierarchy of production oriented towards centre markets serves, in return, as the network by which ultraculture differentially diffuses into the billions of minds which make up human consciousness.

This is not to argue that peripheral–regional cultural production has disappeared. The Indian film industry, for example, draws on regional folklore ('it's what village life would be like if dreams came true') for its predominant themes. Even regionally produced films are so powerful that traditional cultural forms are relegated to an adjunct status – hence puppeteers are forced to mimic the Bombay studios to draw a crowd, while snake charmers serenade with tunes borrowed from musical films (Channel 4, 1984). But in the interaction between centre culture, regional culture and traditional cultures, there can be little doubt which is most dynamic and what direction cultural synthesis takes. The tendency is towards the production of one world mind, one world culture and the consequent disappearance of regional consciousness flowing from the local specificities of the human past.

The people of the periphery are thus subjected to three kinds of domination. Where the forces of production remain undeveloped, they are still the obvious victims of uncontrollable forces of nature. They are increasingly and especially subject to the vagaries of world commodity markets, and their minds are captured by forms of consciousness evolved at the centre of contradiction in the world capitalist system.

Consciousness in the periphery in particular retains its religious nature. But the universalizing religions, especially Christianity linked with Euro-American economic superiority, become ever more universal. Peripheral people thus seek to control local nature by praying to the gods of the powerful centre. This was dramatically exemplified when the Pope visited Papua New Guinea. Villagers in red-feathered head-dresses and painted faces welcomed him in a ceremony that had required abstinence from meat and sex for the previous month, but at the same ceremony they took snapshots with miniature cameras. Meanwhile 'The Pope prayed in pidgin English that the threatening volcanoes on a neighbouring island would be stilled. He told the tribesmen: "May God's peace . . . descend upon your volcanoes" ' (*Guardian*, 8 May 1984, p. 1). Victory of the religious ceremony of the centre over that of the periphery! Mystical consciousness in the periphery is enhanced by the diffusion of modern forms of diversion carried by a persuasive ultraculture. Thus the new gods of entertainment are worshipped around the globe in one big fan-club.

Either way, through formal or informal mysticism, the result is a consciousness alienated from the real and desperate conditions which Third World people must face in the reproduction of their lives.

CONCLUSION

Human history is the history of domination by forces beyond human control. For most of history, the main source of domination was the rest of nature or rather 'Nature' as it was conceived – that is, natural forces endowed with consciousness and will, whose capriciousness alternately favoured and then eliminated human life. The very means by which this form of domination was destroyed (capitalist relations of production) became the main source of new kinds of domination. Thus capitalism likewise capriciously favours or eliminates human life, creates or destroys employment, makes some rich and others poor. What humans have thought about these determining conditions of their existence is the basis of consciousness, in the sense of Marx's general formula that social existence determines consciousness. Likewise the practices and institutions built around experiences of economic domination constitute the core of culture, in the sense of Marx's statement that the mode of production of material life conditions the general process of social, political and intellectual life. Ways of thinking and living have arisen under definite conditions which are not chosen but inherited, not understood but fundamentally mystified, not controlled but controlling. Religion is the main way of understanding events emanating from causes beyond human understanding. Events are interpreted as the material results of an almighty conscious force ('God'). Prayer is the only salvation when the human individual is confronted by an otherwise unalterable force. Human liberation within an accurate structure of consciousness is possible only when these forces are socially and communally controlled. As Marx said:

The religious reflections of the real world can in any case, vanish only when the practical relations of everyday life between man and man, and man and nature, generally present themselves to him in a transparent and rational form. The veil is not removed from the countenance of the social life-process, i.e., the process of material production, until it becomes production by freely associated men, and stands under their conscious and planned control. This, however, requires that society possess a material foundation, or a series of material conditions of existence, which in their turn are the natural and spontaneous product of a long and tormented historical development. (Marx, 1976, p. 173)

This long and tormented development can be shortened by using the fragments of accurate understanding so far achieved to pierce the veil shrouding the countenance of the social life process. This is the essentially political purpose behind radical intellectual labour in the centre and especially in the periphery.

Geography is that part of a whole knowledge which specializes in the relation with the natural environment and relations with others across space. These relations are also characterized by dominance and human servitude, although a transition has occurred from an early domination by local nature to the more recent ('spatial') domination of local events by world-wide forces. What little local control over life there was, at a low level of development of the productive forces, is lost not primarily to centralized control, although elements of this exist, but more profoundly to a lack of control over life mediated by the appearance of central control – that is, a world capitalist system which moves under the power of its unfolding contradictions, and which reduces human action to a series of short-term protective reactions not amounting, even in total, to anything resembling social control. Common ways of misunderstanding this world system, and common forms of escape from its consequences, are the basis of a world culture which overwhelms and destroys local cultures.

It can be argued that all that is lost in the process is a litany of past mysticisms, little deserving lament (see the opening quote from Fanon). But regional and local cultures represent all past experiences under a range of environmental conditions. The selective incorporation of a reinterpreted past into a liberated future cannot occur if the memory of that past has been obliterated, or if its cultural products are known only as museum pieces. This is one danger inherent in the spread of world homogeneous culture. The other is that everyone, including the peoples of the peripheries, becomes caught in a mode of thought unsuited to the solution of the problems thrown up by an inherently contradictory way of life. This is particularly dangerous when the technical devices used to 'solve' problems are capable of widespread physical destruction. Hence the urgent need for a science of society and a revolutionary cultural praxis based on it.

References

Bellah, Roberts N. 1969: Religious evolution. In N. Birnbaum and G. Lenzer (eds), *Sociology and Religion*. Englewood Cliffs, NJ: Prentice-Hall, 67-83.

de Cardona, E. 1975: Multinational television. *Journal of Communication*, 25, 122-7.

Channel 4 (British television) 1984: *There'll Always be Stars in the Sky*. Documentary programme, 25 March.

Enzensberger, H. M. 1974: *The Consciousness Industry: On Literature, Politics and the Media*. New York: Continuum.

Ewan, S. 1976: *Captains of Consciousness: Advertising and the Social Roots of the Consumer Culture*. New York: McGraw-Hill.

Fanon, Frantz 1963: *The Wretched of the Earth*. New York: Grove Press.

Frazer, J. G. 1926: *The Worship of Nature*. London: Macmillan.

Golding, P. 1978: The international media and the political economy of publishing. *Library Trends*, 26, 453-66.

Gould, P. and J. Johnson, 1980: The content and structure of international television flows. *Communication*, 5, 43-63.

Guback, T. H. 1974: Film as international business. *Journal of Communication*, 24, 90-101.

Harris, P. 1974: Hierarchy and concentration in international news flow. *Politics*, 9, 159-65.

Harris, P. 1976: International news media authority and dependence. *Instant Research on Peace and Violence*, 6, 148-59.

Horkheimer, M. and Adorno, T. W., 1972: *Dialectic of Enlightenment*. New York: Herder & Herder.

Hume, D. (ed.) 1968: Origin of religion. In N. Birnbaum and G. Lenzer (eds), *Sociology and Religion*, Englewood Cliffs, NJ: Prentice-Hall, 19-22.

Janus, N. and Roncaglio, R. 1979: Advertising, mass media and dependency. *Development Dialogue*, 1, 81-97.

Marx, K. 1969: The Eighteenth Brumaire of Louis Bonaparte. In K. Marx and F. Engels, *Selected Works*, vol. 1. Moscow: Progress Publishers, 16-30.

Marx, K. 1970: *A Contribution to the Critique of Political Economy*. New York: International Publishers.

Marx, K. 1973: *Grundrisse: Foundations of the Critique of Political Economy*. Harmondsworth: Penguin.

Marx, K. 1976: *Capital*, vol. 1. Harmondsworth: Penguin.

Marx, K. and Engels, F. 1957: *On Religion*. Moscow: Progress Publishers.

Masmondi, M. 1979: The new world information order. *Journal of Communication*, 29, 172-85.

Peet, R. 1980: The consciousness dimension of Fiji's integration into world capitalism. *Pacific Viewpoint*, 21, 91-115.

Peet, R. 1981: Historical forms of the property relation: a reconstruction of Marx's theory. *Antipode*, 13 (3), 13-25.

Peet, R. 1982: International capital, international culture. In M. Taylor and N. Thrift (eds), *The Geography of Multinationals*. London: Croom Helm, 275-302.

Peet, R. 1983: Relations of production and the relocation of United States manufacturing industry since 1960. *Economic Geography*, 59, 112-43.

Peet, R. 1987: *International Capitalism and Industrial Restructuring.*. Boston, MA: Allen & Unwin.

Perez, L. A. 1975: Tourism in the West Indies. *Journal of Communication*, 25, 136–43.

Schiller, H. I. 1969: *Mass Communications and American Empire*. Boston, MA: Beacon Press.

Schiller, H. I. 1978: Decolonization of information: efforts toward a new international order. *Latin American Perspectives*, Special Issue on the Age of Mass Media, 9, 35–48.

Semple, E. C. 1911: *Influences of Geographic Environment on the Basis of Ratzel's System of Anthropo-geography*. New York: Russell & Russell.

Sopher, D. E. 1967: *Geography of Religion*. Englewood Cliffs, NJ: Prentice-Hall.

Timpanaro, S. 1975: *On Materialism*. London: New Left Books.

Varis, T. 1974: Global traffic in television. *Journal of Communications*, 24, 102–9.

Varis, T. 1976: Aspects of the impact of transnational corporations on communication. *International Social Science Journal*, 28, 808–30.

Williams, R. 1980: *Problems in Materialism and Culture*. London: Verso Editions and New Left Books.

Williams, R. 1981: *Culture*. Glasgow: Fontana.

Wittfogel, K. 1929: Geopolitik, geographischer Materialismus und Marxismus. *Unter den Banner des Marxismus*, 3, 17–51; 4, 485–522, 698–735.

Wittfogel, K. 1957: *Oriental Despotism: A Comparative Study of Total Power*. New Haven, CT: Yale University Press.

Wittfogel, K. 1985: Geopolitics, geographical materialism, and Marxism. *Antipode* 17 (1), 21–72.

8

The Individual and the World-Economy

R. J. JOHNSTON

We hold these truths to be self-evident, that all men are created equal, that they are endowed by their Creator with certain unalienable Rights, that among these are Life, Liberty and the pursuit of Happiness.

These ringing phrases, which introduced the American Declaration of Independence, provided the context for framing the Constitution of the United States, a document which became the model for many later exercises in the writing of Constitutions and Bills of Rights. The American Constitution presents an image of a society in which individual control over life and livelihood is paramount. When collective action is taken by the state on behalf of the people, this is done for the general welfare and so for the long-term good of every individual (though, interestingly, in the early years of the United States slaves counted as only 0.5 in the allocation of Congressional seats). The ideology is one of individual freedom and self-control. But what is the reality throughout the world?

This chapter focuses on the position of the individual in the world-economy. Its starting-point is Marx's concept of alienation. Because of alienation, which is inevitable under capitalism, individuals lack control over their own lives – whether they are members of the bourgeoisie or of the proletariat. The empirical reality of this lack of control is the labour market and the maldistribution of economic and political power within society. These become the focus of protest, especially from the relatively powerless. It is countering that protest, creating a consensus within society that legitimizes such an alienating mode of production, which forms a major role of the state under capitalism (Clark and Dear, 1984). The concessions that have been provided include the granting of liberal democracy and human rights. In parts of the world-economy, however, capitalism has been countered by the attempts to create alternative modes of production based on popular rather than liberal democracy.

Democracy and human rights are concepts which are widely accepted in the contemporary world by both capitalist and non-capitalist governments. The United Nations Universal Declaration of Human Rights was unanimously adopted in 1948. It includes the statements (in Article 21), that

Everyone has the right to take part in the government of his country, directly or through freely chosen representatives

and

The will of the people shall be the basis of the authority of government; this will shall be expressed in periodic and genuine elections which shall be by universal and equal suffrage . . .

In this context, Berg (1978, p. 156) has defined *democracy* as

an ideal type of national decision-making system whose members (above some minimum age level) enjoy equality of self-determination

and *self-determination* is defined (p. 167) as

An individual has self-determination to the extent that he is not excluded from making decisions that are relevant to him and to the extent that he makes or effectively participates in the making of such decisions.

Such concepts and definitions are open to a variety of interpretations. In particular, there is a clear division between liberal (or bourgeois) democracy and popular democracy (Hindess, 1980). The former is defined within the context of the capitalist mode of production and the capitalist state. Democracy there is based on a political philosophy regarding the nature of the individual, which when translated into political practice involves competition for the right to manage a capitalist social formation (Macpherson, 1973). Certain parties in such competition may wish to manipulate the social formation towards certain interests (the proletariat's, for example), but, as shown by British and French 'socialist' administrations in recent years, the constraints of the system prevent them steering it far from the short-run, let alone the long-run, interests of capital. (Capitalism as a mode of production is relatively robust, however, and can bring into line local attempts to create alternative modes of production, as in Chile in the early 1970s. This raises major doubts as to the possibility of achieving socialism via liberal democratic means: Hunt, 1980.) Popular democracy, in contrast, can only take place in a classless society produced via the dictatorship of the proletariat. It involves individuals in control of all aspects of their lives, economic and social, the achievement of which leads to the removal of alienation and the withering away of the state (Luard, 1978).

The concepts of liberal democracy are part of the ideology of the capitalist mode of production used by the state to mystify the real meaning of capitalism for the individual, which is alienation. The goal of this chapter is to unravel some of this mystification. It begins with an outline of the concept of alienation. The geography of liberal democracy is then analysed, followed by a discussion of the geography of human rights, contrasting the situation with that achieved under popular democracy. In the final section we look at the right to life itself.

ALIENATION

The concept of alienation was the starting-point of Marx's analysis of the capitalist mode of production. Capitalism alienates the individual from her or his production; individuals are taken over by their own works, which have an independent existence. In a non-alienated condition, the individual is at one with nature, interacting with it in order to reproduce; humanity involves physical commerce with nature, the only influences on which are the individual and the caprices of the environment. In an alienated condition, however, the physical commerce with nature is organized and institutionalized. The individual loses control over it, and therefore over his or her reproduction, and instead can only survive either through the sale of labour power (the proletariat in a capitalist society) or by the purchase of labour (the bourgeoisie); in either case, the individual is no longer a 'being-for-himself' and is depersonalized in the commerce, becoming a 'being-for-another'.

The capitalist mode of production is dehumanizing therefore: instead of the individual controlling the means of production, the means of production control the individual. Those means (land in the traditional trilogy of land, labour and capital) have been captured by the bourgeoisie, and within the proletariat the individual can only survive by agreeing to sell his or her sole resource – labour – on terms very largely dictated by the purchaser in the long term.

Under the capitalist mode of production, therefore, the individual is captured by the system, and becomes subordinate to it. (This is true of individuals in both proletariat and bourgeoisie, for the survival of the latter is just as much dependent on their acceptance of the dictates of the mode of production as is the survival of the former.) Thus as capitalism spread through the world, so its particular form of alienation took over. For most of those involved, there was no alternative. They were forced to give up their pre-capitalist mode of existence, notably through the expropriation of their land resources and the pressures on them to move to urban areas where the commodification of their labour power was complete. Today, as economic power is increasingly centralized, so the processes of alienation are extended.

The causes and consequences of this alienation are readily apparent in capitalist societies. The individual has no control over the basic means of reproduction – food, water, shelter etc. – but can gain access to these only by selling labour and using the proceeds to bargain with those selling what is needed. In selling labour, the individual is usually bargaining in an unfairly structured labour market in which the powers of the individual are constrained by political, institutional and other factors. Similarly, in seeking access to the means of reproduction, individuals operate in environments structured by others.

Alienation involves the removal of self-control from individuals; they are manipulated by others because of the dictates of the mode of production. There is potential within capitalism for alienation to be carried to such extremes that individuals will collectively revolt against it. To counter this threat, and the possibility of the demise of capitalism, the state has been established as a separate (but not independent; Clark and Dear, 1984) body which legitimizes capitalism and maintains social cohesion. One of the strategies that may be pursued is to grant concessions to the proletariat, thereby 'buying' their acceptance of the capitalist system. Such concessions include liberal democracy – the 'apparent' involvement of individuals in the control of the economic system. Liberal democracy is not universally practised within the contemporary capitalist world, however, for reasons outlined in the next section.

Outside the capitalist world, a mode of production has been established which aims to avoid alienation by removing the class differential: if all members of a society are of the same class, then the alienation consequent upon unequal power and exchange cannot ensue. According to Marx, it was the bourgeoisie that should be eliminated, resulting in the dictatorship of the proletariat. That dictatorship was to be achieved, according to Lenin (1917), through the use of the state apparatus to repress the bourgeoisie and achieve a transition to socialism in which the state organized society in the interests of all, as a prior stage to the achievement of communism which would involve the 'withering away of the state' as power is handed over to the people operating in a system of anarchic communities. The transition is still continuing, according to most apologists, involving the dictatorship of the party (and hence of the state, which it controls). Whether it will be achieved, or whether such dictatorship for rather than of the proletariat will turn out to be a permanent feature of such modes of production remains to be seen. At present, in societies promoting a socialist mode of production power is firmly entrenched in a particular form of bourgeoisie and self-control for the majority is absent; alienation is rife there also.

THE GEOGRAPHY OF LIBERAL DEMOCRACY

Democracy is generally conceived as 'government of the people, by the people, and for the people'; all adult individuals have the right to decide what are matters of general concern and how they should be tackled, and the consensus view prevails. This implies that every member of society is an equal participant. In practice, because of the scale problem, it has been replaced by representative democracy, whereby every individual has an equal part to play in both electing a body to organize the society and calling the representatives to account. In popular rhetoric, especially in Western Europe and North America, democracy is equated with individual freedom.

If liberal democracy is defined as government by an elected body (or individual), with that body accountable to the electorate regularly and frequently,

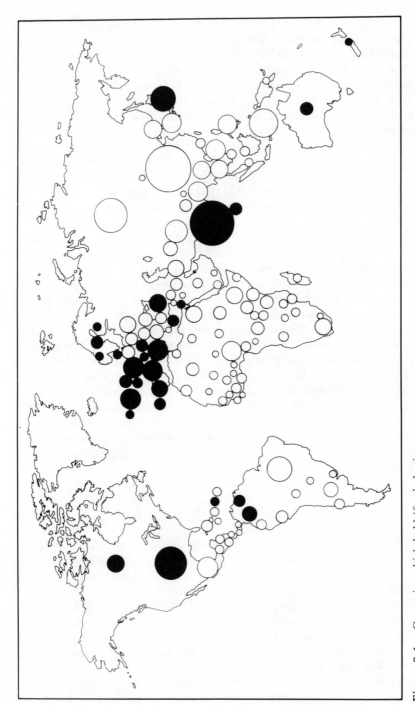

Figure 8.1 Countries which hold 'free' elections

In this and other world maps in this chapter, the size of the symbol for each country is proportional to its population

Source: Butler et al., 1981

with no constraints on the choice of representatives available to the electorate and with no constraints (other than age) on who can vote, then it is a system operated in only a minority of countries at the present time and available to only a minority of the world's population. Figure 8.1 shows all those countries with populations exceeding three million identified by Butler et al. (1981) as satisfying six conditions: (1) universal or near-universal adult franchise; (2) a regular timetable for elections that is adhered to; (3) no substantial group is denied the opportunity to form parties and nominate candidates; (4) all seats in the major legislature can be, and are, contested; (5) campaigns are reasonably fair without violence or intimidation; (6) votes are cast freely and secretly. Only 28 states met these criteria; as the map (figure 8.1) shows, nearly all are in Western Europe.

Democracy is more than holding elections, however. Frequent elections are held in many countries, but the choice available to the electorate is either constrained (certain individuals are prevented from standing as candidates, or certain parties are precluded from fielding candidates) or nil (a one-party state). In others, only a minority are allowed to register as electors and to vote. Dahl (1978) has argued that democracy should give all citizens the opportunity to formulate and signify preferences and to have those preferences weighted equally in the conduct of government. For this to occur, a set of institutional guarantees is required, as laid out in table 8.1.

Study of the geography of liberal democracy should take all these necessary conditions into account. This has been done annually for several years in the publication *Freedom in the World*. Countries have been classified according to their level of political rights into seven categories, from the most free to the least free. The criteria used are as follows (Gastil, 1981, pp. 13–17). Countries in the top category 'have a fully competitive electoral process and those elected clearly rule'. Those in the second category have some constraints on the

Table 8.1   Institutional guarantees for democracy

Necessary condition	Institutional guarantees required
The formulation of preferences	Freedom to form and to join organizations Freedom of expression The availability of alternative information sources The right to vote and to compete for votes
Signifying preferences	All the above plus – Free and fair elections Freedom to stand for public office
Equal weighting of preferences	All the above plus – Institutions which ensure that government policies depend on voting and other popularly expressed preferences

effectiveness of this policy and the equality that it implies; the factors involved include 'extreme economic inequality, illiteracy or intimidating violence' which can weaken effective competition. The next three categories imply increasingly less effective popular control, and most countries in category 5 have no elections. In category 6 no competitive electoral processes are allowed: 'The rulers of states at this level assume that one person or a small group has the right to decide what is best for a nation, and that no one should be allowed to challenge that right', but such rulers do respond to popular desire and operate within the context of local culture. In category 7, the 'least free', 'the political despots at the top appear by their actions to feel little constraint from either public opinion or popular tradition'.

Clearly there is some element of subjectivity in this categorization, although the outcome accords with general beliefs – at least in those countries in category 1, where the ideal of democracy is proclaimed. The map of the 1980 categorization (figure 8.2) confirms those general beliefs. Countries such as the United Kingdom and the United States are in category 1, India, Israel and Jamaica are in category 2, Poland and Yugoslavia are in category 6, and the German Democratic Republic and Haiti are in category 7. (In some of the countries in the lower categories – such as Poland and the German Democratic Republic – the concept of liberal democracy is not accepted; such countries are being classified on criteria that are irrelevant to *their* interpretations of democracy, under which, for example, only one party is necessary to demonstrate the collective will.) Further, the validity of some elections has been criticized by observers, leading to their being depicted as 'demonstration' elections, stage-managed by external, usually US, forces (Herman and Brodhead, 1984).

Liberal democracy implies popularly elected governments. A corollary of this is that governments will only change at the popularly expressed behest. In many countries, however, there are 'irregular' transfers of power without popular consent (figure 8.3). Most of these involve some form of *coup d'état*, and very many (including the only such transfers in Western Europe – France in 1958 and Portugal in 1974) have some form of military involvement. Rule by the military is far from rare (figure 8.4), providing a particular type of government in categories 6 and 7.

Economic Development and Liberal Democracy

The 'geographies of liberal democracy' outlined in figures 8.2–8.4 suggest clear correlations with what we generally term the 'geography of economic development', if the popular democracies are excluded. In broad terms, liberal democracy is associated with the most industrialized and prosperous countries of the world, whereas the absence of political freedom is associated with underdevelopment, especially in the Third World. This suggests, in line with Rostow's (1960) classic, if discredited, model of the processes of economic development, that as countries become more prosperous, so democracy is

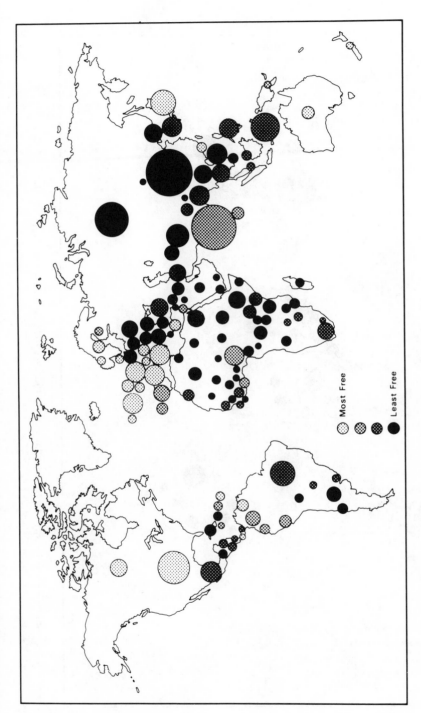

Figure 8.2 Countries according to their level of political rights in 1980
Source: the ratings are taken from Gastil, 1981; categories 3–5 and 6–7 have been combined

Most Free

Least Free

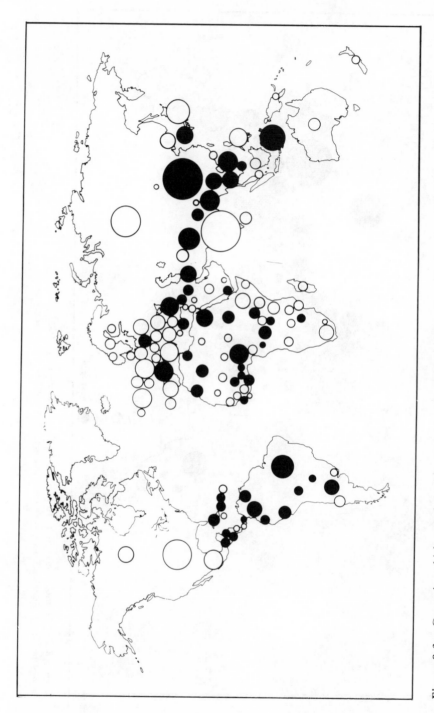

Figure 8.3 Countries which experienced irregular transfers of executive power during the period 1948–1977
Source: the data are taken from Taylor and Jodice, 1983

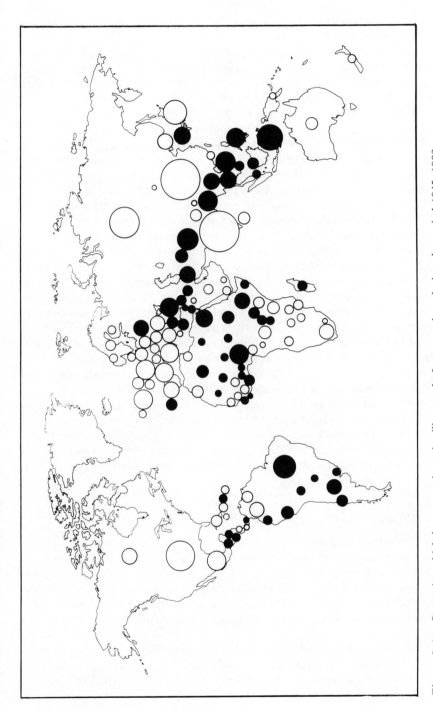

Figure 8.4 Countries which have experienced military rule for some time during the period 1945–1982
Source: the data are taken from Kidron and Smith, 1983

established. Democracy is a condition associated with a particular economic and social condition.

Several authors have tested this proposition. Coulter (1975), for example, regressed indices of liberal democracy against measures of economic development and social mobilization. Three concepts – political competitiveness, participation and public liberties – were combined into an index of liberal democracy. The Netherlands had the highest value (the data refer to the period 1948–69), followed by Belgium, the Federal Republic of Germany and Iceland. The bottom five positions (of the 85 countries studied) were occupied by Pakistan, Kenya, Sudan, Iraq and the United Arab Republic. Social mobilization was measured by variables relating to urbanization, education, communication, industrialization and gross national product. Coulter found that the level of social mobilization accounted for 48 per cent of the variation among countries in their level of liberal democracy, and changes in the level of social mobilization accounted for exactly half of the variation in the level of liberal democracy (see Taylor's, 1985b, 1987, reinterpretations of these data).

A slightly more sophisticated statistical analysis by Bollen (1983) tested the hypothesis that 'economic development . . . increases the likelihood of political democracy' (p. 468), modified to recognize that in the contemporary world-economy 'economic development' in peripheral countries does not bring as much autonomous power to local elites as it does in core countries, because powerful groups in the latter (including multinational companies) have a great deal of influence over those in the former. He used a measure of liberal democracy similar to Coulter's and a single measure of economic development (energy consumption per capita), and his basic equation for a sample of 100 countries accounted for 47 per cent of the variation: the greater the level of development, the greater the level of democracy, but at any level of development the level of democracy was significantly lower in peripheral countries than in the core. (Interestingly, in Coulter's analysis the popular democracies do not stand out as either 'overdemocratized' or 'underdemocratized' relative to their level of social mobilization.)

An Alternative View

Analyses describing the geography of liberal democracy in terms of the geography of economic development provide no understanding of the mechanism whereby such democracy develops. The implication is that it comes about 'naturally' as a consequence of some mechanistic process embodied in economic development: with industrialization, urbanization and prosperity come elections, a wider franchise and freedom of choice. Alternative views suggest that the process is far from automatic, that democratic gains are usually the consequence of long struggles by alienated labour and that victories gained are frequently overturned. To understand why the observed correlations occur, we need a model of the fight for democracy (Therborn, 1977; Taylor, 1987).

Liberal democracy focuses on the state: to understand its operations requires appreciation of the nature and role of the state. In any society, the state has three basic roles – to maintain cohesion, and to promote and to legitimate the mode of production. It is an institution empirically separate from, but totally interlinked with, the powerful class within the mode of production (for full discussions see Johnston, 1982, 1984a,b,c, and Clark and Dear, 1984). By promoting the mode of production it promotes the interests of that class (and so in popular democracies it promotes the interests of all). Under the capitalist mode of production, therefore, the state, by promoting capitalism, promotes the interests of the bourgeoisie; as a consequence, it is acting against the interests of the proletariat. But capitalism is founded on interclass tension. Promotion of capitalism can exacerbate the tension, creating the conditions for conflict which will damage the mode of production and hence the interests of the bourgeoisie (or of capital in general, where it is divorced from individual ownership; Lash and Urry, 1987). Thus the state is also involved in ameliorating that tension by containing proletarian disquiet. It must either obtain support or repress the disquiet.

Support is generally won by granting concessions. But what if it cannot, if legitimation cannot be achieved because the victories would undermine the interests of capital? In such a case coercion is the usual alternative pursued, and legitimation is achieved, not agreed. The capitalist state promotes and legitimates the mode of production. As the latter is built on alienation, then the state must both promote and legitimate it, either by offering concessions to the alienated labour in return for its (grudging) support or by imposing a regime on labour – by coercing it.

These complementary roles are made difficult for the state by general tendencies within capitalist economic systems. Over the last two centuries, capitalist economies have not displayed unbroken economic progress. Rather, they have been characterized by a series of long waves (generally known as Kondratiev cycles) comprising a boom followed by a slump (Taylor, 1985a). Each wave has lasted about 50 years. The slumps are brought about by a combination of factors seemingly endemic to capitalism: investment in productivity increases means that the ability to produce outruns the ability to consume; markets are saturated; sales and profits fall; investment declines; new investments are sought in sectors where profits are more likely and so on (Harvey, 1982, 1985). During such slumps, both classes in capitalist society suffer – the bourgeoisie from a lack of wealth accumulation, and the proletariat from low demand for labour. Of the two, the proletariat suffers most, because it lacks the accumulated wealth on which to live until there is an upturn in the economy and the demand for labour increases.

The state is fully implicated in these crises of capitalism because its twin roles are to promote and sustain the mode of production. It faces three sets of crises (Habermas, 1976). The first are *rationality crises*. The state should promote accumulation, and so is judged to be failing its supporters in the bourgeoisie. The second are *legitimation crises* which focus on the proletariat;

the system legitimated by the state is failing them. Together, these generate *motivation crises*: the state is failing both classes, who will withdraw support (not necessarily immediately and completely).

In order to maintain cohesion within society and avoid motivation crises, the state must ameliorate the impacts of rationality and legitimation crises and create an environment within which they will be removed by a further cycle of prosperity. It must take steps to promote accumulation, without which it has nothing to legitimate. This provides the basis for understanding the changing geography of liberal democracy in the world.

The world-economy comprises three 'zones': core, semi-periphery and periphery. Countries in the core are distinguished by their nodal position within the economic system, in terms of control and the repatriation of profits. Wealth created in the periphery has been expropriated to the core. This has provided the foundation for the major international inequalities, and has allowed the states within the core to legitimate the capitalist mode of production by granting concessions to their proletariats. Such concessions can be 'afforded' because they have been paid for in part by the peripheral countries. One of the concessions has been liberal democracy (Johnston, 1984b). In response to proletarian demands, the state has eventually granted near-universal franchise, allowing widespread participation in government which inevitably means that to some extent the state has granted further concessions to the proletariat. As long as the core country is prosperous, this has been affordable. Further, an ideology of democracy linked to that of capitalism has been promoted by the state. Thus in major motivation crises, the proletariat has largely accepted pro-capitalist arguments in order to counter the rationality crisis (as with the 1931 National Government in the United Kingdom; Taylor, 1983).

The situation in peripheral countries is very different because they lack the accumulated wealth with which to tackle crises. Such countries are almost impotent in their attempts to influence the operations of the mode of production. They are used by outside interests as sources of cheap resources and labour, and because there is an abundance of most resources, certainly of cheap unskilled labour, those outside interests can play off one country against another. Investors want guaranteed cheap commodities plus guaranteed cheap disciplined labour. If the guarantees are not forthcoming, or if they are doubted, the investment will go elsewhere. To win investment, the state operates policies designed to ensure 'order and stability' (Osei-Kwame and Taylor, 1984).

When a state within a peripheral country faces a motivation crisis, it has to tackle a rationality crisis by creating an environment within which capitalist profits can be made, which will probably mean that it cannot afford social democracy; without this, investment will not take place, jobs will not be created and legitimation will be threatened. It can only create such an environment, it feels, by 'disciplining' the proletariat; neither liberal nor social democracy can be afforded (Taylor, 1989). Democracy creates instability – governments may fall or be forced to yield concessions that threaten profits – and so, in the national interest, cannot be afforded. Coercion is necessary to provide the stability that will attract investment and prepare the path to prosperity. Thus non-democratic rule is promoted as being in the national interest, and the state is 'taken over' by groups – often military – pursuing such policies. But what if it fails – relatively, at least? Some investment may be attracted (including, perhaps, some locally generated capital), but it has little multiplier effect, and creates insufficient jobs and prosperity to satisfy the demands of the local proletariat. Their judgement will be that its policies have failed, and this will generate a legitimation crisis. This can be tackled by further coercion and oppression – which is expensive in its demands on well-paid 'safe' manpower (for the police and military apparatus) – or the controllers of the state can yield and 'return the country to democracy'. The former solution is a totalitarian unfree regime. The latter is liable to promote a further rationality crisis. Neither is likely to counter the problems of alienation. This, then, is the politics of failure.

This scenario suggests an ideal-type sequence whereby a peripheral country passes through a continuing cycle of democracy (to counter a legitimation crisis), non-democratic rule (to counter a rationality crisis), democracy again and so on (Johnston, 1984b). A number of countries (e.g. Argentina, Brazil, Chile, Ghana and Nigeria) have recent political histories that display these characteristics. As an ideal-type scenario, it is not a predictive model. Some countries may not change their forms of government: not all states facing legitimation crises have (re)instated liberal democracy, and not all states facing rationality crises have abandoned it. Alternative strategies may be pursued by those in power within the state apparatus, although this may strain their abilities to build coalitions of support (Osei-Kwame and Taylor, 1984; Taylor, 1984); Taylor's (1986) discussion of the particular features of the Indian party system illustrates this. It may be that a strategy works, and that a country rises out of the peripheral status and achieves the level of permanent general prosperity which allows it to afford the luxury of permanent democracy.

And the Core Countries?

Liberal democracy is a form of political organization which, in general, is only possible in the prosperous core countries of the world. Within them, proletarian participation in, and numerical dominance of, government should mean that policies are followed which are very largely in proletarian interests. But this is not the case in absolute terms; however, in relative terms the proletariat is much more prosperous in the core democracies than elsewhere in the world.

Representative government has developed into government by parties. These are not just organizations which allow for some efficiency in government – a party or a coalition of parties with a majority virtually ensures that agreed policies are enacted; they also structure the electoral agenda. The parties can thus determine the issues (Riker, 1983); to the extent that they control the political system, they control the content of political debate. In this way many issues salient to the electors may never become salient within the political system (Schattschneider, 1962).

The strength of this organizational bias depends on the ease with which the electorate can influence the number and nature of the parties. New parties can be established to promote issues ignored by those already existing. They face major problems, however. To win support, they must promote themselves, which requires money and access to the main channels of information – the media. Political campaigning is extremely expensive. To afford lengthy campaigns and permanent organizations, parties must attract funds – either from the proletariat at large (in general very difficult, especially over long periods) or from sectors of the bourgeoisie. Thus only parties, or individuals, which do not threaten the established order have much chance of support. Further, in most countries the media are in private ownership (i.e. they are owned and controlled by the bourgeoisie). They are also firmly committed to the established order, and so are unlikely to give much positive support to new 'threatening' political movements.

This does not mean, of course, that such political movements will not attract support in some situations: the relative successes of the Green Party in the Federal Republic of Germany in the 1980s, Mogens Glistrup in Denmark in the 1970s, Pierre Poujade in France in the 1950s and the Alliance parties in the United Kingdom in 1983 and 1987 illustrate this. But very few are likely to survive for long unless they promote policies acceptable to the established order. The state is not independent of the mode of production, and if parties in control of the state act in ways perceived by the bourgeoisie (either within or outside the country) as harmful to capital interests, then they can be 'disciplined' through the creation of a rationality crisis. A consequence of such rationality crises for many governments has been an inability to service overseas debts, and they have only been 'rescued' from a fiscal crisis by the International Monetary Fund (IMF) on terms favourable to outside capitalist interests and not those of the local proletariat (Johnston, 1982); core countries have also been forced to call on the IMF as a UK Labour Government did in 1976.

Parties structure the political agenda, but are constrained in their actions by the imperatives of the capitalist system. Their degrees of freedom of action are limited by the 'penalties' that can be imposed upon them by capitalist interests. This is more the case now than ever before. The hypermobility of capital and the control of the world-economy by multinational corporations means that rationality crises can be stimulated very rapidly by the threat to withdraw investment. The crisis of the state is potentially deeper than ever before – in the core countries as much as in the periphery: its ability to react positively to the demands of the proletariat is increasingly constrained (Johnston, 1984a).

This situation is apparent in the United Kingdom in the 1980s. The rationality crisis associated with a major depression in the Kondratiev cycle produced the highest levels of unemployment for 50 years. Those in control of the state identified a route out of this crisis as involving a reduction of the democratic luxuries that they believe can no longer be afforded. Thus local government – in some places operated by parties with different prescriptions – has been curtailed by fiscal policies, democratic control over local service provision has been replaced by administrative centralism and participation in elections has been curtailed by raising the deposit to be paid by each candidate ('to discourage frivolous candidates'). Such policies could stimulate a legitimation crisis, but this has in part been forestalled by transferring large sections of the proletariat into a petty property-owning bourgeoisie through the promotion of, for example, owner-occupancy of housing. In this way, the ideology of capitalism – the promotion of individual self-interest – can be promoted as the ideology of the majority, and members of the proletariat are less likely to take action – striking, for example – which would deepen the rationality crisis. (The control of their assets by this transferred proletariat is slight, however. The pension funds established by trades unions are controlled by capitalist interests, which may mean investments that are against the union members' own interests.) The labour force is being disciplined by a process that involves erosion of democratic rights, and yet its support for such authoritarian policies is being 'bought' in other ways (Hall, 1980).

This method of tackling the rationality and legitimation crises has been criticized in the United Kingdom by groups which argue that its potential success is threatened by the existence of a political opposition which, if it gained power, would initiate major shifts in policy. The mere possibility of this happening after the next general election – never more than five years away – creates an unstable environment within which investment in long-term development plans by capitalists is not encouraged (Finer, 1975). The rationality crisis in the United Kingdom is, at least in part, a function of the political system.

Most countries have a range of parties offering alternative policies, but in the United Kingdom there is a much greater chance that, because of the electoral system, a change of government after an election would lead to substantial policy changes within capitalism than is the case in many other countries. The 'first past the post' method means that a party with minority support among the voters can obtain a majority in Parliament, and it magnifies changing preferences within the electorate so that two parties with approximately equal voter support can both have reasonable hopes of electoral victory. If those parties differ on major issues – on support for industry and the welfare state, for example – then outside interests face possible major switches of policy after every election. This, it is claimed, has been the case in the United Kingdom since about 1960 as the two main parties (Conservative and Labour) have become more polarized, creating the unstable situation termed 'adversary politics' (Finer, 1975).

The solution offered by some to the problems of adversary politics and their impact on the rationality crisis that might ensue is electoral reform. An electoral system based on proportional representation would, it is claimed, virtually ensure a permanent 'government of the centre' and remove the possibility of major policy shifts after elections. This would create stability (and so tackle the problems of a rationality crisis) and be more representative of the electorate's views (thus ensuring against a possible legitimation crisis). To some, this would be more democratic. To others it would be less, because it would concentrate power in the parties of the centre who will structure the political agenda in the interests of the bourgeoisie; real choice is diminished and politics has been 'depoliticized' without actually doing away with democracy (Johnston, 1985; Taylor, 1985b; Taylor and Lijphart, 1985).

CIVIL AND HUMAN RIGHTS

Liberal democracy is widely regarded as the foundation of all civil liberties. If individuals lack control of their governments, it is probable that they will be denied many of the other fundamental civil and human rights. In 1948 those rights were defined by the General Assembly of the United Nations which adopted, without a single dissenting vote, a 'Universal Declaration of Human Rights' comprising 30 articles setting out the rights and freedoms to which everyone is entitled. Twenty-eight years later this statement of principles was translated into an 'International Covenant on Civil and Political Rights' which established legal obligations on each state ratifying the Covenant: 77 had ratified it by 1983.

A recent survey assessed countries according to their performance on human rights in the context of the Covenant (Humana, 1984). For 75 countries, it was possible to make a comprehensive assessment, and these were rated on a scale from zero to 100 – the higher the value, the better the human rights record. The average score was 64; the lowest was 17 (Ethiopia) and the highest was 96 (Denmark, Finland and New Zealand). For the countries (49) which had ratified the Covenant, the average was 70. Thirty-two other countries were assessed on a three-point scale only, because of the lack of information: four had a 'fair' human rights record (two had ratified the Covenant – the Dominican Republic and Honduras), 16 had a 'poor' record (including five ratifiers) and 12 had a 'bad' record (including El Salvador, Iran and Libya, which had all ratified the Covenant).

A map showing this categorization of countries according to their human rights records correlates very closely with that of political rights (figure 8.5). A clearer statement of this correlation is provided by comparing Gastil's (1981) sevenfold categorization of countries according to their political rights in 1980 (figure 8.2) with a similar categorization of civil liberties (figure 8.6). For the latter, in countries rated in the top class

> publications are not closed because of the expression of rational political opinion, especially when the intent of the expression is to affect the legitimate political process. No major media are simply conduits for government propaganda. The courts protect the individual; persons are not imprisoned for their opinions; private rights and desires – education, occupation, religion, residence, and so on, are generally respected;

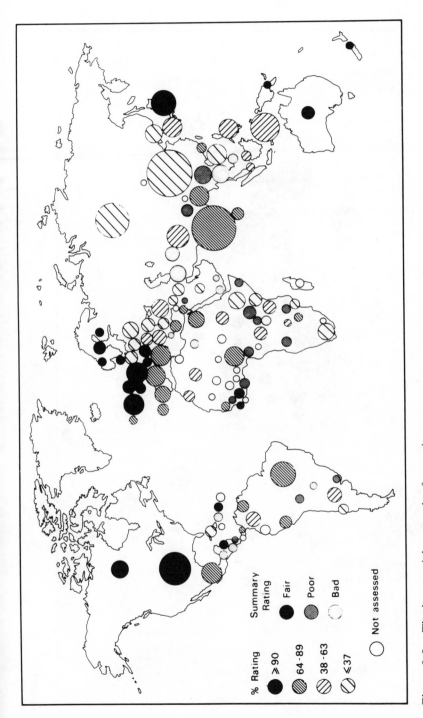

Figure 8.5 The human rights records of countries
Source: the data are from Humana, 1984

Summary
Rating

● Fair

◉ Poor

○ Bad

○ Not assessed

% Rating

● ≥90

◉ 64-89

◎ 38-63

◍ ≤37

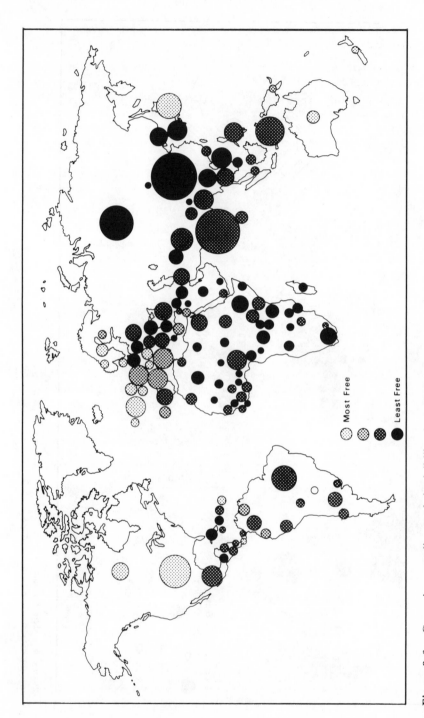

Figure 8.6 Countries according to their civil liberties – 1980

Source: the ratings are taken from Gastil, 1981; categories 3–5 and 6–7 have been combined

Most Free

Least Free

law-abiding persons do not fear for their lives because of their rational political activities. (Gastil, 1981, pp. 17–19)

Moving through the classes, so these freedoms are restricted. Courts are more authoritarian in class 2; political prisoners are common in 3, as is censorship, and torture is practised. By class 6 'the legitimate media are completely under government supervision; there is no right of assembly; and, often, travel, residence and occupation are narrowly restricted' (pp. 19–22). (In some states, those rights are denied because, it is argued, of the existence of 'wartime conditions' which require the exercise of emergency powers. In this sense, the restrictions on freedom imposed by the Sandinistas in Nicaragua against the Contras in the 1980s differ little from those imposed by the UK government in the early 1950s. In both, war was used to justify controls.) By class 7, political terror is common:

> there is pervading fear, little independent expression takes place even in private, almost no public expressions of opposition emerge in the police-state environment, and imprisonment or execution is often swift and sure. (p. 22)

The extent of agreement between the two classifications is shown in table 8.2. Because of different class sizes, complete agreement is not possible. Of the 161 a maximum of 137 could fall on the main diagonal of the table: 92 did. The greatest agreement is in the top left-hand corner: the countries with the best records on political rights also have the best on civil liberties. Towards the lower end of the classifications, there are some disparities: four countries (Bolivia, Central African Republic, Pakistan and Surinam) are in class 5 for

Table 8.2   Cross-classification of countries according to their ratings on political rights and civil liberties

Political rights	Civil liberties							
	1 (most free)	2	3	4	5	6	7 (least free)	Total
1 (most free)	18	5	–	–	–	–	–	23
2	–	14	13	–	–	–	–	27
3	–	–	2	7	–	–	–	9
4	–	–	3	5	1	–	–	9
5	–	–	1	2	21	4	1	29
6	–	–	–	2	8	18	5	33
7 (least free)	–	–	–	–	4	13	14	31
Total	18	19	19	16	34	35	20	161

Source: the ratings are taken from Gastil, 1981

civil liberties but 7 for political rights, whereas one (Syria) rates 7 on civil liberties but 5 on political rights. Overall a good record on political rights is usually consistent with a good record on civil liberties: liberal democracy and human rights go hand in hand.

Even in the countries with good records, however, it must be recognized that there are flaws, that 'these flaws are significant when measured against the standards these states set themselves' (Gastil, 1981, p. 19) and that the interpretation of their standards may be far from ideal. The United States, for example, is given positive ratings by Humana (1984) on the 'freedom to seek information and teach ideas' and on the 'freedom of movement in own country', and racial discrimination is constitutionally outlawed. Yet the interpretation of these freedoms by the courts suggests inequality of treatment (Johnston, 1984c), McCarthyism was a major invasion of freedoms in the 1940s and 1950s, and there is still much discrimination practised (against women, for example; Seager and Olson, 1987). The United Kingdom has similar provisions, but movement between Great Britain and Northern Ireland is politically restricted (Taylor and Johnston, 1984), as is the right to join an independent trade union in certain occupations: in 1984–5, during the National Union of Mineworkers' strike, miners were prevented by the police from travelling to other coalfields on the grounds that they *might* be intending to commit an offence against the anti-picketing laws.

What these data also show is that, according to a definition of human rights that they have accepted, nevertheless the popular democratic states do not have good records. The dictatorship of the party there involves many restrictions. In Romania, for example, Humana (1984) notes the existence of 'Discretionary death sentence for hostility to socialism but usually lengthy prison sentence' and 'Long period of detention or re-education in labour camp' for possession of banned political literature. Romania is given a rating of only 32 per cent and the observation that it operates 'Severe curtailment of human rights despite having signed and ratified the UN covenant' (p. 208). The USSR is given an even lower rating (27 per cent), despite the statement in chapter 6 of its Constitution (Finer 1979) to the effect that (Article 34)

> the equal rights of citizens of the USSR are guaranteed in all fields of economic, political, social, and cultural life.

Again, it was claimed in the early decades of the USSR that a state of warfare existed within the country requiring controls until the 'dictatorship of the proletariat' was achieved. Only in the late 1980s have the actions of President Gorbachev indicated any desire to relax those controls, in the context of the caveat that (Article 39)

> The socialist system ensures enlargement of the rights and freedoms of citizens and continuous improvement of their living standards as social, economic, and cultural development programmes are fulfilled.

Enjoyment by citizens of their rights and freedoms must not be to the detriment of the interests of society or the state, or infringe the rights of other citizens.

In other words, the Communist Party, which has total control of the Soviet state apparatus, has the goal of promoting the advance of human rights as part of its programme of economic and social development; only as that programme is achieved can the rights be fully realized. The arguments are very similar to those applied in countries on the periphery of the capitalist world-economy, where civil rights have to be 'earned' as prosperity is achieved.

THE RIGHT TO LIFE

'Life, liberty and the pursuit of happiness' are the unalienable human rights identified in the American Declaration of Independence. The discussion so far has looked at one definition of liberty (participation in democratic rule) and one of the pursuits of happiness (freedoms of speech, assembly etc.), but has ignored the right to life. In what way is this spatially structured within the contemporary world-economy?

The right to life is defined here as the right to survive without one's life being terminated prematurely by the actions of others. Such premature termination can come about in one of three ways: through abortion prior to birth, through violent death at the hands of others and through death because of ill-health brought about by physical deprivation. Each of these is spatially patterned. There is a clear geography of abortions (Kidron and Segal, 1984), with the greatest freedom to obtain an abortion legally in Europe (including the popular democracies; the USSR has the highest rate of legal abortions; Seager and Olson, 1986). That geography is closely linked in many arguments to the geography of male domination in society. ('The right of women to control their pregnancy is variously abridged by male-dominated law'; Kidron and Segal, 1984. The argument can be extended to male-dominated law – both religious and secular – regarding contraception.) Thus abortion is most readily obtained in countries where women's struggle for equal rights has advanced furthest, though there it may be in conflict with 'pro-life/anti-abortion' groups, many of them religiously based.

With regard to the other two forms of premature death, these are frequently identified as the results of behavioural and structural violence respectively. With *behavioural violence*, death is the product of deliberate acts, of which those involving war in its many forms (Johnston et al., 1987) are the major cause. The geography of war-related deaths is linked to contemporary geopolitics, and is dealt with in detail elsewhere in this book. In most periods that geography shows a very concentrated pattern of premature deaths which are either direct or indirect consequences of war.

Premature deaths from behavioural violence are small in number each year relative to those resulting from *structural violence*, which come about without deliberate actions by individuals intended to cause death. Structural violence has been defined as 'whenever persons are harmed, maimed or killed by poverty and unjust social, political and economic institutions, systems or structures' (Kohler and Alcock, 1976, p. 343). Death from structural violence can be operationally defined in the following way (see also Galtung and Hoivik, 1971). In France in 1985, the life expectancy of a female at birth was 81 years, as a consequence of the system of health care available there, the material affluence that allows the purchase of food, shelter and clothing, and the quality of public utilities (water supply etc.). If it is possible to live for that long in France, it should be similarly possible elsewhere – unless there are genetic and/or environmental reasons to the contrary. However, in Sierra Leone in 1985, the life expectancy for a female at birth was 40 years. Thus on average a Sierra Leonean woman can expect only half of the life-span of a French woman.

This differential between those two extremes (the data are from World Bank, 1987) is not the result of deliberate actions. People in France and similar countries do not want people in Sierra Leone to have shorter lives. (More importantly, they do not want that major differential to come about because many more children die in Sierra Leone than in France – the infant mortality rates in 1985 were 175 per 1000 and 8 per 1000 respectively – for the death of infants is not a desired end.) The differential comes about because of the core–periphery structure of the world-economy, whereby the life chances of the population of the periphery are subordinated to those of the population of the core. Capitalist development is necessarily spatially uneven according to many writers (e.g. Harvey, 1982; Smith, 1984). One aspect of that uneven development is the geography of structural violence.

The nature of that geography can be portrayed in a variety of ways. Figure 8.7 does it very simply by expressing the difference between average life expectancy in France (78 for males and females together) and that in the country in question as a percentage of the French figure. For Sierra Leone this is 49, indicating that the years that the average Sierra Leonean born in 1985 will not live are 49 per cent of the expected life-span of a French person born in that year.

The geography of 'lost years of life' shown in figure 8.7 is a clear depiction of the pattern of uneven development in the capitalist world-economy, and it is also closely correlated with the geographies of liberal democracy and human rights. There are, of course, exceptions: Costa Rica, for example, had an average life expectancy of 74 years in 1985, and in Sri Lanka the figure was 70. Whereas some core countries in the periphery have high life expectancies, however, none in the core have low figures; as with the measures of liberal democracy, the core countries score uniformly highly but there is great variation in the periphery.

Although figure 8.7 clearly demonstrates the geographical variability in life chances, it cannot fully portray the sheer volume of the structural violence

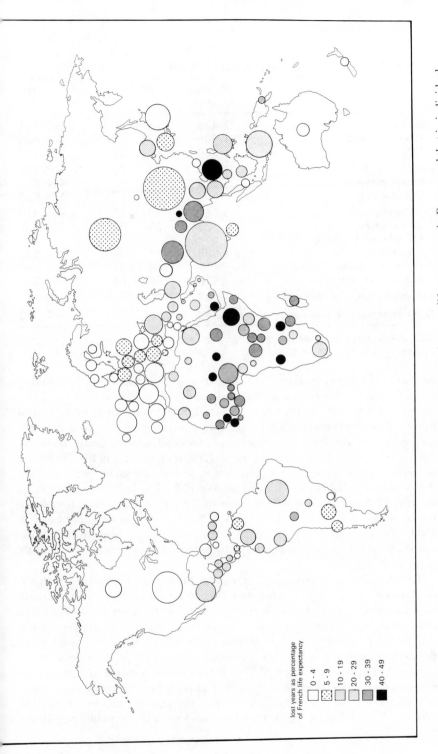

Figure 8.7 The geography of structural violence: the percentage difference between life expectancy in France and that in individual countries in 1985

Source: data from World Bank, 1987

lost years as percentage
of French life expectancy

- 0 - 4
- 5 - 9
- 10 - 19
- 20 - 29
- 30 - 39
- 40 - 49

Table 8.3   The geography of structural violence

	Population (million)	CBR	LE (years)	Years lost (million)
Low income	2,439	29	61	1,202,427
China and India	1,805	24	64	606,480
Other	634	43	52	708,810
Middle income	1,242	32	62	635,904
Lower middle	675	36	58	486,000
Upper middle	567	28	67	174,636
Developing	3,681	30	61	1,877,310
Oil exporters	523	38	58	397,480
Manufactured goods exporters	2,098	24	64	704,928
Highly indebted	554	34	63	282,540
Sub-Sahara Africa	418	48	51	541,728
Industrial market economies	737	13	76	19,162
Non-reporting, non-member	363	19	69	62,073

CBR, crude birth rate per thousand; LE, life expectancy
Source: all data from World Bank, 1987

present in the world-economy. An indication of this is achieved by totalling the lost years. In Sierra Leone, the crude birth rate in 1985 was 48 per 1000, giving 192,000 births in the year. If those 192,000 individuals were born in France, they could expect to live for 78 years on average, giving a total of 14.976 million years of life. With a life expectancy of only 40 in Sierra Leone, however, the total is only 7.68 million years; some 7.296 million years of life will be lost in Sierra Leone as a result of the pattern of structural violence in 1985. Applying that procedure to all countries, using the World Bank's classification, produces the results shown in table 8.3. Because of the pattern of structural violence in the world, 8,520,012 million years of life will be lost by those born in 1985, of which only 0.23 per cent will be lost by those born in the 19 so-called industrial market economies, which house 8.7 per cent of the world's population.

To many observers, reduction of the gross differentials in structural violence displayed here can only be achieved by economic development – in the same way that democracy and civil liberties will be achieved. They favour Rostow's model of economic growth and the various theories of modernization associated with it. What is important to note from figure 8.7 and table 8.3, however, is the low level of structural violence achieved in many of the people's democracies (nine of the largest, including the USSR but not China, form the 'non-reporting, non-member' category in the table). In China, for example, life expectancy in 1985 averaged 69 years and in Vietnam it averaged 65, figures comparable with the 70 for Albania, 72 for the German Democratic Republic and 70 for the USSR.

CONCLUSIONS

Alienation has a variety of forms and appearances. To some analysts, for example, it involves the alienation of individuals from nature, from other individuals, from the product of their hands and minds, and from themselves; to others, it means powerlessness, meaninglessness and self-estrangement. All these conditions stem from the organizational nexus of the mode of production, which requires alienation in order to promote the interests of the ruling class. Under capitalism, those interests involve accumulation by the bourgeoisie, which is achieved through the materialist focus of life, all elements of which are commodified and marketed; even if, as some argue, the bourgeois-proletarian division is obsolete, the subordination of the individual to the dynamic of capitalism remains.

Part of the ideology of capitalism, especially in the core countries of the world-economy, is its association with individual freedom; according to Friedman and Friedman (1980), the market economy of the United States provides freedom and an opportunity for individuals to make the most of their talents, 'through hard work, ingenuity, thrift and luck' (p. 19). Linked to this ideology of economic freedom is the ideology of democracy: according to Friedman and Friedman (p. 21), 'economic freedom is an essential requisite for political freedom', and Macpherson (1973, p. 4) shows that the justification for Western liberal democracy lies in two claims – 'the claim to maximize individual utilities, and the claim to maximize individual powers'. Whereas Friedman and Friedman are passionate advocates of the ideologies, however, Macpherson's conclusion is that 'Neither claim has stood up very well' (p. 6). The present chapter has upheld that conclusion, isolating geographical variations in the provision of liberal democracy but arguing that even where 'freedom and democracy' are supposed to reign this does not imply individual self-control.

Under the capitalist mode of production, individuals are required to behave and organize their lives in certain ways, otherwise their reproduction cannot be guaranteed. (Some societies punish 'deviants'. Others are more tolerant – of hippy communes, for example – in the belief that the market economy will soon 'discipline' them.) Such behaviour and organization are alienating. Some people prosper – absolutely and relatively – but many are degraded. They are treated not as individuals but as commodities, to be used and discarded as the market economy dictates. Many are categorized according to their race, religion, gender or membership of some definable group, and are discriminated against because of what they are rather than who they are.

Resentment against this exploitative alienation has generated political movements which have demanded greater participation by individuals in control over their lives and livelihoods. The liberal democratic system of government has evolved as a response to these demands. The concessions granted to some have, as demonstrated here, largely been won at the cost of others. Liberal democracy and the ability to 'reward' the proletariat is only affordable, it seems, in the core countries of the world-economy; if the concessions are too great, however, crisis will ensue and they must be retracted – even to the extent of reducing, if not removing, democracy. Elsewhere, popular democracy has been introduced to chart an alternative route to material progress. This has also led to restrictions on human rights, as the 'cost' of following the route, in the same way that restrictions have been imposed on the capitalist periphery. But the popular democracies have achieved one major success – a right to life which is nearly that of the capitalist core.

There is a strong correlation between the geography of liberal democracy and the geography of human rights and civil liberties, and also a correlation between both of these and the geography of economic development. This can lead to a crude Rostovian hypothesis that economic development is de-alienating, from which it is inferred that human dignity is best advanced by programmes of capitalist economic development. Four arguments have been assembled here against such a view. First, the Rostow model is wrong and ignores the organization of the world-economy into core and periphery. (Note, however, that the division into core and periphery is not unchanging, and that the geography of liberal democracy has changed markedly in recent decades; over half the present industrial market economies (table 8.3) were not liberal democracies in 1939.) Second, the liberties and freedoms associated with economic development are fragile. Third, there is little evidence that the democratic freedoms of the core countries are de-alienating – though the personal freedoms are clearly much greater with regard to, for example, physical abuse. Finally, the current crisis of the world-economy appears to be increasing alienation, with a reduction of democratic freedoms even in the core countries.

Alienation is an inherent condition of the capitalist mode of production based on materialism. All individuals are subject to the dictates of the market economy and lack control over their own lives as a consequence; some prosper in a material sense, but all are exploited and degraded. The ideology of capitalism has been linked to the ideology of liberal democracy as a form of political organization; this has also brought benefits to some, but the condition of alienation has not been dented and the practice of democracy remains constrained by the dictates of the market-place.

References

Berg, E. 1978: Democracy and self-determination. In P. Birnbaum, J. Lively and G. Parry (eds), *Democracy, Consensus and Social Contract*. London: Sage Publications, 149–72.

Bollen, K. 1983: World system position, dependency and democracy: the cross-national evidence. *American Sociological Review*, 48, 458–79.

Butler, D., Penniman, H. R. and Ranney, A. (eds) 1981: *Democracy at the Polls*. Washington, DC: American Enterprise Institute.

Clark, G. L. and Dear, M. J. 1984: *State Apparatus*. London: George Allen & Unwin.

Coulter, P. 1975: *Social Mobilization and Liberal Democracy*. Lexington, KY: Lexington Books.

Dahl, R. A. 1978: Democracy as polyarchy. In R. D. Gastil (ed.), *Freedom in the World: Political Rights and Civil Liberties 1978*. Boston, MA: G. K. Hall, 134–46.

Finer, S. E. (ed.) 1975: *Adversary Politics and Electoral Reform*. London: Anthony Wigram.

Finer, S. E. 1979: *Five Constitutions*. London: Penguin.

Friedman, M. and Friedman, R. 1980: *Free to Choose*. London: Penguin.

Galtung, J. and Hoivik, T. 1971: Structural and direct violence: a note on operationalization. *Journal of Peace Research* 8 73–6.

Gastil, R. D. 1981: *Freedom in the World: Political Rights and Civil Liberties, 1981*. Oxford: Clio Press.

Habermas, J. 1976: *Legitimation Crisis*. London: Heinemann.

Hall, S. 1980: Popular-democratic vs authoritarian populism. In A. Hunt (ed.), *Marxism and Democracy*. London: Lawrence and Wishart, 157–85.

Harvey, D. 1982: *The Limits to Capital*. Oxford: Blackwell.

Harvey, D. 1985: The geopolitics of capitalism. In D. Gregory and J. Urry (eds), *Social Relations and Spatial Structures*. London: Macmillan.

Herman, E. S. and Brodhead, F. 1984: *Demonstration Elections*. Boston, MA: South End Press.

Hindess, B. 1980: Marxism and parliamentary democracy. In A. Hunt (ed.), *Marxism and Democracy*. London: Lawrence and Wishart, 21–54.

Humana, C. 1984: *World Human Rights Guide*. London: Hutchinson.

Hunt, A. 1980: Introduction: taking democracy seriously. In A. Hunt (ed.), *Marxism and Democracy*. London: Lawrence and Wishart, 7–20.

Johnston, R. J. 1982: *Geography and the State*. London: Macmillan.

Johnston, R. J. 1984a: Marxist political economy, the state and political geography. *Progress in Human Geography*, 8, 473–92.

Johnston, R. J. 1984b: The political geography of electoral geography. In P. J. Taylor and J. W. House (eds), *Agendas in Political Geography*. London: Croom Helm.

Johnston, R. J. 1984c: *Residential Segregation, the State, and Constitutional Conflict in American Urban Areas*. London: Academic Press.

Johnston, R. J. 1985: People, places, parties and parliaments: a geographical perspective in electoral reform in Great Britain. *Geographical Journal*, 151.

Johnston, R. J., O'Loughlin, J. and Taylor, P. J. 1987: The geography of violence and premature death. In R. Vayrynen (ed.), *The Quest for Peace*. London: Sage Publications, 241–95.

Kidron, M. and Segal, R. 1984: *The New State of the World Atlas*. London: Pluto.

Kidron, M. and Smith, D. 1983: *The War Atlas*. London: Pan.

Kohler, G. and Alcock, N. 1976: An empirical table of structural violence. *Journal of Peace Research* 13, 343–56.

Lash, S. and Urry, J. 1987: *The End of Organized Capitalism*. Cambridge: Polity Press.

Lenin, V. I. 1917: *The State and Revolution*. Moscow: Progress Publishers, 1975.

Luard, E. 1978: *Socialism without the State*. London: Macmillan.

Macpherson, C. B. 1973: *Democratic Theory*. Oxford: Clarendon Press.

Osei-Kwame, P. and Taylor, P. J. 1984: A politics of failure: the political geography of Ghanaian elections 1954–1979. *Annals of the Association of American Geographers*, 74.

Riker, W. H. 1983: Political theory and the art of heresthetics. In A. W. Finifer (ed.), *Political Science: the State of the Discipline*. Washington, DC: American Political Science Association, 47–68.

Rostow, W. 1960: *Stages of Economic Growth*. Cambridge: Cambridge University Press.

Schattschneider, E. E. 1962: *The Semi-Sovereign People*. Hinsdale, IL: Dryden.

Seager, J. and Olson, A. 1986: *Women in the World: An International Atlas.* London: Pan.

Smith, N. 1984: *Uneven Development.* Oxford: Blackwell.

Taylor, C. L. and Jodice, D. A. 1983: *World Handbook of Political and Social Indicators,* vol. 2, 3rd edn. New Haven, CT: Yale University Press.

Taylor, P. J. 1983: The changing political map. In R. J. Johnston and J. C. Doornkamp (eds), *The Changing Geography of the United Kingdom.* London: Methuen, 275–90.

Taylor, P. J. 1984: Accumulation, legitimation and the electoral geographies within liberal democracy. In P. J. Taylor and J. W. House (eds), *Agendas in Political Geography.* London: Croom Helm.

Taylor, P. J. 1985a: All organization is bias: the political geography of electoral reform. *Geographical Journal,* 151, 339–43.

Taylor, P. J. 1985b: *Political Geography: World-Economy, Nation-State and Locality.* Harlow: Longman.

Taylor, P. J. 1986: An exploration into world-systems analysis of political parties. *Political Geography Quarterly,* 5, S5–S20.

Taylor, P. 1987: The poverty of international comparisons. *Studies in Comparative International Development,* 22, 12–39.

Taylor, P. J. 1989: Extending the world of electoral geography. In R. J. Johnston, F. M. Shelley and P. J. Taylor (eds), *Developments in Electoral Geography.* London: Routledge.

Taylor, P. J. and Johnston, R. J. 1984: The geography of the British state. In A. M. Kirby and J. R. Short (eds), *Changing Britain.* London: Macmillan, 23–40.

Taylor, P. J. and Lijphart, A. 1985: Proportional tenure vs proportional representation: introducing a new debate. *European Journal of Political Research,* 13, 387–99.

Therborn, G. 1977: The rule of capital and the rise of democracy. *New Left Review,* 103, 3–42.

World Bank 1987: *World Development Report 1987.* Oxford: Oxford University Press.

9

The Question of National Congruence

COLIN H. WILLIAMS

The inherent tension between state nationalism and ethnic nationalism within the world system has already contributed to two major catastrophes this century. At times their interaction has created powerful dynamic socioeconomic structures. At other times the clash of interests represented by these two forces has produced open conflict reflecting a sustained rivalry between striving participants in the developing state system. There is little reason to believe that violence will be eradicated in this relationship, for 'Despite the attempts of capital to tame and rationalise social relations, to subordinate them to its much more coldly destructive logic, violence will always occur' (Shaw, 1984, p. 4).

The pattern of state formation is abundant testimony to the influence of conflict and warfare in the development of national territories. The size and shape of contemporary states are as much a product of international rivalry as they are reflections of the settlement pattern of constituent 'national' populations. Indeed, the quest for national congruence, defined as the attempt to make both national community and territorial state into coextensive entities, has been a major feature of modern history, particularly in Europe. This 'western model' of state formation has been so influential that Williams and Smith (1983, p. 510) claim that the quest for its constituents – authenticity, legitimacy and equality – 'bedevils interstate and intrastate relations all over the world'. The emergence of the 'territorial–bureaucratic state' has had profound consequences for the political organization and structure of the interstate system, especially in those territories carved out of former colonial dynastic rule. For:

The central point . . . of the Western experience for contemporary African and Asian social and political change has been the primacy and dominance of the specialised, territorially defined and coercively monopolistic state, operating within a broader system of similar states bent on fulfilling their dual functions of internal regulation and external defence (or aggression). (Smith, 1983, p. 17).

My intention in this chapter is to examine the quest for national congruence within an interdependent world system and to illustrate the manner in which several political movements have sought to change the system of sovereign states so as to effect a more 'representative' distribution of national states.

Taylor (1982) argues that both statism and nationalism, being expressions of the search for ideological legitimacy, are related to specific epochs in modern capitalist development. I want to examine the relationship between these ideologies and the modern world system, taking Wallerstein's work on the effects of the uneven development of capitalism as representative of the central thread of a world-systems argument.

EXPANSIVE CAPITALISM

Wallerstein's influential analysis of the European-centred world-economy, comprising core, semi-periphery and periphery, is central to the analysis of the modern world-system animated by capitalism. The emerging world-economy encouraged spatial interdependence and a recognizable international division of labour whose profitability was a 'function of the proper functioning of the system as a whole' (Wallerstein, 1979, p. 38). A two-way interaction was initiated between the specialist role of a state's economy within the system and a corresponding set of pressures imposed on domestic political developments by changes within the emerging world-system. Taylor (1982) and Smith (1983) draw attention to the consequences of a state system facilitating the expansion of capitalism. They include the various dynastic and structural changes witnessed in the period 1500–1648, and legitimized in the Treaties of Westphalia, 1648, which established the state system of Europe.

> Since that date, long wars ended by congresses and treaties have become the accepted European norm for state-creation and state-consolidation: witness the treaties of Vienna, Versailles and Yalta. Each new agreement limited the number and extent of new states which could participate in the system; and the later the period, the more did wars and ensuing treaties *create* the recent states. (Smith, 1983, p. 16).

The state system became a framework which facilitated the integrative capacity of capitalism to link previously disparate regions and interest groups into an evolving world-system. Wallerstein's original argument runs as follows:

> Capitalism is based on the constant absorption of economic loss by political entities, while economic gain is distributed to 'private' hands. What I am arguing . . . is that capitalism as an economic mode is based upon the fact that the economic factors operate within an area larger than that which any political entity can totally control. This gives capitalism a freedom

of manoeuvre that is structurally based. It has made possible the constant economic expansion of a world-system, albeit a very skewed distribution of its rewards. (Wallerstein, 1974, p. 348)

The details of how capitalism influenced the development of status groups, national bureaucracies, bourgeois ideology and state boundaries are beyond the scope of this chapter (Rich and Wilson, 1967; Anderson, 1974a; Wallerstein, 1974). But undoubtedly the evolving state structures owed much to regional and transfrontier economic performance. The early integration of Spain, Portugal, France and southern Britain (figure 9.1) stemmed from the superior capacity of the local bourgeoisie and state apparatus to control and influence internal economic arrangements while also pursuing vigorous foreign trade and revenue campaigns. Core states were thus more able than peripheral states to influence the patterns of commodity flow, and hence to realize a greater share of the surplus value (Wallerstein, 1979, p. 292). Wallerstein argues that this initial advantage was translated into a semi-permanent structure wherein

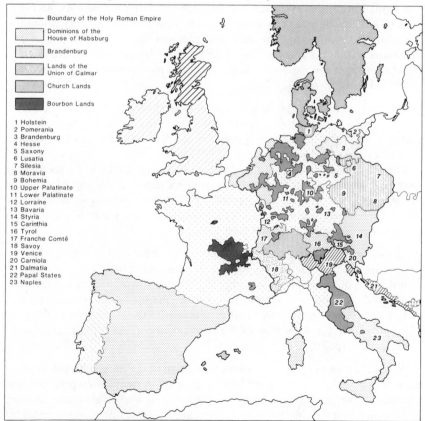

Figure 9.1   Europe in 1519
Source: Treharne and Fullard, 1976, p. 34

the bourgeoisie of the core states were better placed than were their counterparts within peripheral states. This influenced their specific relationship with the core-state proletariat, a relationship which may have been quite different in kind from that between the peripheral bourgeoisie and proletariat. The argument then turns on the differential character of this relationship, mediated through the emergence of the state as the locus of conflict. He writes: 'Since states are the primary arena of political conflict in a capitalist world-economy, and since the functioning of the world-economy is such that national class composition varies widely, it is easy to perceive why the politics of states differentially located in relation to the world-economy should be so dissimilar' (p. 293).

The state is conceived of as a particular kind of social organization which seeks to perpetuate its advantage through intervention, force and economic manipulation. For Wallerstein, this advantage is institutionalized in the concept of a state's 'sovereignty':

a notion of the modern world, is the claim to the monopolization (regulation) of the legitimate use of force within its boundaries, and it is in a relatively strong position to interfere effectively with the flow of factors of production. Obviously also it is possible for particular social groups to alter advantage by altering state boundaries; hence both movements for secession (or autonomy) and movements for annexation (or federation). (p. 292)

While individual states and factions may seek to challenge the existing pattern, it is the state system itself which 'encrusts, enforces, and exaggerates the patterns, and it has regularly required the use of state machinery to revise the pattern of the world-wide division of labour' (p. 292).

Though capitalism is encouraged in part by the regulations of a stable state system, Wallerstein argues that the unequal exchange in the appropriation of its surplus value is spatially differentiated, producing regionally variable effects. At its crudest his argument rests on the inherent unevenness of the patterns of exchange such that, in summary:

Capitalism is a system in which the surplus value of the proletarian is appropriated by the bourgeois. When this proletarian is located in a different country from this bourgeois, one of the mechanisms that has affected the process of appropriation is the manipulation of controlling flows over state boundaries. This results in patterns of 'uneven development' which are *summarized* in the concepts of core, semiperiphery and periphery. (p. 293)

We can accept Wallerstein's characterization of the long sixteenth century as a period within which a multilayered world-system developed, without necessarily accepting his interpretation of this system as a world-empire. Neither need we accept the claim that it was, above all, the economic processes inherent in capitalism which produced this world-system. We should be careful not to promote a historical determinism which argues that, because the division of the world-system into three distinct structural positions is functional to the reproduction of capitalism, the resultant state system was either inevitable or permanent. Wendt (1987) has argued that world-system theorists are prone to interpreting the structure of the world-system as 'given and unproblematic'. In consequence they have tended to reify system structures producing rather static and at times functionalist explanations, primarily because of their earlier preoccupation with the question of structure, treating agencies such as the state or class as 'no more than passive "bearers" of systemic imperatives' (p. 347). Wendt argues that 'without a recognition of the ontological dependence of system structures on state and class agents, Wallerstein is forced into an explanation of that transition (from feudalism to capitalism) in terms of exogenous shocks and the teleological imperatives of an immanent capitalist mode of production' (Wendt, 1987, p. 348). For our purposes the critical feature of this perspective is its recognition of the European state system as the superstructure whose transformations influenced the differential occurrence and subsequent modification of the process of national congruence. Wallerstein's contribution, in articulating these structural transformations, is well recognized by his critics (Skocpol, 1977; Modelski, 1978; Zolberg, 1981), but they would remind us that he pays too little attention to the 'politico-strategic' linkages between parts of the world-system (Zolberg, 1981, p. 262) and underemphasizes non-economic factors in the genesis of early modern states.

Skocpol (1979) is particularly sensitive to an economic reductionist argument that would 'assume that individual nation-states are instruments used by economically dominant groups to pursue world-market orientated development at home and international economic advantages abroad' (p. 22). She argues that the state system, as 'a transnational structure of military competition was not originally created by capitalism' (p. 22), but rather, quoting Hintze, that 'the affairs of the state and of capitalism are inextricably interrelated . . . they are only two sides, or aspects of one and the same historical development' (Hintze, 1975, p. 452, quoted by Skocpol, 1979, p. 299). Her analysis of both the structures of the capitalist world-economy and of individual national responses to that structure is comparative macro-analysis at its best because it allows for the relative autonomy of several layers or scales of analysis in her work. The international state system, for example, 'represents an analytically autonomous level of transnational reality – *interdependent* in its structure and dynamics with world capitalism, but not reducible to it' (p. 22).

In addition to domestic economic performance and comparative international economic position, she recognizes, what Wallerstein underplays, the relevance of factors such as 'state administrative efficiency, political capacities for mass mobilization and international geographical position' (p. 22). In this context the advantage of the world-system approach is that it offers an integrated holistic perspective on an admittedly complex process of global development.

Interstate competition over the past four centuries has animated the ever-fluctuating capitalist system. It has also produced the inexorable integration of diverse culture groups into a 'national population' as part of the process of national congruence. The state has sought to harness the potential of such 'nations' and control their productivity to enhance its own resource base. In consequence, state activity has created a new set of geographies for incorporated peoples, influencing their socio-economic opportunities and political representation. Superordinate 'nations' came to dominate 'unrepresented nationalities' and used the power of the state to buttress their own cultural apparatus as the orthodox legitimized value system. The new opportunities and freedoms were those sanctioned by the state, and woe betide dissident minorities who questioned the verity of state regulations. Indeed, much of the resurgent nationalism of contemporary Europe is but a playing out of minority aspirations unsatisfied during the critical period of state formation.

THE EUROPEAN ORIGINS OF NATIONAL CONGRUENCE

During the two centuries after Charles V's disastrous attempt to revive the Holy Roman Empire as a world monarchy, Europe's fulcrum lurched westward, away from the declining periphery of the eastern territories towards the maritime powers and the promise of the New World beyond. It was a period which saw the political unification of each of three formidable states, Spain, France and Great Britain, the core areas of the expanding European colonial enterprise. State unification necessitated uniform legislation, taxation, conscription, defence and, to an increasing extent, cultural compliance. International exploration and conflict, the search for raw materials and new markets, required internal administrative reform. But domestic stability also required the settlement of religious conflict. The principle of *cuius regio, eius religio* transferred power from the universal claims of pope or emperor to those of monarch or prince, and focused attention on the individuality of emergent states. Further devolution of power into the hands of representative assemblies in much of Protestant Europe during the period also reflected the search for national congruence, anticipated in the eighteenth century and only realized in the nineteenth.

By 1740 maritime Europe had been consolidated into a pattern of core states whose approximate boundaries have survived to the present (figure 9.2). Central Europe, still severely fragmented, witnessed the expansion of Austrian dominions and the consolidation of the Brandenburg–Prussia territories. Eastern Europe, dominated in the past by the regional competition between Sweden and Russia, and between Austria and the Turks, was preoccupied by the drive towards territorial expansion to finance the development of Muscovy's absolute monarchy and Austria's 'defence of Christendom'. For these were frontier regions still, studded with pockets of modernization set in sharp contrast with the overwhelming feudal context of the vast and sprawling multi-ethnic empires.

To the west the Enlightenment in '*Aufklärung* Europe' accelerated the development of national communities, not only through its liberalizing ideology but also through major technical changes in social communication, principally the transmission of the written word. As Anderson (1983) reminds us, print capitalism's search for ever-expanding markets influenced national consciousness in three ways. First, it 'created unified fields of exchange and communications below Latin and above the spoken vernaculars'. Standard literary forms transcended often mutually unintelligible local dialects, and provided, through print, the 'embryos of the nationally imagined community' (p. 47). Second, print capitalism provided a permanent written record, a ready repository of learning, debate and dissent. History could be rewritten, rerecorded and reproduced to correspond with the political needs of the present, particularly in creating, often from mythology, 'that image of antiquity so central to the subjective idea of the nation' (p. 47). Third, it produced new 'languages of power', conferring economic advantage, social position and political privilege. On this basis, Anderson argues that the convergence of capitalism and print technology created the 'possibility of a new form of imagined community, which in its basic morphology set the stage for the modern nation' (p. 49). Called

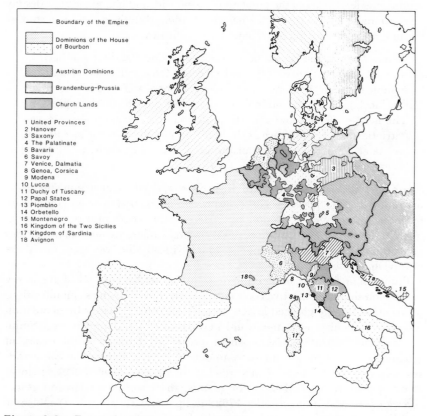

Figure 9.2   Europe in 1740
Source: Treharne and Fullard, 1976, p. 54

unto liberty by the power of the printed word, whole peoples could set their political aspirations beyond the confines of class, region and even history itself, if the potency of history could now be channelled to serve the will of the masses.

It is now commonplace to attribute the genesis of political nationalism to the French Revolution's insistence on human freedom and the selfhood of both state and citizen. But it was the violent passionate export of these ideas, via the French revolutionary armies, which, often by default, spawned reactive nationalism in Europe. Snyder (1978, p. 58) put it thus: 'When Napoleon decided to spread his ideas of the French Revolution as he interpreted them into other countries by military force, he invariably ignited the fires of nationalism'.

While many acknowledge that the force of democratic ideas crystallized in the French Revolution (Ronen, 1979), others have argued that the Revolution symbolized a people's conflict which had a profound effect on French collective identity. Warfare was central to this process, because it emphasized the centralizing tendencies of the previous two centuries and bound people's destiny together more firmly than ever before. Anthony Smith argues that conflict became a mobilizing agent of national significance:

> Within France, the Revolution reduced the long-standing gulf between state and army, and society, and so could release the tide of patriotic sentiment. In the great *fêtes*, the tricolour, anthem and *levées en masse* under the Jacobins, French national consciousness gained classic expression. The need to defend and export the Revolution threw up a whole new patriotic imagery of France as the home of republican virtue and liberty, and later, under Napoleon, of the great nation in arms, an imperial liberator and civiliser. (Smith, 1981b, p. 386)

Smith does not claim that warfare, by itself, shapes ethnic or national identity, but rather that the 'historical consciousness', fundamental to ethnic community reproduction, is often a 'product of warfare or the recurrent threat therefore' (p. 379). Imminent fear for collective survival maximizes the internal similarities of a people and serves to distinguish more critically the 'we' from the 'they' in any shared territory.

Preparation for war and resultant post-war settlements also have a key bearing on shaping collective consciousness: the former by mass mobilization, propaganda and heightened loyalty to the state, and the latter by population transfer, boundary adjustment and new resource development. But Smith observes that warfare is also a salient test of the character of the emergent bureaucratic state to act efficiently to defend its own interests in the world-system. The cutting edge of this trend is the development of a professional army, especially after 1792. Prior to this, the effect of warfare on ethnic consciousness was indirect; after 1792, and particularly after the Napoleonic era, 'by strengthening a stable network of territorial states, the professional army indirectly fostered a loyalty and cohesion in the demarcated population, which helped to erode local allegiances and encouraged the belief in an all-powerful bureaucratic state and its standardised culture' (pp. 385–6).

The medium-sized states of the Atlantic seaboard used their military might, together with their mass taxation systems, developing educational institutions and social communication networks, to enforce their effective sovereignty throughout their territory. Ethnocentrism was forged in the citizenry as a response to repeated and incessant warfare. Each successive generation was charged with the solemn duty to protect 'the state' and to honour the sacrifice of previous generations' yielding to the call of war in defence of their national liberties. Over the centuries the process, whereby ethnic consciousness is mobilized through interstate conflict, has had profound effects: it

> tended to 'harden' national space, to eliminate all twilight zones and interstices between compact, clearly bounded national states, and so to present to other comparable units a fully mobilized and nationally conscious population wielding military power through its political apparatus. These wars have determined not only the extent and shape of a national territory; they have often hammered the resident population into a community and created potent associations between it and the territory it came to occupy. (Williams and Smith, 1983, p. 515)

Nowhere was this process more evident than in the great unification nationalisms of nineteenth-century Germany and Italy. Under the twin influences of the romantic idealists and the aggressive military and economic impulses of the Prussian elite, German speakers were to be partially united under the umbrella of organic nationhood. Snyder (1978, p. 59), among others, has argued that the special romanticism so prevalent in Germany during the last century was a desperate attempt to hide the ignominy and shame of Napoleonic defeats by focusing on a mythical folk tradition lost since the Middle Ages. The stress on land, language, lore and loyalty was an expression of original historicism; an attempt to reuse traditional values in a changing context against the universal principles, inspired by the French Revolution, which sought to negate the distinctiveness of folk communities in favour of rational individualism. This great clash of ideals, between revolutionary liberalism and conservative nationalism, was to forge a dynamic political tension which reverberated in the domestic socio-economic issues and factions of most European societies.

If, in general, the nineteenth century can be characterized as a century of national unification, reaching three peaks in 1848, the 1870s and the 1890s, it was also a period of enormous changes in the political boundaries of several European states. Connor, commenting on this wholesale change, says, 'In the one hundred and thirty year period separating the Napoleonic Wars from the end of the World War II, all but three of Europe's states had either lost extensive territory and population because of ethnonational movements or were themselves newly created self-determined states' (Connor, 1981, p. 210; the three in question were Portugal, Spain and Switzerland). Comparison of figures 9.3 and

9.4 reveals the effect of the wholesale dismantling of multinational empires in Central and Eastern Europe under the impulse of nationalism. Most historians are agreed that the motivations underlying nationalist movements are a complex interplay of religious and economic factors, expressing deep-rooted social discontent. In Central and Eastern Europe these elements were skilfully welded primarily in defence of language rights, whose denial was interpreted as a symbol of oppression, historic domination and conquest. Language agitation proved extremely virulent in its capacity to mobilize previously disparate poeples.

Under the banner of liberal nationalism (which so often turned into a conservative force by the end of the century), the struggling nationalities of the east revolted against their former rulers. Some clung to their institutional distinctiveness and sought to reassert their former independence, as did Hungary. Others, such as the Croats, fearful of Magyar intolerance towards

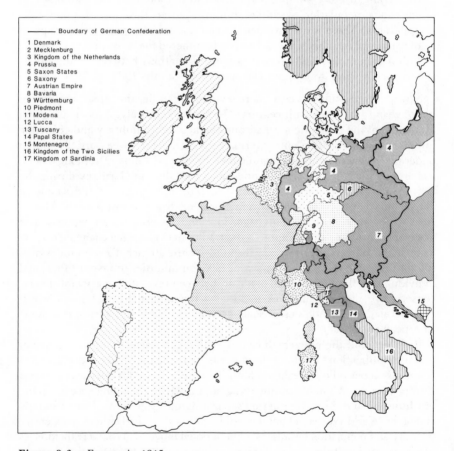

Figure 9.3   Europe in 1815
Source: Treharne and Fullard, 1976, p. 64

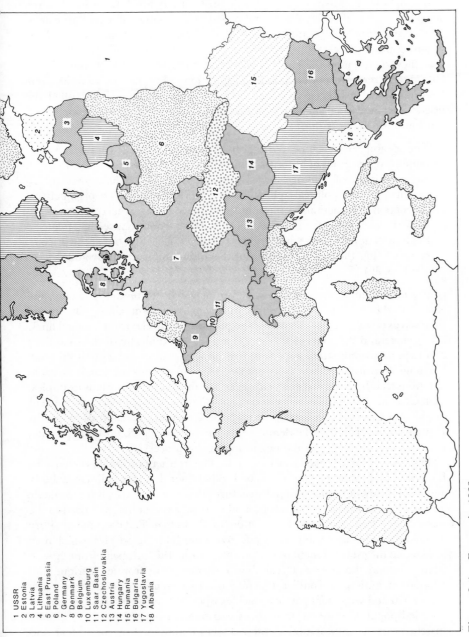

1 USSR
2 Estonia
3 Latvia
4 Lithuania
5 East Prussia
6 Poland
7 Germany
8 Denmark
9 Belgium
10 Luxemburg
11 Saar Basin
12 Czechoslovakia
13 Austria
14 Hungary
15 Rumania
16 Bulgaria
17 Yugoslavia
18 Albania

Figure 9.4 Europe in 1923

Source: Treharne and Fullard, 1976, p. 84

minorities, stressed their special institutional and legal relationships, guaranteed by the Nagodba concession of 1868. Still others, deprived of institutional or legal distinctiveness, sought their legitimacy in the 'natural rights' of 'the people' and adopted a Fichtean version of a linguistically defined nationhood as the supreme locus of their loyalty, as happened in Romania and Czechoslovakia.

The attraction of language as a badge of nationality was twofold. First it gave authenticity to the putative claims of the nationalist intelligentsia that they represented the real conscience and destiny of their people, i.e. it legitimized their corporate political aspirations. But in the view of some historians this insistence on a conservative organic nationalism stifled the liberalizing individualism derived from transatlantic ideological developments. Kitromilides suggests that:

> The idea of Herder appeared especially relevant to these concerns and reinforced the trends of historicism and folklorism that became the primary characteristics of Southeastern European thought in the nineteenth century. In the political sphere, these trends sustained an organicist conception of the national community that allowed an eventual compromise with the corporatism of traditional culture – a compromise that made modern nationalism acceptable to entrenched social elites but at the price of sacrificing the liberal impulse of the Enlightenment. (Kitromilides, 1983, p. 59)

Second, language nationalism gave an inclusivity to the target group and excluded the nobility and 'foreign intrusive' merchant class, who found themselves categorized as enemies of the people on both class and national lines. As Orridge and Williams (1982, p. 29) argue, an ethnically distinct landlord and capital-owning strata superimposed on a peasantry speaking a different language is a situation especially conducive to autonomist nationalism, as in late-nineteenth-century Wales. However, we must also cite variant examples, as in Croatia, where the nobility and the intelligentsia, who formed the nucleus of the nationalist movement, both spoke Croatian, in contrast with the Hungarian aristocracy, who demonstrated greater loyalty to the emperor than to Hungary, especially at the beginning of the process.

Two further factors need consideration. The frontier position of the Eastern European multinational empires so bedevilled their drive for political stability and establishment of institutional structures that few resources were available for the administrative centralization and cultural assimilation practised in the west (p. 29). The struggle with the Ottoman Turks to defend disputed territory delayed the implementation of any effective homogenizing policies, which may have led to the establishment of nation-states along the European frontier with Islam. Second, because of Great Power rivalry, cross-frontier intrigue and religious and linguistic conflict, this region witnessed a continuously changing population and settlement pattern which exasperated any attempts at socializing a settled population into a permanent patriotic citizenry. In short, the state structures of the multinational empires were maintained by force, fear and foreign ordinances. In consequence the region lagged behind the economic, social and political developments of Western Europe and preserved feudal elements, such as the 'second serfdom', which reinforced already existing cultural divisions (p. 32). More importantly the south-east was subject to the first impulses of westernization, whereby in resisting the foreign domination of empires nationalist leaders sought to imitate the equally foreign model of the nation-state, developed on the Atlantic seabord and offered as an epitome of all that was rational, democratic and uplifting in modern civilization. In consequence, we can recognize here many of the classic features of a European periphery identified earlier by Wallerstein: the east was fragmented, dependent and 'underdeveloped', and its economic and political rhythms were reactions to impulses generated either by the core of the modern world-system or by the last vestiges of medieval Christendom and Islam.

Poland's Unequal Struggle for National Survival

Illustration of the impact of nationalism and warfare on national sovereignty and state boundaries in this region can best be undertaken by reference to Poland and its neighbours over the past two centuries.

We have seen that a country's geopolitical context is a vital factor in shaping its 'national history'. Nowhere is this more evident than in modern Poland. The search for Polish national congruence is one of the most tragic elements of European history, and still suggests resonant lessons for aspiring nationalists elsewhere in the world. Since the eighteenth century at least, great-power intervention, annexation and forced assimilation have characterized Polish political life. Initially sparked off by Catherine II's concern for the treatment of its Orthodox population, the three partitions of Poland – in 1772, 1793 and 1795 – were a grand display of great-power prerogative. By these partitions Poland was divided between the Russian Empire, the Hapsburg Monarchy and the Prussian Kingdom. Polish hopes of liberation were realized within 11 years, when in 1806 Napoleon's army entered Poland *en route* for the Russian frontier. The Treaty of Tilsit, July 1807, had re-established a semi-autonomous Polish state as the Grand Duchy of Warsaw, placed under the sovereignty of the King of Saxony (Seton-Watson, 1977, p. 123), and neutralized *pro tempore* the threat of the Russians who now busied themselves in a half-hearted continental blockade. In 1809 the Duchy's territory was further increased after Napoleon had, once again, subdued the Austrians and ceded many of their former territories as tribute to the fidelity of the many thousand Polish volunteers in his armies. Poland's characteristically brief period of liberty ended on the defeat of Napoleon in 1812, when the partitions were restored, with Russia the net beneficiary of Poland's demise.

For the remainder of the century, though Poles were active in European nationalist movements and revolutionary societies as far flung as Paris, Sicily, Piedmont and Hungary, at home they failed to win any more than a limited regional autonomy for Galicia and representation in the Viennese *Reichstrat* from their most sympathetic occupying power, Austria (p. 128). Policies of systematic Russification and, after 1880, of forced Germanization, as part of Bismarck's *Kulturkampf*, brought limited assimilation and a more general strengthening of Polish national consciousness. However, the international climate did not favour mass mobilization in the cause of freedom, as it had done to the south and east, and therefore various factions sought to negotiate with the occupying powers with the aim of isolating and removing one of them from Polish territory. Russia was the most likely victim (despite the efforts of national democrats to seal an alliance) and Austria the willing instrument for the eviction, and on the eve of the First World War Pilsudski's Polish Patriots mustered to the Austrian ranks and in due course expelled the Russians.

But centuries of ethnic intermingling and post-war refugee movements were to confound the rational logic of the Wilsonian principle of 'national self-determination' embodied in the Versailles Treaty of 1919. Here, as Seton-Watson observes, 'the difference between the Polish nation and the historical Polish state emerged in acute form', as indeed it was to do in the boundary delimitation of most Central European states legitimized by the Versailles deliberations (see figure 9.4). The three common problems of establishing a sovereign territory, respecting the rights of newly disenfranchised ethnic minorities and creating a buffer zone betwen rival continental powers took on extra significance as the West sought to overcompensate for Poland's suffering at the hands of voracious neighbours.

> In the West, the victorious great powers gave the benefit of the doubt to the Poles at the expense of the Germans in drawing the frontiers in Pomerania and Silesia – though they gave a good deal less than Polish nationalists would have liked, and created an awkward problem through the separation of East Prussia from Brandenburg by a belt of Polish territory, the so-called 'Polish Corridor'. (Seton-Watson, 1977, p. 129).

Post-war readjustments provided Poland with several gains, specified in the 1919 peace treaty, e.g. the industrialized coalfield of Upper Silesia, Posen's rich agricultural hinterland, a Baltic coastline, a well-developed communication system and a primary-educated population in the new territories (Tägil, 1977, p. 138). However, long-established trading patterns and communication links were abruptly broken in the new areas as they transferred from the pre-existent east–west patterns of trade and commerce and stagnated before a comprehensive north–south rail and road link could be instituted. Old markets, resource areas and rules of trade were abandoned wholesale. This fragmentation was compounded by the mass emigration of German-speaking professionals, urban artisans and rural workers, only some of whom were replaced by Polish migrants from Galicia and Congress Poland (p. 139).

Both sides of the new German–Polish borders contained substantial minorities, an inflammatory element in subsequent Nazi propaganda and a lasting grievance for irredentist organizations determined to undo the post-Versailles settlement. To the east, the government determined to assimilate its Belorussian and Ukrainian minorities, but met with only limited success in the interwar period. Resistance to state integration had been a characteristic feature of Eastern European society for well over 50 years, and the 'unsatisfied nationalism' of the Ukraine in particular was not about to yield its cause, especially as the principle of self-determination had the most radical effect on the dismantling of empires, producing a mosaic of nation-states for almost all the Ukraine's neighbours and confederates in the nationalist struggle (see figure 9.4).

The clearest evidence of the influence of warfare on delimiting national boundaries is provided by the several territorial changes in Eastern Europe following the Second World War. The question of Polish boundary adjustments had been determined at the Yalta Conference as part of the agreement on administering occupied German territory. At the Potsdam Conference 17 July–2 August 1945, the Allies sought to minimize the extent of the Soviet zone of occupation in Germany by reluctantly acknowledging that the disputed western

Figure 9.5   Territorial changes in Eastern Europe following the Second World War
Source: Day, 1982, p. 31

frontiers, portions of East Prussia not occupied by the Soviet Union and the Free City of Danzig should be under the administration of the Polish state, pending a final peace settlement. By a strange paradox of fate Poland had reverted to occupying territory it had not held for nearly two centuries (figure 9.5), but the restoration, like its previous dismemberment, was largely determined by external forces, not by an application of national will in the form of self-determination.

CONTEMPORARY CURRENTS OF ETHNICITY

In many ways the boundary problems of Poland and its neighbours reflect the invention of nation-ness, for, despite their putative ethnic origin, successive populations were socialized *in situ* to identify themselves as Poles. The architects of national consciousness were the intelligentsia (who spread their ideological hold over the masses) and reluctant nobility alike, for the inclusivity of nationalism would brook no class antagonism in its rhetoric if not always in its actions. Literacy in the vernacular was the key to the unleashing of popular support. Throughout Eastern Europe independence movements preached the sovereignty of the people, with all its principled implications for the abolition of serfdom, the emancipation of the proletariat and the drive towards an ethnically determined 'universal suffrage' for all co-nationals. Liberation from the shackles of the colonial empires promised a glorious future for peoples who barely two generations earlier had no right to expect a future at all as a nation among nations. Their liberation came via conflict and warfare, for the dismantling of empires in the east was the product of revolution and post-war settlements producing a different European order after both 1919 and 1945.

In the west, however, a different pattern emerged. Previously strong nation-states emerged from the cataclysmic conflict to face both domestic reconstruction and the challenge of decolonization of their far-flung empires. The 'descent from empire' not only produced a new set of international actors but also a questioning of the very basis of state legitimacy and popular representation at home. The myriad forms of discontent prompted new approaches to old problems. Government commissions inquired into popular democracy, central–peripheral relations, and the form and function of government itself. Alternatives to class-related politics were canvassed throughout Western Europe, chief of which were the ethnically inspired inheritors of nineteenth-century liberal nationalism.

The post-war 'ethnic revival' in Western Europe has been interpreted in at least four different ways: first, as the continuation of a historic struggle between superordinate 'states' and subordinate 'nations' (Foueré, 1980); second, as a response to the decolonization processes and the visible success of liberation movements in overthrowing their imperial shackles; third, as a realization on behalf of local ethnic activists that their socio-economic position is a direct result of systematic discrimination by the core majority – such discrimination structures the individual's life chances in employment, educational opportunity and general standards of living, and also determines the limits of political activity superimposed upon the peripheral minority by the host state's majority (Hechter, 1975); last, as a direct response to the uneven spread of capitalism, operating as a transcendent force in the world-system. For Nairn (1981), capitalism's uneven development is necessarily 'nationalism producing' because development always comes to the less advanced peoples within the 'fetters of the more advanced nations', as with France and England (Smith, 1981a, p. 37). The latter two interpretations of ethno-nationalism have aroused great interest, particularly among social scientists and ethnic activists concerned to portray their cause within both historic and an academically respectable framework. The force of Nairn's account is that he sets the nationalist reaction within the operation of a global division of labour and provides a holistic interpretation for the scope and the timing of the current round of nationalist resurgence.

Central to this relationship between uneven capitalist development and reactive nationalism is exploitation at the national, regional, collective and individual level. In other words, it is an all-encompassing relationship. Orthodox modernization theory would argue that it is only by being transformed to resemble 'modern' majority cultures that minority 'dependent' cultures can escape the pejorative connotations of this unequal relationship. But as Glyn Williams forcefully stresses, in so doing they cease to exist, so that the 'ideological or hegemonic nature of this argument should be evident' (Williams, 1980, p. 364).

Smith argues that Nairn is correct when he suggests a close relationship between foreign domination and nationalism, and has shown the significance of conflict and warfare in heightening national consciousness (Smith, 1983). But conflict and exploitation have been the stuff of intergroup relations since time immemorial; why should capitalism *per se* generate a particularly ethnic, as opposed to a class, religious or territorial reaction? The most revealing answer is Smith's explanation of the rise of the bureaucratic state and its role in the formation of 'national culture' (Smith, 1981a). Because the state apparatus and traditional institutions are barred to them, the ethnic intelligentsia resort to the one resource they have access to – the people and their own unique history, which they can rewrite in accordance with a nationalist view of global history and disseminate through print capitalism (Anderson, 1983). In stressing the historicism of the intelligentsia, scholars such as Smith, Berry (1981) and G. A. Williams (1982) point to the utility of language, religion, territory, oppression and myths as agencies for group mobilization. Critically, these are factors independent of the state, whose legitimacy is internally derived and not externally imposed, and which may often be the only 'political' resource not fully controlled by a powerful state machinery.

In an earlier phase of capitalist development, though the peripheral regions of Western Europe may have been characterized by an imbalanced economic structure, there was nevertheless sufficient mobility and growth in the system to promote state integration. For ethnic aspirants in colonial homelands, blocked in the upper echelons of government and commerce, there was always the release valve of serving in the colonial territories (Smith, 1982). But with the run-down of empire many of the repatriated ethnic professional classes failed to secure positions commensurate with their talent and expectations. In consequence 'exclusion breeds "failed assimilation" and reawakens an ethnic consciousness among the professional elites, at exactly the moment when the intellectuals are beginning to explore the historic roots of the community' (p. 31). When historicism and reawakening combine, argues Smith, the ethnic revival can 'blossom and assume full political form' (p. 32). This trend may be exacerbated by regional economic decline within the ethnic homelands. For as capital investment transfers from lagging industrial regions to new growth areas, the ethnic intelligentsia interpret such structural manifestations as the inability or unwillingness of the central state agencies to redirect growth into the periphery. Nationalist leaders conclude that the only means of redressing this economic imbalance is by wresting sovereign political control, or at least substantial elements of autonomy, from the constituent state. Very few nationalists have been able to make a strong *a priori* case for an economic revolution and a substantial increase in living standards for their homeland on independence. But most conceive of foreign rule as the greatest barrier to national development and make autarky (self-sufficiency) the economic counterpart and basis for sovereignty (Williams and Smith, 1983, p. 509).

Ethnic Resurgence in Western Europe: Violence and Reform

Given the overall context of state integration in Europe it is remarkable that current ethnic unrest should be so virulently manifested in the three European states with the longest history of unification and consolidation. France, Spain and the United Kingdom all possess minorities within minorities who seek greater autonomy. Their style of resisting centralization and assimilation varies from the extra-constitutional activities of violent movements in Ulster, Euskadi and Corsica, through non-violent resistance in Catalonia, Wales and Brittany, to party political opposition in all of the above plus Scotland, Galicia and Alsace (figure 9.6). Obviously we should be wary of generalizing about such movements as if they reflected a single concern for decentralization, for a plethora of factors account for their initial emergence and subsequent developments (Anderson, 1978; Foster, 1980; Krejci and Velimsky, 1981; Breuilly, 1982). Here I concentrate on two case studies which reflect much of the ethnic discontent in contemporary Europe. I have refrained from discussing the nationalist problems of the UK state, preferring to concentrate on the less well understood examples of the Basque country and Corsica.

Basque nationalism The Basque case illustrates many of the classic features of nationalist opposition to state stability. The Basques enjoyed a long period of autonomy prior to their incorporation into the Spanish state; elements of an institutional framework, the *fueros*, survived until fairly recently and were used

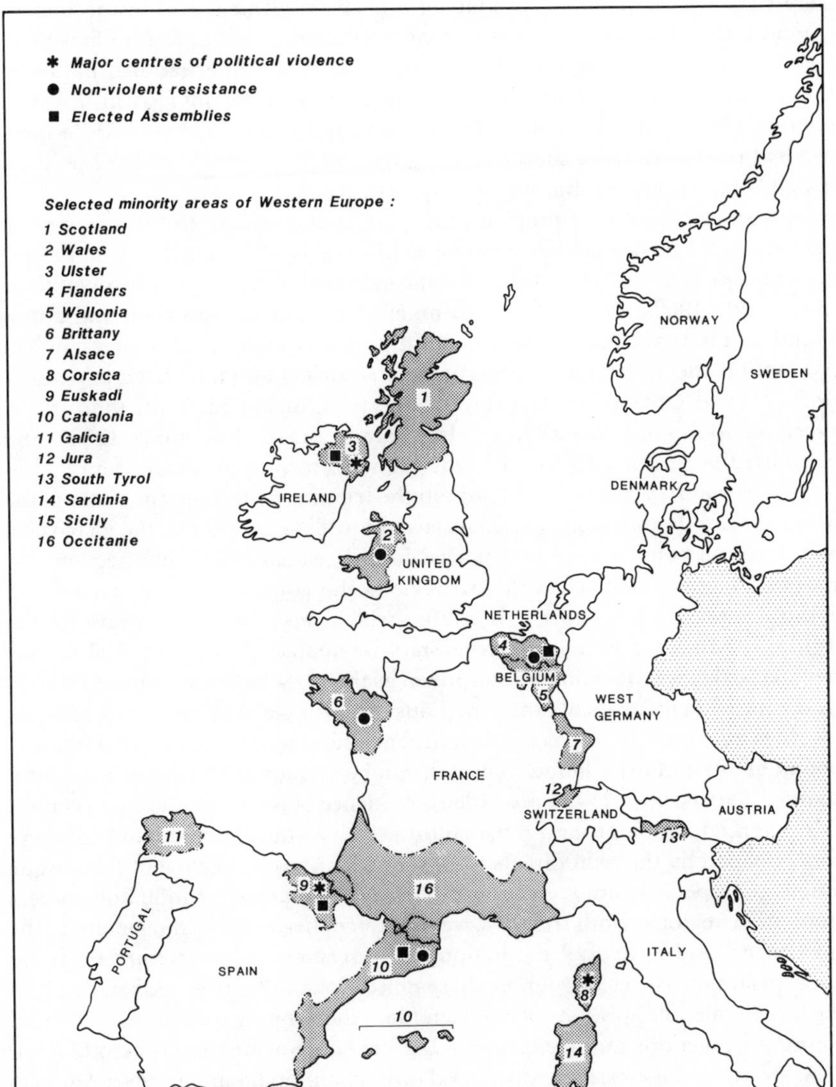
✱ Major centres of political violence
● Non-violent resistance
■ Elected Assemblies
Selected minority areas of Western Europe :
1 Scotland
2 Wales
3 Ulster
4 Flanders
5 Wallonia
6 Brittany
7 Alsace
8 Corsica
9 Euskadi
10 Catalonia
11 Galicia
12 Jura
13 South Tyrol
14 Sardinia
15 Sicily
16 Occitanie
NORWAY
SWEDEN
DENMARK
IRELAND
UNITED KINGDOM
NETHERLANDS
BELGIUM
WEST GERMANY
FRANCE
SWITZERLAND
AUSTRIA
ITALY
PORTUGAL
SPAIN

Figure 9.6 Selected minority areas of Western Europe

as evidence of a prior claim to legitimate statehood. Their language and culture were deemed unique, among the oldest surviving elements of civilization in Europe, but, under the impress of state-building, non-Spanish elements were ruthlessly eradicated, producing a deep resentment within the Basque community. State oppression was confirmed during the Spanish Civil War, when many of the Basques, though conscious of the international nature of the conflict against fascism, believed themselves to be engaged in a war of national liberation. Thereafter violence and oppression characterized Basque–Madrid relationships as the Franco regime sought to eradicate local political dissent and to destroy Basque cultural identity.

The amalgamation of previously disparate movements to form *Euskadi ta Askatasuna* (ETA; Basque Homeland and Liberty) in 1957 reinvigorated Basque nationalism. A strategic switch to mount spectacular urban-guerilla operations in the early 1960s produced a government backlash of repression and a new round of 'internal colonial domination' by a police state. Such repression had a profound effect on large sections of Basque society and mobilized previously uncommitted citizens to the cause of liberty. Outside Euskadi, the turmoil precipitated major political crises elsewhere in Spain (Medhurst, 1982), and heralded the return of a socialist government in the post-Franco era.

The crucial struggle revolved around the legitimacy and the character of the Spanish state. Both Basque nationalists and socialists questioned the ideological and material control exercised from Madrid, which effectively negated the Basque identity. Central to this struggle is the control over the agencies of cultural reproduction, since it is culture which serves as the medium for the legitimization of power and structured inequality (Williams, 1980). For nationalists, state education is interpreted as the basis for social control through its legitimizing ideological function. Thus the struggle over Euskera education involves a struggle for ideological control at whose heart lies the revolutionary potential of a minority language which can be employed to transmit a radical counter-state position (Williams, 1980). Evidence of the state's desire to control this potential and expropriate the cultural role within the dominant ideology was provided by the twin threats to Basque identity: the decline of the Basque language, especially among young people, and the post-war influx of workers from other regions. Both trends serve to reduce the ethnic homogeneity of the 'provinces' and have split the nationalist movement in its attempt to devise appropriate measures to counter these deleterious influences. A key result of the large-scale inmigration was to strengthen the support base for socialist and communist factions at the expense of the Partido Nacionalista Vasco (PNV).

As working-class resistance increased throughout Spain in the 1960s, workers from the Basque provinces were in the vanguard of the illegal strike movement. ETA was particularly disturbed by this trend, for sections of its elite were Marxist, and their attempt to combine class and ethnic appeals confused the largely inmigrant population of urban industrial areas (Clark, 1981; Medhurst, 1982), leading to vacillation on key principles of autonomy. Further divisions concerned the appropriateness of employing violence as a mobilizing strategy to expose the 'real' character of the Spanish state in its overt repression. The details of ETA's violent campaign are too well known to bear repetition here and it is tendentious to enquire whether recent reforms would have been inaugurated had it not been for this persistent threat to the stability of the state. However, it is only since 1975 that the government has acknowledged several of the traditional demands of the Basque activists. In May of that year it issued a decree allowing the teaching of Euskera on an optional basis, and in October it issued a second decree providing protection for several regional languages, but took the opportunity to reaffirm that Spanish was still the only official language of governance (Clark, 1981). These concessions were strengthened in May 1979 when the Ministry of Education issued its long-awaited decrees accepting responsibility for programmes of instruction in Euskera at all levels, and providing financial assistance to the private Euskeran *ikastolas* (schools).

The culmination of Sr Suarez's determination to institute liberal democratic principles in Madrid–Basque relations was the approval of the Basque Autonomy Statute in 1979. The first Basque parliament was elected in March 1980. Its earlier deliberations strengthened the autonomous domain of Euskera, a relatively problem-free task compared with the more daunting exercise of improving the economic position of Euskadi *vis-à-vis* both Spain and Europe. Unlike Brittany or Corsica, a history of rural underdevelopment cannot be invoked as a catalyst for Basque nationalist developments. Quite the contrary, it has long been a leading industrial sector, more akin to South Wales and Central Scotland historically. What is at issue though is whether the formal trappings of autonomy can effect real structural changes in Euskadi's economic performance. If they cannot, will this promote separatism or give credence to the socialist argument that a healthy Euskadi can only be realized with increased socio-economic integration between all constituent parts of Spain?

Whatever the outcome, violence against the state apparatus continues to animate Basque resistance. Clark's (1984) analysis of ETA's strategy examines a number of recent 'explanations' for the persistence of conflict, but concludes that, despite the relevance of socio-political factors in inducing a violent political culture, insurgent violence is self-sustaining, 'Almost apart from the wishes of the participants or the objective conditions that led to the insurgency' (p. 278).

This observation must be tempered by two significant changes. First, the Basque resistance is now directed towards a socialist government which has shown an astonishing readiness to accommodate many of the autonomist demands, yielding concessions which Francoists would have considered tantamount to a denial of the indivisibility of the Spanish state. Second, the extradition of wanted insurgents from their relatively safe exile in France casts doubts on the capacity of ETA to sustain a large-scale violent opposition to the current regime, especially when that regime exercises a great deal of leniency in its formal dealings with former 'terrorists' who return voluntarily. Spain's successful membership of the European Economic Community (EEC) and the recent co-operation between socialist regimes in France and Spain can also work as a counterbalance to 'regionalist' threats within France itself.

Extradition orders are a clear signal to neighbours that, though decidedly more liberal in the treatment of minorities than previous governments, the current French administration does not countenance centres of violent opposition on French soil.

Nationalism within France Napoleon's inheritance and the Jacobin centralist tradition continue to influence the manner in which the French state negotiates with its constituent 'dissident' nationalities. Previous attempts at determining a specific regional role for areas such as Brittany, Corsica and Alsace in relation to the needs of the French economy have been increasingly questioned since 1945. Clear economic differentials between core and periphery, a failure to devise appropriate regional development policies and a continued stigmatization of 'traditional cultures' are cited as preconditioning grievances for the emergence of reactive nationalist movements. The root of these periodic disturbances is economic exploitation and external control.

In Corsica, after decades of neglect, the French state sought both to colonize the island and to develop its natural resources. Kofman (1982) demonstrates the results of state and capitalist penetration for the period 1950–80. Developments in the three most important employment sectors – agriculture, construction and tourism – produced two forms of marginalization: first, a spatial polarization between coastal development and interior neglect; second, a social polarization consisting of a tripartite stratification. Key positions were reserved for French mainlanders and foreigners, while the Corsicans were squeezed between these spiralists and the influx of North Africans and Iberians imported as semi-skilled labour. Corsicans resented the development of their territory by 'outsiders', and a colonizer–colonized mentality was intensified with the repatriation of some 15,000–17,000 settlers from Algeria and an increase of non-Corsican-born inhabitants from 10 per cent of the population in 1954 to 45 per cent in 1975 (Kofman, 1982, p. 305).

Numerous resistance organizations were formed and re-formed, the strongest of which were the Action Régionaliste Corse dating from 1967 and the more vitriolic Front de Libération Nationale de la Corse formed in 1976. The socialist government's *statut particulier* (a devolutionist reform recognizing Corsica as a territory with a regionally elected assembly), announced by Gaston Defferre, the Socialist Minister of the Interior and 'Decentralization', in August 1981, has not assuaged the sporadic violence associated with Corsican autonomists (Kofman, 1982, pp. 309–10). But the new interdependent structure of councils, the state–regional employment committee and new agencies for transport, hydroelectric power, agriculture and regional development do provide an innovative institutional framework wherein grievances can be voiced. However, as Mény has demonstrated, if the reforms are limited to institutional changes they will be ineffective; the real transformation would accompany 'the establishment of the rule of law and a respect for universal suffrage' (*Le Monde*, 20 November 1983, quoted by Mény, 1984, p. 74). The Defferre reforms seek to increase local democracy and involvement and have gone a long way to stifling the opposition cries that a diminution of centralist control would threaten the viability of the French state.

Government pronouncements that its decentralist measures will reduce the source of core–periphery conflict in France have been judged premature and overoptimistic. Ményi (p. 75) observes that the very creation of new organs of government and the establishment of new relationships may create problems, for they 'constitute a disruption of the system'. But in resolving such problems attention will have been lifted from the activities of the central government to competition at the local level for the exercise of regional power (Hirsch, 1981; Ményi, 1984). After reviewing the experience of Mitterrand's regionalist accommodation programme, Ményi concludes that in offering territorial minorities 'a right to roots' as well as a 'right to *options*' (Urwin, 1985), 'the State has laid the groundwork for a consensual integration of far greater efficacy than earlier, more authoritarian attempts' (Ményi, 1987, p. 60). Such reforms may very well strengthen the central state apparatus in time, by making it an arbiter of local and regional conflicts and by distancing it from events in Brittany, Corsica and Occitania, so 'objectifying' and 'depersonalizing' the role of the state that its legitimacy increases.

The break-up of the state? The French and Spanish examples were selected to reflect the wider structural discontent in contemporary Europe. The expansion of the national territorial state ideal, in the past two centuries in particular, has strained the basic resources which constitute the building blocks of nation formation. Yet while materialist approaches often view the state as the unwitting midwife of global capitalism, unable to control the very processes it brought into being, idealist approaches root much of the threat to global stability in the aggrandizing process inherent in the competitive state system itself. Idealist alternatives focus on a return to small historical communities as a panacea for the problems created by the emergence of a world-economy and its concomitant state system.

Leopold Kohr argues that aggression and warfare are a product of a near-critical mass of power which is periodically unleashed as open conflict because of the failure of countervailing pressures. Great size and power, rather than being a positive benefit as theories derived from economies of scale would have us believe, are in fact dangerous attributes producing widespread social misery. Kohr's antidote to this danger is to dismember large states, for he concludes:

> If wars are due to the accumulation of the critical mass of power, and the critical mass of power can accumulate only in social organisms of critical *size*, the problems of aggression, like those of atrocity, can clearly again be solved in only one way – through the reduction of these organisms that have outgrown the proportions of human control. (Kohr, 1957, p. 54)

Small may not necessarily be beautiful, but it is certainly safe, or safer than superpower rivalry! To reduce Europe's states to their constituent historic nations, or, in Kohr's terms, 'Europe's natural and original landscape', would require massive structural changes and would produce a plethora of new polities (figure 9.7) but he claims that the effort would make Europe more peaceful for it would reduce the problem of contested border areas and non-state national minorities; 'With all states small, they would cease to be mere border regions of ambitious neighbours. Each would be too big to be devoured by the other. The entire system would thus function as an automatic stabilizer' (p. 59). He does not claim that wars would be eliminated in a reconstituted Europe, but they could be contained and made more bearable as periodic events unlikely to bring the global system to the brink of a world war. Localized peace just as much as localized war is 'divisible', and Kohr's thesis echoes the principles of the Middle Ages where warfare was divisible not only in space but also in time according to the controlling institution of *Treuga Dei*, the Truce of God (p. 63).

Restoration of Europe's old nations in a European federation smacks of romantic idealism from the perspective of power politics, *realpolitik* or integration theory, and the fact that Kohr's ideas have lain neglected by mainstream political science for three decades may prove conclusive to the cynic. Analogous proposals have had an interesting, if chequered, career and have deeply influenced both regionalists and federalists. But the works of Kohr (1957), Heraud (1963) and Foueré (1980) now seem both more cogent and also more respectable than they did when such ideas were first mooted. Of particular significance is the attempt by several European institutions to legitimize and channel the growth of ethno-national awareness so as to promote it as a positive force and deflect the potentially disastrous effects of neglect or further repression. Evidence of this can be found in the Council of Europe's proposal to establish a Charter of European Regional and Minority Languages and its long-term promotion of regional-authority responsibility in as many aspects of daily life as possible. Further evidence is the establishment of both public and private agencies, such as the Bureau of Unrepresented Nations in Brussels, the European Bureau of Lesser Used Languages in Dublin and the Federal Union of European Nationalities, while transnational and transfrontier co-operative ventures continue apace, and are particularly fruitful in the Alto Adige and Friuli–Venezia–Giulia regions.

The parallel development of a set of European superstructural and substructural relationships will inevitably place territorial considerations at the centre stage of regional international politics (Sharpe, 1987; Williams, 1988) for the freedom of labour mobility guaranteed in the New European Order post-1992 will create more interethnic tensions and language conflict, and that in a fourfold manner. First, we shall probably witness the unquestioned supremacy of English as the *lingua franca* of the new Community. Functionally this will cause other international languages, such as French, Spanish, Italian and German, to jockey for position in a secondary role within the educational, legal and commercial domains of a restructured Europe. Second, autochthonous European language minorities, such as the Irish, Frisian and Welsh, will be further marginalized unless they can significantly influence national/regional

Figure 9.7   Europe's constituent ethnic territories
Source: Kohr, 1957, p. 233

patterns of stable bilingualism with a much reduced language switching by domain than hitherto. Third, there will be increased pressure on local authorities to provide mother-tongue education and public services to the children of mobile workers and their families, especially in the already multicultural cities such as London, Brussels, Paris, Berlin and Amsterdam. Finally, there will be increased pressure at the Community level to provide resources for the instruction, absorption and occupational integration of the children and dependants of non-European migrant workers, such as the Turkish *gastarbeiters*, and of the religiously distinctive descendants of North African and Asian migrants of the past three decades. Each of these major trends will have key social, economic and political ramifications which may enhance the productive capacity of the European economy but will also strain the public agencies of an 'integrated Europe'. In time, of course, such trends could serve to replace the nineteenth-century nation-state pattern of statehood by a more functionally oriented system of territories. However, it would be foolish in the extreme to expect such a transition to occur without sustained political socialization and a concomitant vigorous rearguard defence of the status quo by a plethora of interest groups and agencies acting in the name of the nation.

The international state system, if not its constituent parts, remains intact, virulent and eminently capable of defending itself. That Europe has not witnessed a cataclysmic continent-wide eruption of violence in this generation, contrary to the experience of every generation since 1648, gives cause for much rejoicing. But, as Shaw has warned, we should not grow complacent in our attitude to periodic warfare:

> In the face of the immense evidence that periods of peace are so often periods of preparation for the next war, there are still those who are able to ignore or 'normalise' the relationship. The overwhelming majority of writings about modern capitalism grossly underplay the role of the Second World War in creating the system which has subsequently developed, and underestimate the likelihood that this whole period of capitalist development will culminate in nuclear catastrophe. (Shaw, 1984, p. 3)

National Congruence in Former Colonial Territories

In contrast with the long experience of state formation and national congruence in Europe, Third World societies convinced of the value of the European-derived 'nation-state' have a seemingly impossible task of reconciling divergent interests in the pursuit of state stability and economic development. In many respects nationalism was more virulent in the colonial context, even though the traditional factors which were conducive to nationhood were often but a pale reflection of their European origins. Indeed, the very notion of a nation itself, linked intimately to its own state structure for full political expression, was an alien concept to all but a few of the western-educated elite. Let us illustrate the interplay of nationalism, conflict and state formation by reference to the

Nigerian experience, conscious, of course, that we are selecting but one of a large number of possible cases.

History permits us to interpret colonialism as a fascinating example of nineteenth-century liberal notions superimposed on the *anciens régimes* of Africa, Asia and Latin America. High ideals, and even higher profit levels, determined the extension of European influence over the newly conquered territories of the far-flung empire, regulating the interaction between a world-system of core, semi-periphery and peripheries. The 'accidents' of imperial Balkanization produced a diverse pattern of multi-ethnic colonies whose transition to independence is one of the most squalid and tragic episodes in world history. Not that the call to liberty was in itself a tragedy, but the passage from servitude to statehood was so often marred by widespread conflict, warfare and subsequently new forms of domination that their effects largely determine the future role and direction of many newly independent states.

The basic political crisis is legitimacy. Principles of national unity are employed in the most unpromising circumstances to shape a population largely incorporated forcibly into a nation served by a strong state apparatus. Under the drive of nationalism, post-independence developments have strengthened the process of filling power vacuums, extending the reaches of the bureaucratic--military elite to the furthest periphery. Central to this European-style pattern of nation-building is identity formation. As we have seen, national integration is a difficult enough process in Europe; it is well-nigh impossible in many African contexts. The basic building blocks are absent from the national construction of a unified society. Reflect on Azikiwe's statement in 1945: 'Nigeria is not a nation. It is a mere geographic expression. There are no Nigerians in the same sense as there are English, Welsh or French. The word ''Nigerian'' is merely a distinctive appellation to distinguish those who live within the boundaries of Nigeria from those who do not' (Sklar, 1963, p. 233, quoted by Oberschall, 1973, p. 91).

The interdependent relationship between the professional intelligentsia and varieties of nationalism is crucial in the post-colonial state (Smith, 1983, pp. 90–4). In the search for national pride, economic development, state unity and international recognition, the intelligentsia are uniquely placed to interpret local and global events, and to analyse their effects on the 'nation' which they have helped to forge into a self-conscious political community. However, as Smith demonstrates: 'it follows that the chief political struggles in Africa today, including ethnic ones, are at root factional conflicts within the intelligentsia – civilian versus military, liberal versus marxist, regional or ethnic conflicts – and that any involvement on the part of the other strata or classes is at the invitation or behest of one or other faction within the ruling stratum of the intelligentsia' (p. 90). In Nigeria, the scale of the intelligentsia's problems in transmitting these various schemes for the state's future was daunting – a task compounded abroad by the constraints of neo-colonialism and at home by the vast ethnic and regional disparities inherited on independence.

It is now commonplace to attribute to Nigeria's ethnic diversity the seeds of the eventual civil war. But many commentators on the period 1950–66 have argued that certain historical factors, not inevitable elsewhere, have exacerbated these differences. Three are of prime importance: the legacy of colonial administration – direct rule in the south and indirect rule in the north; the internal dynamics of the Nigerian military and of civil–military relations; federal–regional rivalry with each of the major political parties trying to outmanoeuvre its opponents in gaining access to power, patronage and privilege.

The period prior to independence (1951–9) and just after (1960–6) was one of intense regional and ethnic bargaining between competitive elites and their respective ethnic voting blocks (for details see Luckham, 1971; Himmelstrand, 1973; Young, 1976). It was difficult to develop a sense of national identity, and there were constant threats of secession from disaffected regions and unrepresented minorities. The salient issues were in effect mediated by the regional governments which were controlled by political parties reflecting the majority ethnic group in each of the three regions: the Northern People's congress (NPC) in the Hausa–Fulani Emirates of the north, the Action Group, led by Chief Akintola, in the west and the National Council of Nigerian Citizens (NCNC) based upon Ibo dominance in the east. Luckham (1971, p. 209) argues that one reason for the consolidation of political power in the regions was to promote each group's bargaining position at the centre. A number of themes dominated the regional bargaining process, including (a) the division of power between the federal core and the constituent regions, (b) the distribution of revenue allocations and management of scarce economic and natural resources, (c) the allocation of key regional and federal posts in patronage politics, (d) the timing of independence, (e) the status of Lagos as either the chief city of the Yoruba region or the proposed new federal capital, (f) the struggle for political power within the federation, (g) the constitutional provision to be made for ensuring minority-group rights and (h) the vexing question of the creation of new states after independence to accord more equitably with Nigeria's ethnic groups (Luckham, 1971).

On the eve of independence (1 October 1960) the various expressions of anti-colonial nationalism and regional power were far from being harmonized in the quest for national congruence and statehood. Indeed, the great diversity of available options was itself evidence both of the fragility of the state idea and of the clash of rival sectional groups. Himmelstrand (1973) has outlined five contending nationalist options, each of which was present to some degree and none of which was sufficiently powerful to command total loyalty when the integrity of the state was threatened. The first was Nigerian nationalism, devoted to the territorial sovereignty of an indivisible Nigeria capable of withstanding regional challenges. The second was regionalism, a policy whereby the federal core is reduced to the status of a manipulated distributor of patronage to the regional ruling elites. This reflected a competition described by Luckham (1971, p. 210) as 'an oligopolistic bargaining' between the three principal parties, two of whom sought to exclude the third from coalition power. The third represents an underlying grievance, ethnic nationalism within multi-ethnic regions. Here Himmelstrand has in mind the demands for redrawing the administrative boundaries to reflect 'ethno-cultural realities', demands for the creation of new states from disaffected minorities and attempts to build up local ethnic associations tied neither to the dominant regional group nor to the largesse of the federal state. The fourth option is ethnic nationalism between regions, i.e. interregional irredentism, which was particularly attractive for Ibos in the mid-west, Yorubas in the north and the middle-belt peoples generally. The final option, secession, was often threatened by the north in its attempt to produce an independence settlement which was entirely favourable to the preservation of the status quo and hence to its dominant position. But it was the Ibo military leadership which in the event precipitated a Nigerian civil war by declaring secession and establishing the Republic of Biafra on 30 May 1967 (Himmelstrand, 1973).

I do not wish to recount the details of Biafran separatism, but merely to illustrate that much can be learned from this episode of the improbabilities of ethno-regional secession. It is scarcely in any state's interest to legitimize a wholesale disintegration of the multi-ethnic state structure; Bangladesh is the only successful separatist attempt of late and provides ample evidence of the international strictures involved. However, it may be useful to analyse the Biafran case in terms of preconditioning factors and triggering events to highlight the fluidity of political competition.

Among the many preconditioning factors none was more powerful than the periodic exclusion from power of the Ibo elite after independence. Among the most ardent of Nigerian nationalists, contributing a disproportionate input to both the civil service and the military, the Ibos saw themselves in the vanguard of Nigerian modernization – a viewpoint inherited from their position as colonial intermediaries, and vigorously put into practice for the first four years of self-government.

However, in 1964–5 the struggle for political power and patronage went outside the established electoral process (Luckham, 1971). The federal elections of 30 December 1964 produced a vitriolic and unconstitutional campaign. The results weakened the position of the NCNC in the coalition, which was now dominated by an alliance between the NPC and the Nigerian National Democratic Party (NNDP). Realizing that they were heading for the same degree of political exclusion suffered by the Action Group in 1962, the leaders of the east openly threatened secession. Unsupported by any other political party, the east's grievances festered, to be compounded by the increasing conviction that their isolation and insecurity was the result of northern discrimination and western manipulation of the coalition. This was confirmed by the character of the Western Region's elections of October 1965 which were 'openly rigged by the NNDP' (Luckham, 1971, p. 218). The results were widely challenged and the Action Group refused to recognize them. Outbursts of collective violence characterized the period from October 1965 to January 1966, when the military intervened to restore order. The first *coup* sought to end corruption, nepotism, inefficiency and 'regionalism'. General Ironsi, though head of a unified military structure, sought to maintain a quasi-federal system with military governors replacing regional administrators. But it was clear to most observers that the trend towards a unitary form of government would accelerate after Ironsi's speech on 28 January in which he declared 'All Nigerians want an end to regionalism. Tribal loyalties and activities which promote tribal consciousness and sectional interests must give way to the urgent task of national reconstruction. The Federal Military Government will preserve Nigeria as one strong nation' (quoted by Panter-Brick, 1970, p. 16). In the event Ironsi issued a decree establishing a unitary state on 24 May, which was followed by widespread rioting in the north. In the context of inter-regional rivalry, Luckham argues that the January *coup* was but one more instance of this 'periodic shifting power between the regions' – isolating the north from the political core – 'though the method of transferring power was new' (Luckham, 1971, p. 219). The July *coup*, when Yakubu Gowon emerged to succeed the assassinated Ironsi, restored the north's pre-eminent position and denied the east access to the centre of political power. It confirmed the awareness of political isolation felt so keenly by the eastern civilian and military elite. Added to this were the fears for survival engendered by the massacres of May

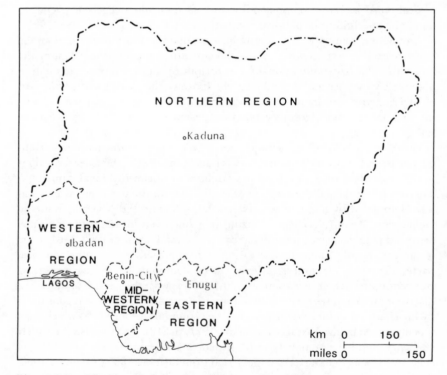

Figure 9.8   Nigeria as a federation of four regions, 1966
Source: Panter-Brick, 1970, p. x

and September–October 1966, both of which were directed at Ibos living in the north. In consequence 1.2 million Ibo refugees left, crowding into the eastern heartland of Biafra and bringing with them stories of atrocities, fears of genocide and rumours of imminent northern invasion (Panter-Brick, 1970; Luckham, 1971; Oberschall, 1973). Federal oil revenues were retained in the east for the relief of refugees, a development which troubled non-Ibo tribes who feared being subjugated by an Ibo majority in an independent Biafra.

But if the east were allowed to secede it was feared that the west would also, for it could not withstand the north in an emasculated Nigeria (Oberschall, 1973, p. 99). To forestall such developments the federal military government declared a state of emergency on 27 May 1967 and divided Nigeria's four regions into 12 states (figures 9.8 and 9.9). This effectively abolished the three-party oligopoly dominated by the north and allowed the stronger minority groups to play a more independent role (Luckham, 1971). The Ibos interpreted this reform as an attempt to deny them access to their oil supplies and revenues now under the jurisdiction of the Rivers state (figure 9.10). They also discerned an appeal to non-Ibo easterners in the Ogojo, Calabar and Rivers area to resist Ibo influence with federal support (Oberschall, 1973, p. 100).

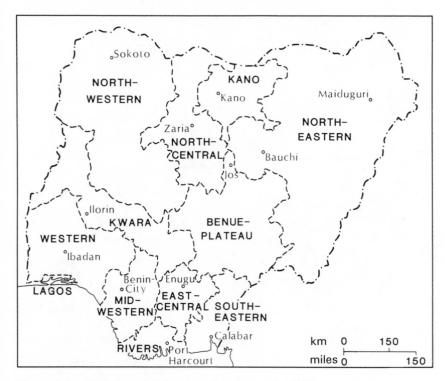

Figure 9.9   Nigeria as a federation of twelve states, 1967
Source: Panter-Brick, 1970, p. x

Underlying all this, of course, was the key role of the military, which is examined superbly by Luckham (1971). The triggering factors leading to the secession of the Eastern Region on 30 May 1967 relate as much to the internal military conflict as they do to the wider issues of federalism and ethno-regionalism. The key conflict derived from the fragmentation of the army after July 1966 and the challenge to a hierarchical authority structure induced by two military *coups*. The initial disagreement between Lt-Colonel Gowon, who declared himself Supreme Commander, and Lt-Colonel Ojukwu, leader of the army in the east, was over the rules of precedence and legitimate continuity in the army hierarchy. Subsequently Ojukwu enunciated that the key to legitimacy was effective control and leadership, qualities he claimed he represented over and above any other contender. In consequence, once the army lost its cohesion 'the nation was on the verge of disintegration' (Luckham, 1971, pp. 298, 147–50).

Thirty months of warfare demonstrated that secession would not be tolerated. The civil war ended in January 1970 with the lessons of conflict and the 'unvindictive peace' apparent to all African states (Kirk-Greene and Rimmer, 1981). Gowon's twin challenges of post-war reconstruction and the return to

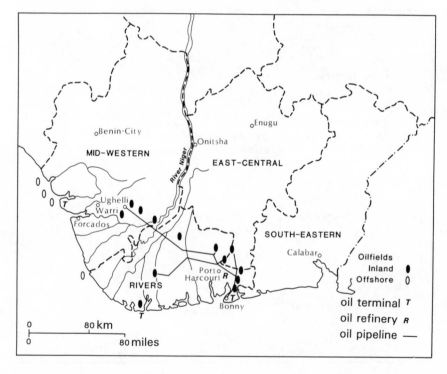

Figure 9.10   Location of oilfields, terminals and pipelines, 1969
Source: Panter-Brick, 1970, p. xi

civilian rule were mediated through a comprehensive reform programme designed to induce a 'period of lasting peace and political stability' (p. 4). The Nigerian experience suggests that, though nationalism may be a potent force in the anti-colonial struggle, it by no means offers a blueprint for governance on independence. The constant oscillation between federal core and ethnic–regional homeland called into question the effectiveness of pre-independence negotiations in harmonizing interethnic and inter-regional disparities. Economic autarky was hindered by the harsh reality of under-development and of the limited scope allowed to Nigeria in its tightly structured role in the international division of labour. Above all, it was the fragility of the state ideal itself, which strained deep-rooted and often bitter rivalries, as citizens fought to defend their new-found freedoms symbolized in the paradoxical, but nevertheless powerful, sentiments of an indivisible Nigerian 'territorial state'.

CONCLUSION

We have seen that nationalism has become an increasingly influential factor in the emergence of an interstate system. Its enigmatic character is displayed in its utility as an instrument both for nation-building and for state disintegration at various junctures. However, our emphasis on the question of national congruence has sought to highlight an abiding dilemma common to many contemporary societies. In their attempt to represent political and economic developments as essentially populist and communal, the leaders of modern states have to appeal to the 'common ground' of popular involvement in the territorial state. But this appeal presupposes a solidarity of interest and involvement among the populace, which is often lacking, especially in former colonial territories. In order to strengthen this identification states often resort to centralist bureaucratic institutions to buttress the state apparatus. In turn this reinforces the relative advantage of groups already able to control the direction of future state activity and economic development. In consequence, marginalized minorities perceive the state apparatus as the monopoly institution of an opposition group, both unwilling and unable to recognize their legitimate grievances within the framework of state activities. Thus in seeking to incorporate the masses in the national socialization processes, the bureaucratic state often re-emphasizes the very lines of conflict which discriminate against subordinate groups.

Attempts to modify the state structure have often led to conflict, and civil or international warfare is far too common an outcome to treat the issues of this chapter as ephemeral products of an expanding world system. In the mid-1970s commentators spoke of the demise of the nation-state in certain western democratic societies; more than a decade on we can conclude that the salutary experience of being challenged by ethnic–regional demands (among others) has made a number of states far more sensitive to the needs of their constituent ethnic minorities, while simultaneously strengthening their central agencies of government. Nationalism and ethnic antagonism are far too prevalent in such diverse settings as Serbia, Sri Lanka, southern Sudan and Azerbaijan to allow us to treat questions of identity and resource competition in a complacent manner (Williams and Kofman, 1988). We cannot and should not treat nationalism as an epiphenomenal diversion from an evolutionary state-building programme, but neither should we unquestioningly accept the primacy of the nation as the most salient form of political and cultural identification. It is but a product of human invention and is quite capable of being replaced by other forms of social organization. However, there is little imminent fear of the collapse of the nation-state system, though caution should be expressed as to the nature and direction of national identification in selected multi-ethnic societies as we enter a new period of economic and cultural readjustment and redefinition of identity.

We have much to learn before we answer the question: 'Why do the nations rage asunder?'. If indeed the concerns of this chapter are powerful contributors to a world in crisis, the best we can perhaps hope for is that we recognize it as a permanent crisis, and one that demands permanent tolerance for fear of something worse.

Acknowledgements

I owe a particular debt to the encouragement of A. D. Smith, Ron Johnston and Peter Taylor. I also wish to thank a number of friends who read the manuscript and offered constructive criticism: Wilbur Zelinsky, Sven Tägil, Kerstin Nyström, Ralf Rönnquist and Rune Johansson. Jane Williams, former cartographer at the Staffordshire Polytechnic, drew the figures with her customary skill and patience.

References and Bibliography

Anderson, B. 1983: *Imagined Communities*. London: Verso Editions.
Anderson, M. 1978: The renaissance of territorial minorities in western Europe. *West European Politics*, 1, 128–43.
Anderson, P. 1974a: *Lineages of the Absolutist State*. London: New Left Books.
Anderson, P. 1974b: *Passages from Antiquity to Feudalism*. London: New Left Books.
Berry, C. J. 1981: Nations and norms. *The Review of Politics*, 20, 75–87.
Best, G. 1980: *Humanity in Warfare*. London: Weidenfeld & Nicolson.
Boal, F. W. and Douglas, N. J. (eds) 1982: *Integration and Division: Geographical Perspectives on the Northern Ireland Problem*, London: Academic Press.
Boyce, G. 1982: *Nationalism in Ireland*. London: Croom Helm.
Breuilly, J. 1982: *Nationalism and the State*. Manchester: Manchester University Press.
Clark, R. P. 1979: *The Basques: The Franco Years and Beyond*. Reno, NV: University of Nevada Press.

Clark, R. P. 1981: Language and politics in Spain's Basque provinces. *West European Politics*, 4, 85–103.

Clark, R. P. 1984: *The Basque Insurgents, ETA, 1952–1980*. Madison, WI: University of Wisconsin Press.

Connor, W. 1981: Nationalism and political illegitimacy. *Canadian Review of Studies in Nationalism*, 8, 201–28.

Day, A. J. (ed.) 1982: *Border and Territorial Disputes*. Harlow: Longman.

Foster, C. R. (ed.) 1980: *Nations Without a State*. New York: Praeger.

Foueré, Y. 1980: *Towards a Federal Europe*. Swansea: Christopher Davies.

Gilbert, F. (ed.) 1975: *The Historical Essays of Otto Hintze*. New York: Oxford University Press.

Hamers, J. F. and Blanc, M. 1984: *Bilingualité et Bilingualisme*. Brussels: Pierre Mardaga.

Haugen, E., McClure, J. D. and Thomson, D. S.: 1981: *Minority Languages Today*. Edinburgh: Edinburgh University Press.

Hechter, M. 1975: *Internal Colonialism: The Celtic Fringe in British National Development*. London: Routledge & Kegan Paul.

Heraud, G. 1963: *L'Europe des ethnies*. Paris: Presse d'Europe.

Himmelstrand, U. 1973: Tribalism, regionalism, nationalism and secession in Nigeria. In S. N. Eisenstadt and S. Rokkan (eds) *Building States and Nations*, vol. 2. London: Sage Publications, 427–67.

Hintze, O. 1975: Economics and politics in the age of modern capitalism. In Gilbert (ed.), *The Historical Essays of Otto Hintze*. New York: Oxford University Press.

Hirsch, J. 1981: The apparatus of the state, the reproduction of capital and urban conflicts. In M. Dear and A. J. Scott (eds) *Urbanisation and Urban planning in Capitalist Society*. London: Methuen, 593–607.

Kirk-Greene, A. and Rimmer, D. (eds) 1981: *Nigeria Since 1970*. London: Hodder & Stoughton.

Kitromilides, P. 1983: The Enlightenment east and west: a comparative perspective on the ideological origins of the Balkan political traditions. *Canadian Review of Studies in Nationalism*, 10, 51–70.

Kofman, E. 1982: Differential modernisation, social conflicts and ethno-regionalism in Corsica. *Ethnic and Racial Studies*, 5, 300–13.

Kohr, L. 1957: *The Breakdown of Nations*. Swansea: Christopher Davies.

Krejci, J. and Velimsky, V. 1981: *Ethnic and Political Nations in Europe*. London: Croom Helm.

Lancaster, T. D. 1987: Comparative nationalism: the Basques in Spain and France. *European Journal of Political Research*, 15, 561–90.

Luckham, R. 1971: *The Nigerian Military*. Cambridge: Cambridge University Press.

Medhurst, K. 1982: Basques and Basque nationalism. In C. H. Williams (ed.), *National Separatism*. Cardiff: University of Wales Press, 235–61.

Mény, Y. 1984: Decentralisation in socialist France: the politics of pragmatism. *West European Politics*, 7, 66–79.

Mény, Y. 1987: France: the construction and reconstruction of the centre, 1945–86. *West European Politics*, 10, 52–69.

Modelski, G. 1978: The long cycle of global politics and the nation state. *Comparative Studies in Society and History*, 20, 214–35.

Nairn, T. 1981: *The Break-Up of Britain*. London: Verso.

Oberschall, A. 1973: *Social Conflict and Social Movements*. Englewood Cliffs, NJ: Prentice-Hall.

Orridge, A. W. 1981: Uneven development and nationalism. *Political Studies*, 29, 1-15, 181-90.

Orridge, A. W. and Williams, C. H. 1982: Autonomist nationalism: a theoretical framework for spatial variations in its genesis and development. *Political Geography Quarterly*, 1, 19-39.

Panter-Brick, K. S. (ed.) 1970: *Nigerian Politics and Military Rule: Prelude to the Civil War*. London: Athlone Press.

Panter-Brick, K. S. (ed.) 1978: *Soldiers and Oil*. London: Frank Cass.

Pettman, R. 1979: *State and Class*. London: Croom Helm.

Reece, J. E. 1977: *The Bretons Against France*. Chapel Hill, NC: University of North Carolina Press.

Rich, E. E. and Wilson, C. H. (eds) 1967: *The Economy of Expanding Europe in the 16th and 17th Centuries*. Cambridge: Cambridge University Press.

Rokkan, S. and Urwin, D. W. (eds) 1982: *The Politics of Territorial Identity*. London: Sage Publications.

Ronen, D. 1979: *The Quest for Self-Determination*. New Haven, CT: Yale University Press.

Rosi Garcia, M. and Strubelli Trueta, M. (eds) 1984: Catalan sociolinguistics. *International Journal for the Sociology of Language*, 47.

Rudolph, J. 1982: Belgium: controlling separatist tendencies in a multinational state. In C. H. Williams (ed.), *National Separatism*. Cardiff: University of Wales Press. 263-97.

Seton-Watson, H. 1977: *Nations and States*. Boulder, CO: Westview Press.

Sharpe, L. J. 1987: The West European state: the territorial dimension. *West European Politics*, 10, 146-67.

Shaw, M. 1984: War: the end of the dialectic? *Journal of Area Studies*, 9, 3-6.

Sklar, R. 1963: *Nigerian Political Parties*. Princeton, NJ: Princeton University Press.

Skocpol, T. 1977: Wallerstein's world capitalist system: a theoretical and historical critique. *American Journal of Sociology*, 82, 1075-90.

Skocpol, T. 1979: *States and Social Revolutions*. Cambridge: Cambridge University Press.

Smith, A. D. 1981a: *The Ethnic Revival*. Cambridge: Cambridge University Press.

Smith, A. D. 1981b: War and ethnicity: the role of warfare in the formation, self-images and cohesion of ethnic communities. *Ethnic and Racial Studies*, 4, 375-97.

Smith, A. D. 1982: Nationalism, ethnic separatism and the intelligentsia. In C. H. Williams (ed.), *National Separatism*. Cardiff: University of Wales Press, 17-41.

Smith, A. D. 1983: *State and Nation in the Third World*. Brighton: Wheatsheaf Books.

Snyder, L. L. 1978: *Roots of German Nationalism*. Bloomington, IN: Indiana University Press.

Tägil, S. (ed.) 1977: *Studying Boundary Conflicts*. Lund: Lund Studies in International History, Esselte Studium.

Tägil, S. 1982: The question of border regions in western Europe: an historical background. *West European Politics*, 5, 18-33.

Taylor, P. J. 1981: Political geography and the world-economy. In A. Burnett and P. J. Taylor (eds), *Political Studies from Spatial Perspectives*. Chichester: Wiley.

Taylor, P. J. 1982: A materialist framework for political geography. *Transactions, Institute of British Geographers*. NS 7, 15-34.

Treharne, R. F. and Fullard, H. 1976: *Muir's Historical Atlas*. London: George Philip.

Urwin, D. 1985: The price of a kingdom: territory, identity and the centre–periphery dimension in Western Europe. In Y. Mény and V. Wright (eds), *Centre–Periphery Relations in Western Europe*. London: Allen & Unwin, 151-70.

Wallerstein, I. 1974: *The Modern World System*. New York: Academic Press.

Wallerstein, I. 1979: *The Capitalist World Economy*. Cambridge: Cambridge University Press.

Wendt, A. E. 1987: The agent–structure problem in international relations theory. *International Organization*, 41, 334–70.

Williams, C. H. (ed.) 1982: *National Separatism*. Cardiff: University of Wales Press.

Williams, C. H. 1984: Ideology and the interpretation of minority cultures. *Political Geography Quarterly*, 3, 105–25.

Williams, C. H. (ed.) 1988: *Language in Geographic Context*. Clevedon: Multilingual Matters.

Williams, C. H. and Kofman, E. (eds) 1988: *Community Conflict, Partition and Nationalism*. London: Routledge.

Williams, C. H. and Smith, A. D. 1983: The national construction of social space. *Progress in Human Geography*, 7, 502–18.

Williams, G. 1980: Review of E. Allardt's Implications of the Ethnic Revival in Modern Industrial Society . *Journal of Multilingual and Multicultural Development*, 1, 363–70.

Williams, G. A. 1982: *The Welsh in their History*. London: Croom Helm.

Wolf, E. 1971: *Peasant Wars of the Twentieth Century*. London: Faber & Faber.

Young, C. 1976: *The Politics of Cultural Pluralism*. Madison, WI: University of Wisconsin Press.

Zolberg, A. R. 1980: Strategic interactions and the formation of modern states: France and England. *International Social Science Journal*, 32, 687–716.

Zolberg, A. R. 1981: Origins of the modern world system: a missing link. *World Politics*, 33, 253–81.

10

The New Geopolitics: The Dynamics of Geopolitical Disorder

JOHN AGNEW and STUART CORBRIDGE

Old intellectual controversies die hard. In this chapter we touch on two of them in our examination of the nature of geopolitics in the 1980s. The first concerns the dispute between those who see the world as immutably divided into nation-states that are the sole actors in international relations and those who see an emergent world-economy that is increasingly independent of states (see chapter 12). Early in this century the controversy was labelled that between 'imperialists' and 'superimperialists'. The second concerns the choice of synchronic versus diachronic conceptions of causation. Most writing on geopolitics takes the synchronic position in arguing that certain fixed geographical features of the world (such as the disposition of the oceans and the continents) or fixed processes of economic expansion–military strength determine the strategic interests and foreign policy choices of dominant states.

In this chapter we depart from conventional wisdom in both viewing 'superimperialism' as an emerging context for the redefinition of sovereign powers and seeing geopolitics as a dynamic (hence diachronic) phenomenon changing over time as the world is changed. In these respects the chapter differs from much recent writing on geopolitics, including work by Gilpin (1987), Kennedy (1987) and Huntington (1987–8). There is agreement with these authors, however, in seeing the 1980s as a critical period of transition from a *geopolitical order* (or prevailing system of power) of superpower blocs dominated by the United States and the USSR to a geopolitical order as yet unknown.

Since the late 1960s the basic bipolar power balance between the United States and the USSR that has existed since 1945 has altered to the detriment of both. A multipolar world now exists at the economic level. Japan's gross national product (GNP) has overtaken that of the USSR, while the GNP of the European Economic Community (EEC) has overtaken that of the United States. What remains to be seen is whether a multipolar system of five clusters of power (the United States, the USSR, Japan, China and the EEC) will emerge at the political–military level as well. At the present time the world is in a period of geopolitical disorder between the order of the post-war US–Soviet conflict and a new order that has not yet emerged. As Gramsci once observed of periods of change: 'the crisis consists precisely in the fact that the old is dying and the new cannot be born; in this interregnum a great variety of morbid symptoms occurs' (Gramsci, 1971, p. 276).

There are numerous indicators of the contemporary disorder. These include the inability of either the United States or the USSR successfully to defend unpopular client governments (as in Vietnam and Afghanistan respectively), the failure of both superpowers to translate their pre-eminence as nuclear powers into effective control over their immediate spheres of influence (as in Central America and Eastern Europe respectively), and the weakening of American and Soviet industrial and technological strength relative to Japan and the Federal Republic of Germany.

There is some controversy over the cause of the geopolitical disorder. Some commentators propose one version or another of the 'imperial overstretch thesis' made popular by Paul Kennedy (1987). This focuses on the overextension of military commitments relative to declining state resources. Others suggest that the progressive globalization of the world economy under American influence has produced a mismatch between the political and economic pretensions of dominant states (and their apologists) on the one hand, and their ability to control and enhance their economies, on the other.

We propose that to understand this new situation requires a new view of geopolitics, which can no longer be seen in terms of the impact of *fixed* geographical conditions (heartlands/rimlands, lifelines, choke-points, critical strategic zones, domino effects etc.) upon the activities of the Great Powers. Rather, today *geopolitical economy* is replacing classical geopolitics as the fundamental context for the constitution of foreign policy. There is a lag in recognition of this, however, as some political leaders remain wedded to particular geopolitical ideas even as they become outmoded. This reflects the fact that a geopolitical order involves a particular mode of reasoning and gains expression in a discourse about the division and control of global political space. Geopolitics, both old and new, is an active process of constituting the world order rather than an accounting of permanent geographical constraints. The 1980s are a time of crisis in geopolitics precisely because as an old order is dying a new one has not yet been born.

We begin this chapter with an argument for a dynamic conception of geopolitics as opposed to the usual static ones. A review of some of the morbid symptoms of disorder is provided in the second section. Finally, the present geopolitical disorder is discussed in terms of the imperial overstretch and globalization theses.

THE NEW GEOPOLITICS

In common usage geopolitics is used to refer to either a fixed and objective geography constraining and directing the activities of statesmen or the writings of certain key geopoliticians such as Mahan, Mackinder and Haushofer whose theories of geopolitical relationships are held to have influenced US naval policy, Nazi expansionism and the American policy of 'containing' the USSR. In either case geography is seen as a set of immutable facts of existence, rather after the fashion of current attempts in the United States to bring geography 'back' into the curriculum of elementary schools.

Spykman clearly articulated the dominant conception of geopolitics as a determining geography when he wrote:

> Geography is the most fundamental factor in the foreign policy of states because it is the most permanent. Ministers come and go, even dictators die, but mountain ranges stand unperturbed. (Spykman, 1942, p. 41)

From this point of view geography is a metaphysics, always present and always exerting its influence in prescribed ways upon the conduct of international politics.

From the perspective of the geopolitician, the geographical is more akin to a constraint on action. To Mackinder (1904), for example, 'man and not nature initiates; but nature in large measure controls'. By this, he appears to mean that there is a possibility of human agency but only if the constraints are appreciated and understood. This was the basis for the geopolitician's entry into the domain of statecraft to urge what was 'necessary' and 'possible' in foreign policy upon the leadership of the state.

These fixed-form views of geopolitics have been joined in recent years by attempts to define geopolitics in terms of fixed processes of one kind or another. Adding distance as a variable to models of international conflict, for example, has become increasingly popular (e.g. Starr and Most, 1985; O'Loughlin, 1986; O'Sullivan, 1986). The imperial overstretch thesis discussed in a later section is another example. The tendency of some world-systems theorists to hypostatize the world into cores and peripheries that are defined today by essentially the same processes of labour exploitation that characterized them 400 years ago is perhaps the most egregious example of changelessness (Agnew, 1982; Corbridge, 1986).

The apparent disintegration of the post-war bipolar division of the world, what we can call the post-war geopolitical order, provides an appropriate setting in which to challenge established views of geopolitics. To use Berman's apt phrase – following Marx – we are living in a period in which 'all that is solid melts into air' (Berman, 1984).

What is needed is a framework for defining geopolitics which incorporates both processes of economic and political change and the rhetorical understanding that gives a geopolitical order its appeal and acceptability. We are in a period of geopolitical disorder in which no dominant geopolitical discourse has yet emerged to match changed circumstances. The constitution of a dominant geopolitical discourse is achieved by a widely accepted division of the world into spaces of greater or lesser importance from the point of view of the contemporary Great Powers. Geopolitics, therefore, involves a process or reductive reasoning by which practitioners of statecraft constitute and read places (continents, regions, states etc.) in terms of their utility and importance to regional and global objectives and interests. States and regions are thus read geopolitically as 'clients', 'buffer zones', 'spheres of influence' or 'strategic locations'. It is a process of representation by which political leaders (and experts) define the world and fill it with subjects, narratives and scenarios with little sense of the contradictions, ironies and pluralities, or even the histories, of particular states or regions (Agnew and O'Tuathail, 1987). It is the representation of dominant powers, perhaps especially the United States after 1945, that 'script' or write the agenda for the world community at large. With the disintegration of the post-war geopolitical order we are as yet without a new script to match.

This does not mean that the post-war geopolitical order has been without contradictions and tensions. There have certainly been discontinuities and periods of lessened geopolitical confrontation, such as the *détente* of the 1970s (between two periods of cold war; Halliday, 1983; Crockatt, 1987). The following statement by Cyrus Vance not only provides a clear recognition of the tensions, but also attempts a tortuous resolution:

> When the two concepts – human rights and national security – are uttered in the same breath, it is often to express an unavoidable conflict, a fundamental tension between the pursuit of the good and the pursuit of the practical. I strongly reject the idea that there is a fundamental incompatability between the pursuit of human rights and the pursuit of self-interest. By this I do not mean to say that there can never be a conflict between human rights concerns and security concerns. We must weigh constantly how best to encourage the advancement of human rights while maintaining our ability to conduct essential business with governments, even unpopular ones, in countries where we have important security interests. (Vance, speech delivered 27 March 1980)

The post-war geopolitical order was also based upon tacit agreement between the Great Powers over shared interests and zones of vital strategic interest (Franck and Weisband, 1972; Keal, 1983). This is what made it 'hegemonic' (Cox, 1983). Southern Africa, Europe and the Middle East, for example, have been regarded by the United States and the USSR as vital regions of conflict, partly because of their resources but also because of their position *between* agreed spheres of influence. There have also been dissenters to such geopolitical 'agreements', but regionally (for example, contemporary Iran and Gaullist France) and globally (for example, Nazi Germany and Japan in the 1930s and early 1940s). Revisionist powers, by definition, attempt to introduce a new geopolitical order and to practice alternative geopolitical discourses. Historically, this has led to major wars and the later establishment or normalization of new geopolitical orders.

If a geopolitical order finds room for contesting voices, so also does it embody a temporality not captured in synchronic accounts of geopolitics. Successive discourses of geopolitics have taken shape against a background of world-economic development based on a recurrent rhythm of expansionary boom, crisis formation and attempted restructuring. We are currently in the middle of the most recent of these crisis-induced restructurings – one signified by the term globalization. This is the fourth in a sequence which stretches back to include the Great Depression, the Long Depression of the nineteenth century and the political-economic crisis period of 1830–48.

In each of these periods a significantly different geopolitical order with distinct geopolitical rivalries took shape: Britain–France in the first (1830–48), Britain–Germany–USA in the second, USA–USSR in the third and multiple rivalries with an as yet unknown outcome in the fourth. The Wallerstein version of world-economic hegemony would add seventeenth-century Netherlands as a kindred case (Netherlands–Britain; Wallerstein, 1980; Chase-Dunn, 1982), and Modelski (1978, 1982) would regard the US-based post-war regime as the fifth replication of a 'long cycle' of world politics that includes sixteenth-century Portugal and eighteenth-century as well as nineteenth-century Great Britain (see chapter 11; for a good review of long-cycle theories, see Rosecrance, 1987). For us, it is important to stress that a recurrent rhythm of boom, crisis and restructuring is not a function of a fixed process of capital accumulation or hegemonic rivalry. The current trend to globalization, for example, is a qualitatively different mode of restructuring based upon the emergence of a new regime of accumulation which has been called global Fordism. This involves a generalization of mass production and mass consumption to the world scale, where the marriage is consummated through extended circuits of international trade and investment and credit creation. We further believe that each geopolitical order is involved with a particular geopolitical discourse about the spatial division of the world and its geographical orientation. Until the order is established or normalized the previous geopolitical discourse tends to persist, even in the face of apparent disorder linked to shifts in the regime of accumulation.

This is true today. Although there are voices in the United States which argue, for instance, that the United States lost out in preparing for the Third World War with the USSR while its allies (read Japan) were mobilizing their people for trade wars, hard-line cold war 'toughness' still prevails in many quarters. The Reagan administration based its foreign and defence policies upon the aim of restoring American power and pre-eminence *vis-à-vis* the USSR after the humiliations of Vietnam, Watergate and the Iran hostage crisis. The Nixon and Carter administrations had in different ways acknowledged that American foreign policy could no longer count on ready acquiescence elsewhere. The Reagan administration, in contrast, saw loss of primacy as a failure of policy-making rather than a long-term secular trend in the world-system (Williams, 1987).

The signs of geopolitical disorder discussed later suggest that the attempts to reassert US dominance have failed. Yet much of US foreign policy is still based upon the continuing existence of a simple bipolar world (Dalby, 1988).

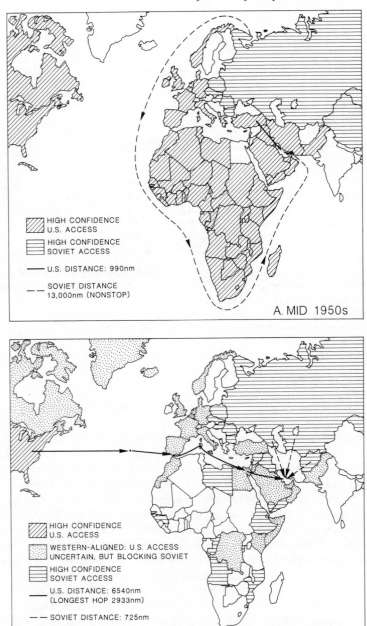

Figure 10.1　How far is it to the Persian Gulf? Trends in US and Soviet access to airfields or air space: (a) mid-1950s; (b) 1987
Source: Iklé and Wohlstetter, 1988

Table 10.1   The dance of the dinosaurs: US and Soviet procurement of major
weapons systems, 1977–1986

	United States		USSR	
	Pentagon	Gervasi	Pentagon	Gervasi
ICBMs/SLBMs	850	850	3,000	1,198
IRBMs/MRBMs	200	3,496	1,000	880
SAMs[a]	16,200	84,000	140,000	140,000
Long- and intermediate-range bombers	28	28	375	310
Fighters[b]	3,450	3,450	7,150	2,948
Military helicopters	1,750	2,043	4,650	1,450
Submarines	43	45	90	59
Major surface combatants[c]	89	89	81	44
Tanks	7,100	12,655	24,400	9,370
Artillery	2,750	3,750	28,200	5,625

ICBM, intercontinental ballistic missile; SLBM, submarine-launched ballistic missile; IRBM, intermediate-range ballistic missile; MRBM, medium-range ballistic missile; SAM, surface-to-air missile
[a]Includes naval SAMs
[b]Excludes anti-submarine warfare and combat trainers
[c]Excludes auxiliaries
Source: Gervasi, 1988, p. 121; Gervasi's estimates come from publicly available sources, especially Jane's annual reference volumes on military equipment and Central Intelligence Agency (CIA) annual reports to Congress

For example, in January 1988 a group of leading American experts on military and foreign affairs produced a report laying out a grand strategy for the United States over the next 30 years (Iklé and Wohlstetter, 1988). The report restricts itself to a discussion of military hardware and assumes a largely bipolar world. Moreover, it engages in massive statistical and cartographic legerdemain to demonstrate American vulnerability to Soviet threats (usually in the vicinity of the USSR) and US economic pre-eminence relative to all other countries both today and in 2010 (Kennedy, 1988; see also figure 10.1). This victory for ideology over reality is also a feature of the Pentagon's publication Soviet Military Power. Gervasi shows in his annotated edition that Soviet capabilities are massively overstated (see table 10.1). He explains this tendency to engage in statistical sophistry as a result of the needs of the military–industrial complex to invent 'missile gaps' and 'bomber gaps' in order to keep up military spending (Gervasi, 1988; see also Posen, 1988). It could also be that the 'experts', having grown up in a bipolar world to which they have become accustomed, are not capable of adjusting their ideas to a changing world.

The imagery of bipolarity also persists because, as yet, there is nothing to replace it. Possible emergent geopolitical orders are only dimly outlined. One, popularized by Gore Vidal (1986), has the declining superpowers joining together to fend off potential usurpers. Other less fanciful speculations accept the reality of a multipolar world and the need, along the lines argued by Kissinger (1968), to design a geopolitical order based around it. In some circles in the United States there is discussion of the need to expand capabilities for foreign military intervention and direct rule in an imperial fashion (Klare, 1988). The economic costs of this move towards a North–South rather than an East–West military strategy, however, are not likely to aid in competing economically with Japan and the EEC (Thompson and Zuk, 1986).

All these images maintain a state-centrism and an assumption of absolute empty space between dominant powers that may no longer be valid. This is not once again to annouce the imminent demise of the territorial state (Herz, 1957). It is to insist that the superpowers are territorial states in a world of trading states and that whatever new geopolitical order does emerge will have to recognize this fact (Rosecrance, 1986). Indeed, this suggests that the present period is one which might produce a more liberal international order (Keohane, 1984) and in which discourses questioning the definition of security in statist–military terms can flourish (Halliday, 1983; Mendlovitz and Walker, 1987). Contemplating these alternatives, however, challenges one of the key claims of much recent writing on international relations and geopolitics – i.e. that the world needs a hegemonic power to provide stability and leadership (Krasner, 1978; Gilpin, 1987; see also Haggard and Simmons, 1987; Smith, 1987; Yarbrough and Yarbrough, 1987; Corbridge, 1988a; Hall, 1988). We shall return to this in the conclusion.

SIGNS OF GEOPOLITICAL DISORDER

The signs of geopolitical disorder in the 1980s can be classified into a number of categories that together symptomatize an era of change and challenge. First of all, the nation-states that are the building blocks of classical and post-war geopolitics are less and less 'full societies' (Williams, 1983). They are at once too large and too small for a wide range of social and economic purposes. They are often too large to develop full social identities and real, as opposed to artificial, national interests. This can be seen in the political and economic division between North and South in a supposedly unified 'England' (despite increased capacities for state surveillance; Giddens, 1985). It is also evident in the numerous separatist movements around the world (see chapter 9). At the same time it is obvious that for many economic purposes most states are too small. They are increasingly 'market sectors' in an intensely competitive and unstable capitalist world-economy. There is, then, a profound contradiction between the social claims and military pretensions of many states and the economic realities that they now face. This is manifested in the severe monetary and fiscal crises that many states, including the United States, the USSR and much of Latin America, now face in maintaining their appeal to capital investment at the same time as they provide social welfare investment and high levels of military spending (O'Connor, 1973; Offe, 1985).

A second sign is the erosion of Great Power leadership capabilities and the emergence of a more plural international system. In particular, the hegemonic leadership of the United States has entered into decline. Because US leadership rested upon superordinate economic capabilities that translated into political and military power (and then back again), it has been relative economic decline that has received most attention. Rupert and Rapkin (1985) show that US shares of global resources and capabilities (e.g. the ratio of US GNP to global GNP) have declined in a more or less linear fashion since 1950, but that indicators of interdependence or susceptibility to 'external shocks' (e.g. the ratio of US trade to the GNP) have increased dramatically since 1970 (figures 10.2 and 10.3). They borrow the phrase 'scissors effect' from Bergsten to describe the joint consequences of a declining capability share and increased interdependence. In Bergsten's terms

> The United States has simultaneously become much more dependent on the world economy and much less able to dictate the course of international economic events. The global economic environment is more critical for the United States and is less susceptible to its influence. (Bergsten, 1982, p. 13)

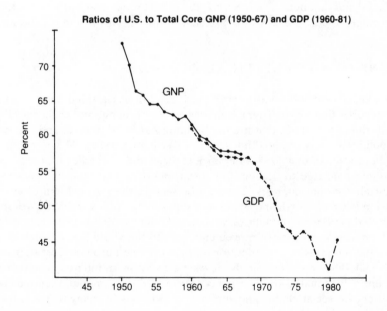

Ratios of U.S. to Total Core GNP (1950-67) and GDP (1960-81)

Figure 10.2   Ratios of US to total core gross national product (1950–1967) and gross domestic product (1960–1981)

The total core consists of the United States, the United Kingdom, the Federal Republic of Germany, France, Italy, Canada and Japan

Source: Rupert and Rapkin, 1985, p. 162 (uses OECD National Accounts)

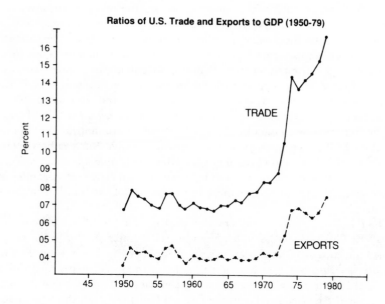

Figure 10.3   Ratios of US trade and exports to gross domestic product (1950–1979)
Source: Rupert and Rapkin, 1985, p. 172 (uses *Economic Report of the President*, 1983)

Most recently, the United States, the creator of the post-war geopolitical order and home of the world's key currency, has become the largest debtor in history. Japan, thought of as a developing country only 30 years ago, has become its largest creditor. Previously unthinkable questions arise as a consequence. Can the United States continue to lead alliances, especially those with Japan and the North Atlantic Treaty Organization (NATO), if it is increasingly in debt to its erstwhile followers? Can it continue to dictate military policy to countries which have a greater degree of economic autonomy?

Meanwhile, the USSR, the other great pillar of post-war geopolitics, is in the throes of its greatest period of crisis since 1917. Its problems are also economic but of a different variety. In the Soviet case the domestic economy stagnated during the Brezhnev years (1964–80). Consequently, the current government, through its much publicized policies of *perestroika* (restructuring), is attempting, so far without much success, to stimulate the economy by removing controls and regulation and opening up the country to joint ventures and international trade (McGwire, 1988).

Thus far the troubles of the superpowers have led to a more plural world-system rather than a new challenger or bloc. Leaders of both the United States and the USSR are now under pressure to take into consideration the demands of both friends and foes (Calleo, 1987; Dawisha, 1988). They can no longer act as if they are two 'dinosaurs' alone in the world (Shulman, 1987–8). However, each country is still heavily militarized. This is important because there are those in each with a stake in seeing the other superpower as an immutable enemy, irrespective of the problems that each faces. It is also important because the persisting militarism has generated questions about the need for ever-newer weapon systems in the face of economic retrenchment elsewhere in government budgets. Political movements for change (such as the Nuclear Freeze Movement in the United States, the Campaign for Nuclear Disarmament (CND) in the United Kingdom and European Nuclear Disarmament (END) in Europe) have arisen to challenge the notion of a world divided into two permanently warring blocs allied around the United States and the USSR. The signing of the Intermediate-range Nuclear Forces (INF) Treaty at the Washington Summit of December 1987 may not mark a decisive reversal in the nuclear arms race between the superpowers (Bromley and Rosenberg, 1988), but it does mark a recognition of the limits each faces in a more plural world.

The INF Treaty also provides modest evidence for a third sign of geopolitical disorder. This is that the military–industrial complexes that have been so powerful within the United States and the USSR are no longer as predominant as was once the case. For one thing, many of their very expensive products have failed to perform when used in actual combat situations (e.g. Soviet missiles in Syria and American radar in the Persian Gulf). Second, the technological sophistication of the superpowers has not helped them win conventional wars in which they have been involved. Third, other countries, including France, the United Kingdom and Brazil, have increasingly replaced the United States and the USSR as arms suppliers around the world. This reflects judgements about the quality of weapons as well as declarations of independence from superpowers increasingly distracted by domestic woes. Fourth, both superpowers have called into question the strategic doctrine of massive retaliation upon which their nuclear weapons policies have been based. Each is involved in expensive anti-ballistic missile (ABM) research and development, although the American 'Star Wars' or Strategic Defense Initiative (SDI) is best known. This offers the prospect of further destabilizing the arms race and decoupling Western Europe from American nuclear weapons policy. Finally, in both the United States and the USSR military spending is now seen as a burden rather than a benefit (although it is more a boom for some regions than others in the United States; O'hUallacháin, 1987). This is a change from the period in the 1950s and 1960s when spending on military hardware was seen as a mechanism for stimulating the US economy (Mintz and Hicks, 1984).

A fourth and final sign of geopolitical disorder is the inability of the Great Powers to impose their will over many previous allies and clients, including some in their immediate vicinity. Although there are some 25 wars going on around the world at present, many of them fuelled if not sponsored by the United States and the USSR, few if any will lead to determinate victories for either superpower. Their most famous direct interventions (in Vietnam and Afghanistan) were military and political disasters. Within their spheres of influence (the United States in Latin American and Western Europe, and the

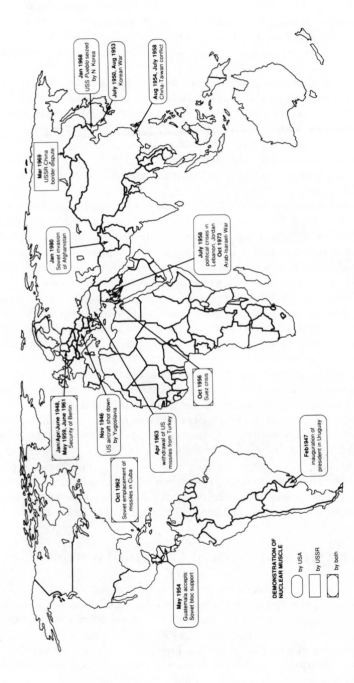

Figure 10.4 The declining use of nuclear threats
Source: Freedman, 1985, pp. 110–11

USSR in Eastern Europe and parts of Africa) they are massively unpopular and increasingly unheeded. According to one 1986 study, Soviet 'influence' is now substantial in only 11 per cent of the world's countries, down from a 1950s peak of 15 per cent (Lappé et al., 1988). Nuclear power does not translate into easy domination (figure 10.4).

The 1980s predicament of the superpowers is illustrated by Nicaragua. It is not just American policy that has failed in Nicaragua (i.e. to bring the Contras to power). The Soviet patrons of the Sandinista government have been saddled with yet another economic burden (Cuba, Vietnam, . . . the list goes on) at a time when they can ill afford such a commitment. Nicaragua, devastated by earthquakes, war and now hurricanes over the past 20 years, continues to pay a price for the pretensions of geopoliticians in both Washington and Moscow (Slater, 1987; Pastor, 1988).

Balky unreliable allies and conflicts that do not redound to the advantage of either the United States or the USSR pale in significance next to the emergence of entire world regions, most nearly the Middle East, but also perhaps South Asia and East Asia, in which the writ of the global superpowers is challenged by such regional powers as Iran, India and China. Since 1970, with the oil price shock of 1973–74, the opening-up of China after 1972 and the fatal challenge to the US use of regional surrogates in Iran in 1979, 'worlds' previously subordinated to the imperatives of the US–Soviet geopolitical order of 'containment' and 'balance' have emerged as regional geopolitical arenas in their own right. The Kissinger missions of the early 1970s have no credible counterpart in the late 1980s.

In most world regions US opposition to political radicalism has aligned the United States with forces which proved to be losers in the competition for state power, perhaps most clearly in the cases of China and Iran. The United States has survived such debacles, however, because foreign insurgencies have rarely challenged US military security or even investments by US corporations. The real fear of US policy-makers has been any suggestion that the US model of economic development is not the last word in 'democracy' and 'freedom'. Any demonstration that the choice is not just between US-style capitalism and Soviet-style state socialism would undermine the legitimacy of the simple bipolar world and American centrality to it (Calleo, 1970; Keal, 1983; Ross, 1987).

THE DYNAMICS OF GEOPOLITICAL DISORDER

From 1945 until the late 1960s the global geopolitical order was based upon a bipolar balance between the United States and the USSR. Recently, a number of writers have argued that this bipolar balance is in the process of unravelling because of the decline of the superpowers, especially the United States, relative to competitors for hegemony, i.e. China, Japan and perhaps a united Europe (Gilpin 1987; Huntington, 1987–8; Kennedy, 1987); a good review of this literature is provided by Schmeisser (1988).

The arguments of Kennedy (1987) have probably been the most widely quoted and influential. His narrative moves from the primacy of Spain in the sixteenth-century, through that of France under Louis XIV and Britain in the eighteenth- and nineteenth-centuries, to end with the rise of Russia and the United States whose antagonism formed the bipolar world system after 1945. His last chapter surveys evidence for the decline of the post-war cold war order in terms of the process of 'imperial overstretch' that he argues undermined previous geopolitical orders. This process is defined as follows:

> The relative strengths of the leading nations in world affairs never remain constant, principally because of the uneven rate of growth among different societies and of the technological and organizational breakthroughs which bring a greater advantage to one society than to another . . . If, however too large a proportion of the state's resources is diverted from wealth creation and allocated instead to military purposes, then that is likely to lead to weakening of national power over the longer term. (Kennedy, 1987, pp. xv–xvi)

According to this thesis, decline has an inevitability about it. Ruling and defending far-flung territorial possessions and national interests is fundamentally distracting. At a certain point in time 'the political benefits from external expansion may be outweighed by the great expense of it all – a dilemma which becomes acute if the nation concerned has entered a period of relative economic decline' (p. xiv). Moreover, to withdraw from positions once acquired is to create a vacuum into which other forces might flow and perhaps to start a retreat that ends as a rout.

Seen in this perspective the United States and the USSR are successor states to previous European hegemonies – the United States to Britain's maritime empire, and the USSR to French and German attempts to assemble an overwhelming continental bloc. These can be identified, although they are not by Kennedy, with the 'outer or insular crescent' and the 'heartland' of Sir Halford Mackinder's (1904) essay 'The geographical pivot of history'. By extension, recent challenges could be framed in similar terms.

However, the imperial overstretch thesis ignores four points about the contemporary situation that set it apart from earlier periods of transition (and it is not clear that the overstretch thesis fits them either). For a start, it is unlikely that the United States will fall victim to the challenge of a single competitor as has happened to dominant powers in the past. The United States does not face a threat from any state with both strong military and economic assets. The USSR continues to fit one category and Japan now fits the other, but nowhere have they been combined as in past eras.

Second, a country as large, populous and rich as the United States is not likely to be diminished by the rise of any other power in a way that the participants in the old European state system were diminished by the rise of the United States and the USSR. Even if the United States were to decline economically by several orders of magnitude, its technological and resource base, its rich consumer markets and its 'capitalist culture' would ensure its economic significance *vis-à-vis* the rest of the world for many years to come: it is worth noting that the GDP of the United States in 1985 was almost exactly three times that of its closest rival, Japan (World Bank, 1987). In any case, some of America's recent economic decline can be traced to domestic policy choices, such as those detailed by Stockman (1986) and Greider (1987), rather than to foreign competition. The United States is at the very least *primus inter pares* economically (Parboni, 1986; Strange, 1987).

Third, American foreign policy and associated military commitments in the 1950s and 1960s were incredibly ambitious even for so wealthy a country. However, the experience of failure in Vietnam and reversals elsewhere have led to a much more modest foreign policy, despite presidential rhetoric, in the 1980s. Relative to the Carter years there have been some new territorial commitments – for example, in the Caribbean and the Persian Gulf – but not on the same scale as the 1960s.

Fourth, and most importantly, the modern interdependent world economy, itself largely a product of American involvement, has given the purely economic dimension of power greater importance (Gilpin, 1975). Global free trade and access to markets in industrial countries have been crucial to the economic success of Japan and the newly industrializing countries of East Asia (Taiwan, South Korea etc.; see chapter 2). China now seems attracted by the same prospect. Moreover, Japan would prejudice its leading economic role if it set about re-arming on a massive scale in pursuit of a territorial empire. Japan can ill-afford to revive the suspicion of widespread ambitions throughout Asia, especially now that the rest of Asia is better organized politically and economically than it was in the past.

This last point of contention with the imperial overstretch thesis as applied to the contemporary situation directs attention to an alternative conception of the decline of the US–Soviet geopolitical order. The links between economic expansion, military power and political empire upon which the imperial overstretch thesis relies appear to have been cut. In particular, possession of military superiority is no longer readily convertible into enhanced political status. Nuclear weapons have made war between Great Powers almost too destructive to contemplate, never mind practice. This suggests that enhancement of purely economic power is now of primary importance.

Economic power, however, is no longer a simple attribute of states that have more or less of it. The world today is characterized by a global economy in which no single national economy has a dominating role and in which many of the major actors, such as banks and multinational corporations, are global in outlook and orientation. It is this globalization of the world-economy that provides the context for the relative decline of the US economy and hence for explaining in large part the decline of the cold war geopolitical order.

What is the evidence for this? In the first place, there has been a major revival of world trade over the past 30 years that has created a system of trading links in which most of the world's national economies, including the heretofore

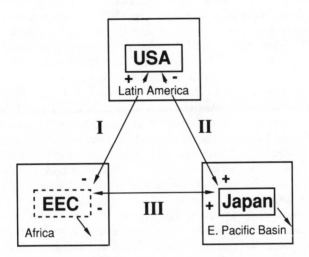

Figure 10.5   The circuits of world trade
Source: Davis, 1985

isolated and insulated American economy, are deeply involved (figure 10.5). In the 1970s the main circuits of this system were as follows: (1) a US manufacturing trade deficit with East Asia (cars, electronics, apparel) balanced by a surplus ($20 billion in 1980) with Europe (computers, aerospace); (2) a net US trade surplus with Latin America offset by an outflow of US direct investment, especially to Brazil and Mexico; (3) European manufacturing deficits with the United States and East Asia balanced by surpluses with Africa; (4) Japan's surpluses with Europe and the United States balanced by payments for energy and raw materials and the export of capital to other parts of East Asia and elsewhere (Davis, 1985).

Since 1980 this system has destabilized. Japan and other East Asian countries have boomed through increasing their exports to the United States and Europe. However, the continuation of this imbalance is in doubt as long as the other circuits of the world trade system remain depressed and encumbered by debt. Within this system the long-term growth of one party is largely determined by the growth of others. Large-scale trade places a premium upon open trading and commerce rather than territorial expansion and military superiority (Rosecrance, 1986).

Second, American multinational firms have been major agents of the transformation of the world-economy (Brett, 1985; Agnew, 1987). Approximately 30 per cent of the US trade deficit with East Asia (including Japan) is due to American corporations' making things there and selling them in the

United States. For example, of the four largest exporters from Taiwan, one is RCA and another is AT&T; the largest exporter of computers from Japan is IBM. According to Lipsey and Kravis (1987), US corporations accounted for the same proportion of world exports (18 per cent) in 1984 as they did in 1966. At the same time, the US economy's share of world exports dropped by a quarter. Today about half the total exports of US corporations originates outside the United States. Much of this trade is intra-industry as well as intrafirm; similar products are being shipped back and forth (Browne, 1986).

Third, even a protectionist economy such as that of Japan is increasingly internationalized and subject to stresses generated abroad (Higashi and Lauter, 1987). This can be seen in the increased dependence of Japan upon one major national market – the United States (Miyazaki, 1987; Lincoln, 1988). In 1981, 25.4 per cent of Japan's exports went to the United States; in 1986 this share was 38.5 per cent. It is further apparent in the increased scale and diversification of Japanese-held foreign assets. In 1987 Japanese foreign assets totalled close to $1 trillion compared with the $1.1 trillion of the United States (Uchitelle, 1988). Japanese investment has also diversified away from bonds and portfolio investment towards direct investment (Thrift and Leyshon, 1988). Much of this is no longer oriented towards the product cycle (i.e. investment in countries with comparative advantages in production) but involves overseas production for reasons of product diversification and macro-economic advantage (e.g. avoiding tariffs and quotas). Japanese institutions also own much of the short-term debt of the United States, as well as much prime real estate in that country (including part of the Citicorp Tower in New York City). The increased openness of the Japanese economy is also evident in recent pressures upon Japan to reform the conditions of entry to its stock exchanges and major financial markets, in the growing role played by the yen in international trade and investment, and in the ties being forged by Japanese capital with the economies of East Asia, the Pacific Rim and the USSR (Fujita and Ishigaki, 1985; Dicken, 1987).

Fourth, there is an increasingly internationalized world financial system (chapter 2; see also Hamilton, 1986; Strange, 1986; Wachtel, 1986). The growth of institutional investors such as pension funds and insurance companies has encouraged financial innovation and and deregulation. Many financial markets (e.g. foreign exchange, futures and Eurobonds) operate around the clock. They sell global products to global markets. This has led to vastly increased interdependence between previously insulated national financial systems. This was clearly illustrated by the rapid diffusion of the shock from the Wall Street Crash on 19 October 1987 around the world's other stock exchanges. It has also redefined the space for national sovereignty with regard to macro-economic management. Heightened fears of capital flight, together with easy access to offshore markets, has constrained the governments of industrial countries to tight counter-inflationary policies in the 1980s. Even in the United States the option of national economic development based upon competitive devaluation or low interest rates is diminished by the realities of international economic competition and by the fear of a ballooning balance-of-trade deficit (Parboni, 1988). The disciplines of finance capital have been felt more strongly still in Mitterrand's France and throughout the indebted Third World (Lipietz, 1987).

However, while making the practice of conventional state-oriented foreign policy increasingly difficult for the Great Powers, globalization has not as yet had much effect upon its formulation, except in terms of American calls for increased 'burden-sharing' of military expenditures among US allies and the progressive emergence of trading blocs such as the EEC and the Canada–US Trade Agreement. In part, this is because globalization is a trend rather than a *fait accompli*. States have not disappeared as units of policy-making or as instruments for political mobilization (Parboni, 1986; Gordon, 1988). In part, this is because the space for globalization was created by and for dominant interests in particular nation-states: for example, finance capital in the United Kingdom (Harvey, 1982; Coakley and Harris, 1983). More pertinently, states tend to institutionalize the moments of their success and the ideas associated with them. Thus, in the United States the Reagan administration's attempt to restore American primacy involved increases in military spending and some foreign intervention (in Lebanon, Central America etc.) that could have been sustained in the 1950s but in the 1980s weakened the national economy and disrupted rather than maintained connections with allies (Gill and Law, 1987; Reich, 1987–8; Williams, 1987). It seems today that, given a certain military threshold, a Great Power's relative influence is determined largely by its ability to compete in world markets, to provide markets and finance for others, and to co-ordinate its economic policies with other states (Rosecrance, 1986; Putnam and Bayne, 1987). While acknowledging the diseconomies of military spending identified by the imperial overstretch thesis (though note the important revisionism of Browne, 1988), it is surely more important to consider the profound restructuring of the world economy in recent years as the necessary context for explaining the disintegration of the post-war geopolitical order.

CONCLUSION

The globalization of trade and investment is creating new capabilities for the nation-state and makes new and varied demands of it. As we move to the *fin de siècle* it is clear that the long-standing equation of economic, military and political power is being eroded. The possibility of hegemony in turn is being redefined. The United States remains hegemonic in the military alliances of the capitalist powers, as does the USSR within its bloc (although nuclear proliferation is a challenge to the bipolar order). At the level of the economy, matters are less straightforward. The internationalization of the US economy post-1945 laid the foundations for a system of global Fordism in which the main dynamic shifted from the United States to Western Europe and thence to Japan and the Pacific Rim. The internationalization of capital has also reshaped – and on balance weakened – the possibility of national macro-economic management. Nation-states have delivered to the 'markets' some of the power to set interest rates and exchange rates and so, ultimately, levels and rates of employment and inflation. These powers can be reclaimed, but it is a distant prospect. The best that can be hoped for is the construction of a *geopolitical and economic order* which recognizes a dispersed-power principle appropriate to an age of complex interdependence, and which promotes a decentralized system of governance. In its absence we remain prisoners of a new 'order' wedded to the anarchy and disorder of the markets. The debt crisis is one sign of geopolitical disorder founded on a North–South confrontation and linked to the unregulated expansion of credit monies (Corbridge, 1988b). It is far removed from the 'classical' conflict of East and West.

Meanwhile, the discourse of the post-war geopolitical order plods on. Geopolitics in general is an ordering of spatial inclusion and exclusion (Kearns, 1984). It is about delimiting spheres of influence and defining strategic zones. It is a discourse about geographically defined interests rather than a set of immutable facts about the world. However, any specific discourse of geopolitics has its own timetable, dictated only in part by the timetables of geopolitical order and succession it claims to describe. In the 1980s the dominant bipolar geopolitical order of the post-war years has disintegrated, but no clear alternative has emerged to replace it. So it is in geopolitics as discourse. Synchronic conceptions remain the norm, with the result that disorder in practice is compounded by disorder in representation. At stake are issues of concern to us all. Mikhail Gorbachev recently expressed the hope that the United States and the USSR will not appear to future historians as 'two dinosaurs circling each other in the sands of nuclear confrontation' (Gorbachev, quoted by Shulman, 1987–8, p. 494). We hope so too: that would indeed be a case of old ideas dying hard.

References

Agnew, J. A. 1982: Sociologizing the geographical imagination: spatial concepts in the world-system perspective, *Political Geography Quarterly*, 1, 159–166.

Agnew, J. A. 1987: *The United States in the World-Economy: A Regional Geography*. Cambridge: Cambridge University Press.

Agnew, J. A. and O'Tuathail, G. 1987: Geopolitical order and domesticated space: towards a critical historiography of American Geopolitics. Presented at the *International Studies Association Washington, DC, April 1987*.

Bergsten, C. F. 1982: The United States and the world economy. *Annals of the American Academy of Political and Social Science*, 460, 11–20.

Berman, M. 1984: *All That is Solid Melts Into Air*. New York: Basic Books.

Brett, E. 1985: *The World Economy Since The War: The Politics of Uneven Development*. London: Macmillan.

Bromley, S. and Rosenberg, J. 1988: After exterminism. *New Left Review*, 168, 66–94.

Browne, L. 1986: High technology industry in the world marketplace. *New England Economic Review*, May–June, 21–5.

Browne, L. 1988: Defense spending and high technology development: national and state issues. *New England Economic Review*, September–October, 3–22.

Calleo, D. 1970: *The Atlantic Fantasy: The U.S., NATO and Europe*. Baltimore, MD: Johns Hopkins University Press.

Calleo, D. 1987: *Beyond American Hegemony*. New York: Basic Books.

Chase-Dunn, C. 1982: International economic policy in a declining core state. In W. P. Avery and D. P. Rapkin (eds), *America in a Changing World Political Economy*. New York: Longmans.

Coakley, J. and Harris, L. 1983: *The City of Capital*. Oxford: Blackwell.

Corbridge, S. 1986: *Capitalist World Development: A Critique of Radical Development Geography*. London: Macmillan.

Corbridge, S. 1988a: The asymmetry of interdependence: the United States and the geo-politics of international financial relations. *Studies in Comparative International Development*, 23, 3–29.

Corbridge, S. 1988b: The debt crisis and the crisis of global regulation. *Geoforum*, 19, 109–30.

Cox, R. 1983: Gramsci, hegemony and international relations: an essay in method. *Millenium*, 12, 162–75.

Crockatt, R. 1987: The cold war past and present. In R. Crockatt and S. Smith (eds), *The Cold War Past and Present*. London: Allen & Unwin.

Dalby, W. S. 1988: *American Geopolitics and the Soviet Threat*. Ph.D. dissertation, Simon Fraser University, Burnaby, BC, Canada (unpublished).

Davis, M. 1985: Reaganomics' magical mystery tour. *New Left Review*, 149, 45–65.

Dawisha, K. 1988: *Eastern Europe, Gorbachev and Reform: The Great Challenge*. Cambridge: Cambridge University Press.

Dicken, P. 1987: Japanese penetration of the European automobile industry: the arrival of Nissan in the United Kingdom. *Tijdschrift voor Economische en Sociale Geographie*, 78, 94–107.

Franck, T. M. and Weisband, E. 1972: *Word Politics: Verbal Strategy Among the Superpowers*. New York: Oxford University Press.

Freedman, L. 1985: *The Atlas of Global Strategy*. Oxford: Equinox.

Fujita, M. and Ishigaki, K. 1985: The internationalisation of Japanese banking. In M. J. Taylor and N. J. Thrift (eds), *Multinationals and the Restructuring of the World Economy*. London: Croom Helm.

Gervasi, T. 1988: *Soviet Military Power: The Pentagon's Propaganda Document Annotated and Corrected*. New York: Vintage.

Giddens, A. 1985: *The Nation-State and Violence*. Cambridge: Polity Press.

Gill, S. and Law, D. 1987: Reflections on military–industrial rivalry in the global political economy. *Millenium*, 16, 73–86.

Gilpin, R. 1975: *U.S. Power and the Multinational Corporation*. New York: Basic Books.

Gilpin, R. 1987: *The Political Economy of International Relations*. Princeton, NJ: Princeton University Press.

Gordon, D. M. 1988: The global economy: new edifice or crumbling foundations? *New Left Review*, 168, 24–64.

Gramsci, A. 1971: *Selections From the Prison Notebooks* (edited and translated by Q. Hoare and G. N. Smith). New York: International Publishers.

Greider, W. 1987: *Secrets of the Temple: How the Federal Reserve Runs the Country*. New York: Simon & Schuster.

Haggard, S. and Simmons, B. 1987: Theories of international regimes. *International Organization*, 41, 491–517.

Hall, J. A. 1988: American interests and international political economy. *States and Social Structures Newsletter*, 7, 1–5.

Halliday, F. 1983: *The Making of the Second Cold War*. London: Verso.

Hamilton, A. 1986: *The Financial Revolution*, Harmondsworth: Penguin.

Harvey, D. 1982: *The Limits to Capital*. Oxford: Blackwell.

Herz, J. 1957: The rise and demise of the territorial state. *World Politics*, 9, 473–93.

Higashi, C. and Lauter, G. P. 1987: *The Internationalization of the Japanese Economy*. Boston: Kluwer.

Huntington, S. P. 1987–8: Coping with the Lippmann gap. *Foreign Affairs*, 66, 453–77.

Iklé, F. C. and Wohlstetter, A. 1988: *Discriminate Deterrence: Report of the Commission on Integrated Long-Term Strategy*. Washington, DC: US Government Printing Office.

Keal, P. 1983: *Unspoken Rules and Superpower Dominance*. New York: St Martin's Press.

Kearns, G. 1984: Closed space and political practice: Frederick Jackson Turner and Halford Mackinder. *Environment and Planning D*, 2, 23–34.

Kennedy, P. 1987: *The Rise and Fall of the Great Powers: Economic Change and Military Conflict From 1500 to 2000*. New York: Random House.

Kennedy, P. 1988: Not so grand strategy. *New York Review of Books*, 35 (May 12), 5–8.

Keohane, R. O. 1984: *After Hegemony: Cooperation and Discord in the World Political Community*. Princeton, NJ: Princeton University Press.

Kissinger, H. A. 1968: The central issues of American foreign policy. In *Agenda for the Nation*. Washington, DC: Brookings Institution.

Klare, M. T. 1988: Policing the Third World: a blueprint for endless intervention. *Nation*, 247 (July 30–August 6), 77, 95–8.

Krasner, S. D. 1978: *Defending the National Interest: Raw Materials Investments and Foreign Policy*. Princeton, NJ: Princeton University Press.

Lappé, F. M., Danaher, K. and Schurman, R. 1988: *Betraying the National Interest*. San Francisco, CA: Grove Press.

Lincoln, E. 1988: *Japan: Facing Economic Maturity*. Washington, DC: Brookings Institution.

Lipietz, A. 1987: *Mirages and Miracles: The Crises of Global Fordism*. London: Verso.

Lipsey, R. and Kravis, J. 1987: The competitiveness and comparative advantage of U.S. multinationals, 1957–1984. *Banca Nazionale del Lavoro Quarterly Review*, 161, 147–65.

Mackinder, H. J. 1904: The geographical pivot of history. *Geographical Journal*, 23, 421–42.

McGwire, M. 1988: *Perestroika and Soviet National Security*. Washington, DC: Brookings Institution.

Mendlovitz, S. and Walker, R. (eds) 1987: *Towards a Just World Peace: Perspectives from Social Movements*. London: Butterworth.

Mintz, A. and Hicks, A. 1984: Military Keynesianism in the United States, 1949–1976. *American Journal of Sociology*, 90, 411–17.

Miyazaki, Y. 1987: Debtor America and creditor Japan: will there be a hegemony change? *Japanese Economic Studies*, Spring, 58–96.

Modelski, G. 1978: The long-cycle of global politics and the nation-state. *Comparative Studies in Society and History*, 20, 214–35.

Modelski, G. 1982: Long-cycles and the strategy of U.S. international economic policy. In W. P. Avery and D. P. Rapkin (eds), *America in a Changing World Political Economy*. New York: Longmans.

O'Connor, J. 1973: *The Fiscal Crisis of the State*. New York: St Martin's Press.

Offe, C. 1985: *Disorganised Capitalism*. Cambridge: Polity Press.

O'hUallacháin, B. 1987: Regional and technological implications of the recent buildup in American defense spending. *Annals of the Association of American Geographers*, 77, 208–23.

O'Loughlin, J. 1986: Spatial models of international conflict: extending current theories of war behaviour. *Annals of the Association of American Geographers*, 76, 63–80.

O'Sullivan, P. 1986: *Geopolitics*. New York: St Martin's Press.

Parboni, R. 1986: The dollar weapon: from Nixon to Reagan. *New Left Review*, 158, 3–18.

Parboni, R. 1988: U.S. economic strategies against Western Europe. *Geoforum*, 19, 45–54.

Pastor, R. A. 1988: *Condemned to Repetition: The United States and Nicaragua*. Princeton, NJ: Princeton University Press.

Posen, B. R. 1988: Is NATO decisively outnumbered? *International Security*, 12, 186–202.

Putnam, R. D. and Bayne, N. 1987: *Hanging Together: Cooperation and Conflict in the Seven-Power Summits*. Cambridge, MA: Harvard University Press.

Reich, R. B. 1987–8: The economics of illusion and the illusion of economics, *Foreign Affairs*, 66, 516–28.

Rosecrance, R. 1986: *The Rise of the Trading State: Commerce and Conquest in the Modern World*. New York: Basic Books.

Rosecrance, R. 1987: Long cycle theory and international relations. *International Organization*, 41, 283–301.

Ross, A. 1987: Containing culture in the cold war. *Cultural Studies*, 1, 328–48.

Rupert, M. and Rapkin, D. P. 1985: The erosion of U.S. leadership capabilities. In P. M. Johnson and W. R. Thompson (eds), *Rhythms in Politics and Economics*. New York: Praeger.

Schmeisser, P. 1988: Taking stock: is America in decline? *New York Times Magazine*, 17 April, 24–27, 66–68, 96.

Shulman, M. D. 1987–8: The superpowers: dance of the dinosaurs. *Foreign Affairs*, 66, 494–515.

Slater, J. 1987: Dominos in Central America: Will they fall? Does it matter? *International Security*, 12, 105–34.

Smith, R. K. 1987: Explaining the non-proliferation regime: anomalies for contemporary international relations theory. *International Organization*, 41, 253–81.

Spykman, N. 1942: *America's Strategy in World Politics*. New York: Harcourt Brace.

Starr, H. and Most, B. A. 1985: Contagion and border effects on contemporary African conflicts. *Comparative Political Studies*, 16, 92–117.

Stockman, D. 1986: *The Triumph of Politics: The Inside Story of the Reagan Revolution*. New York: Harper & Row.

Strange, S. 1986: *Casino Capitalism*. Oxford: Blackwell.

Strange, S. 1987: The persistent myth of lost hegemony. *International Organization*, 41, 551–74.

Thompson, W. R. and Zuk, G. 1986: World power and the strategic trap of territorial commitments. *International Studies Quarterly*, 30, 249–67.

Thrift, N. and Leyshon, A. 1988: "The gambling propensity": banks developing country debt exposures and the new international financial system. *Geoforum*, 19, 55–69.

Uchitelle, L. 1988: When the world lacks a leader. *New York Times*, 31 January, Business Section, pp. 1, 6.

Vidal, G. 1986: Requiem for the American Empire. *Nation* 11 January 1, 15–19.

Wachtel, H. 1986: *The Money Mandarins: The Making of a Supranational Economic Order*. New York: Pantheon.

Wallerstein, I. 1980: *The Modern World System II: Mercantilism and the Consolidation of the Capitalist World-Economy, 1600-1750*. New York: Academic Press.

Williams, P. 1987: The limits of American power: from Nixon to Reagan. *International Affairs*, 63, 575-587.

Williams, R. 1983: *The Year 2000: A Radical Look at the Future and What We Can Do to Change It*. New York: Pantheon.

World Bank 1987: *World Development Report, 1987*. New York: Oxford University Press.

Yarbrough, B. V. and Yarbrough, R. M. 1987: Cooperation in the liberalization of international trade: after hegemony, what? *International Organization*, 41, 1-26.

11

World-Power Competition and Local Conflicts in the Third World

JOHN O'LOUGHLIN

Let's not delude ourselves. The Soviet Union underlies all the unrest that is going on. If they weren't engaged in this game of dominoes, there wouldn't be any hot spots in the world. (Ronald Reagan)

Wars are needed only by imperialists to seize the territories of others. American imperialists lay claim to the whole world living under their heel and threaten humanity with a rocket and nuclear war. (Nikita Khruschchev)

In 1977 a simmering and long-standing border dispute between Somalia and Ethiopia erupted into full-scale war, claiming over 25,000 lives by 1980 and resulting in hundreds of thousands of refugees in crowded and miserable conditions along the border of the two countries. The outbreak of the war was only one of a series of disasters and conflicts that have plagued the Horn of Africa since 1968. The great famine of the early 1970s, causing at least 400,000 deaths, the civil wars in the Ethiopian provinces of Eritrea and Tigre (see figure 11.3), the military coups in Somalia in 1969 and Ethiopia in 1974, and the recurrence of famine throughout the region in the 1980s have been only the visible elements of endemic economic and political difficulties in the area. Though barely noticed in the West, preoccupied with its own economic difficulties and the US–Soviet confrontation of a second cold war, the crisis in the Horn of Africa reveals much about the state of international relations in the late twentieth century.

The United States and the USSR became heavily involved in the Somali–Ethiopian conflict as a result of their global rivalry. The second cold war between the superpowers, beginning about 1979 and ending in 1985–6, differed fundamentally from the first cold war (1947–53) in the following ways: the shift of the regional focus of competition from Europe to the 'Arc of Crisis' (Kenya through the Arabian Peninsula to Pakistan); a diminution of the ideological basis of the differences between the superpowers, replaced by an emphasis on their struggle for influence throughout the world; approximate parity in nuclear weapons now accompanied by a renewed emphasis on the feasibility of using conventional forces in scattered locations around the globe; a narrowing of the military superiority of the United States over the USSR; a questioning of American foreign policy and military strategy by its Western allies; perhaps most importantly, a focus on local contests in the Third World (Halliday, 1986). Regional disagreements between the superpowers often have a negative impact on the state of their direct relations. The regional competition is predicated on three background developments: the early post-war zone of competition (in Europe) has stabilized or even ossified; the USSR has developed the capability to aid Third World allies since the 1960s, and its conventional power projection forces are now on a par with those of the United States (McFarland, 1987); the Third World contains the most unstable regimes, anxious to acquire protection and aid for regional and global allies (Litwak and Wells, 1988).

Though we now seem to have entered an era of improved United States–USSR relations, named the third *Détente*, this development must be viewed in the light of continued superpower efforts to improve their positions in the world's regions. Since 1960, 81 major wars, almost all in the Third World, have resulted in the deaths of about 12.5 million people. World military outlays of $930 billion in 1987 equalled the income of 2.6 billion of the world's poorest populations (Sivard, 1988). American foreign policy shifted from multilateralism to unilateralism during the late Carter years and showed few signs of deviating from that course in the Reagan presidency. Unilateralism, the 'America First' syndrome, is the determination to act in world affairs without regard for alliances, international institutions or international law: 'The unilateralist fallacy is that the United States knows the interests of other nations better than they know their interests' (Schlesinger, 1988, p. 77). Not only did Reagan wish to 'roll back' the revolutions in many Third World states but his administration wished to re-establish a clear global power and income hierarchy.

Wallerstein (1987) claims that the Reagan attempt to turn back the tides of history was a failure. Only in Grenada did a clear result, favourable to US interests, appear. However, US aid to anti-government forces in Afghanistan, Angola and Nicaragua, and direct US military intervention in Libya, Lebanon and the Persian Gulf have sent a clear message to the parties involved. Short of a troop commitment that would result in substantial American loss of life, the Reagan presidency made a major, and mostly successful, effort to put pro-Soviet governments on the defensive and to reassert American power in the world's regions. Reagan's inability to persuade American domestic opinion to support a full-scale military involvement (the administration even spent $400,000 on a national survey before deciding that American public opinion would support the air attacks on Libya in April 1986; Mayer and McManus, 1988) precluded the fulfilment of a full restoration of American global power as it was in the 1950s. Though the conjuncture of superpower relations has now swung into the positive mode, the fundamental differences, based on very different social, economic and historical formations, have not eroded and are likely to result in another round of confrontation and arms races at some point within the next decade.

By some accounts, the contemporary Third World is like the Balkans of the 1900–14 period as Great Power tensions are expressed by alliances with local political forces and by delivering large amounts of military and economic aid to their allies (Brock, 1982). In this scenario, just as tensions in the Balkans helped create the climate of suspicion, fear and rearmament as well as the spark that set off the First World War, so, too, the deep and growing involvement of the two superpowers in Third World disputes could set off the Third World War. The crisis in the Horn of Africa, currently less evident than in 1978 but no closer to resolution, is only one of dozens of regional disputes in which the superpowers have taken sides (Chaliand and Rageau, 1983). Among the major tensions, those in Afghanistan, between Israel and its neighbours, in the Horn of Africa, between Vietnam and its neighbours, in Southern African (South Africa and neighbouring Angola, Mozambique and Namibia), in Central America, in the Persian Gulf and in the Korean Peninsula spring readily to mind as cases of potential escalation to hemispheric or global conflict. It is difficult to measure whether the world of 1988 is a more dangerous place than the world of 1950, or even of 1910, but widespread publicity about the nuclear arms impasse between the United States and the USSR, the near-total abdication of the 1953–78 recognition of spheres of influence and the aggressive foreign policy of the Reagan administrations, the independent and growing importance of Third World states in a bipolar world, the continued downturn in the global economy and the widening income gap between rich and poor nations have generated a pervasive sense of global crisis.

In this chapter, I will focus on superpower roles and strategies in the Third World as an important aspect of their hegemonic struggle. Hegemony, as the term is used in this chapter, is a combination of economic and military superiority achieved by the leading global state. Hegemony is usually measured by economic indicators (total gross national product (GNP), percentage of world trade, industrial strength etc.) and military indexes (number and type of nuclear weapons, number of troops, military expenditures and equipment etc.). Measures of hegemonic status and historical developments are described by Goldstein (1988) and Modelski and Thompson (1988). Political influence throughout the world is viewed as a logical extension of economic and military pre-eminence. Most observers agree that the United States emerged from the Second World War as the hegemonic state. America's hegemony was evident until the late 1960s, but since that date the USSR on the political–military side and other capitalist states in the West and East Asia on the economic side have challenged America's hegemony. As a result of the global economic and political changes of the past two decades, the bipolar post-war world on which the foreign policies, geopolitical theories and strategies of states are based has been undermined, and this deconcentration of power has demanded a new geopolitics and new foreign policies (see chapter 10).

Rather than describing the details of the behaviour of the global powers in the 40 or so contemporary wars, I will examine only the conflicts in the Horn of Africa, Afghanistan and Central America to illustrate the general range of commitment, strategy, doctrine and choice faced by the superpowers in areas perceived vital to both (the Horn) and in their own 'backyards'. The three research perspectives in the fields of international relations and geopolitics are reviewed first for their potential theoretical insights into one of the most critical issues of our day – international conflicts involving major powers. The major theme of the chapter, the hegemonic decline of the United States in the late twentieth century and the efforts by the Reagan administration to reverse it, will be developed from one of the three theoretical perspectives. This theory, with a focus on superpower hegemonic struggle in the Third World, is applied to three geopolitical struggles. The conflict in the Horn, now involving both superpowers, and the particular threats to Soviet interests in Afghanistan and to American interests in the Central American isthmus by indigenous popular liberation movements are examined. Some conclusions on superpower conflict in the context of a rapidly changing global arrangement of political and economic power and possible geopolitical analyses complete the chapter.

THEORY IN INTERNATIONAL RELATIONS

The field of international relations is characterized by a bewildering array of approaches, models, ideologies, policies, advocacy positions, disciplinary emphases and publication outlets. The global disaster of the First World War promoted an interest in international relations in Europe and North America. Early research was predominantly of an 'idealist' nature, attempting to understand the causes of international disputes and suggesting policies for their solution. After the Second World War a 'realist' or power politics approach, viewing international behaviour as essentially in the national interest and not necessarily motivated by an ideology, replaced early idealist work. All nations, regardless of location, history, size, political orientations, leadership style, government form, and military and economic strength, were motivated by the same goal – maintenance of world order according to the logic of power. Forming alliances, balancing a strong power by a coalition of weaker states, viewing war as an instrument of foreign policy and using international organizations for national goals were considered part of the normal game of power politics. Diplomatic history was the key discipline, teaching modern practitioners of the art of international politics the rules of the game, the strategies for successes and the lessons of failure.

We can view *Geopolitik* (the school of geopolitics) as the geographic contribution to this *traditionalist school* of international relations, both idealist and realist. A state's location, in both absolute and relative space, was seen as a strong motivating force in its external relations. Going beyond the historical geography of Mackinder's 'heartland–rimland' model, the German geopoliticians in the period following the First World War advocated a national policy of aggression and conquest to achieve regional and eventual world dominance (Kristof, 1960). This development points to one of the major problems with the traditional power politics approach to international relations. Most of its practitioners are strongly motivated by national concerns, adopting a state-centred or ethnocentric policy focus. Henry Kissinger could be considered a classic example in his early academic writings, in his government service in the Nixon and Ford administrations, and in his memoirs. The approach has a strong policy focus, which is not a problem in itself but, in practice, leads to developing a nationalist, often biased, perspective on international conflicts (Agnew, 1983; Paterson, 1987). In the recent work by Cline (1980) and Gray (1988), we can see the geographical role in foreign affairs raised to new heights. An alliance strategy is advocated for the United States by taking each nation's power status, adding up the totals for friends and adversaries (the Soviet Bloc), identifying key global choke-points, such as the Straits of Malacca, the Cape of Good Hope, and the Bab-el-Mandeb, and urging an American commitment to maintaining an aggressive foreign and military policy to ensure the 'Pax Americana'.

A combination of dissatisfaction with the state-centred or unique view of world politics of the traditionalist school and a social-scientific trend towards positivist research in the 1950s and 1960s, with its use of hypothesis testing, acquisition of large data sets and widespread use of statistical analysis, led to the 'hegemony' of the *behavioural approach* in the field of international relations. Currently about 70 per cent of the researchers and key works in world politics in North America are classified as belonging to this group (Alker and Biersteker, 1984). Less than 20 per cent are from the traditional school, and the remaining 10 per cent are from the 'dialectical' school (see below). Behaviouralists share a 'neutral' analytical and empirical perspective on interstate relations. Although specific theories and tests for detailed relationships, such as relating external behaviour to internal domestic politics and the outbreak of war to imperial competition between major powers, have been developed, the field has been labelled as 'non-theoretical'. One of the major behavioural research efforts, the Correlates of War (COW) Project led by J. David Singer, has accumulated a massive data bank on interstate, colonial and civil wars since 1815. In responding to critics, who castigate the project's almost exclusive data orientation, Singer (1981) has defended the work by stating that, because of the complexity of the causes of conflict and the weak knowledge base of the international relations discipline, the best prospects for the field lie in a 'brush clearing operation of some magnitude'. Consequently, thousands of correlation analyses of attributes derived from *realpolitik* (power politics), such as alliance behaviour, geographical contiguity, military and economic capacity and crisis behaviour, and conflict indices, such as war severity or length of conflict, have been produced but no systematic explanation of the causes of conflict seem to be emerging from this inductive approach. By inserting the geographical location factor (measuring the neighbouring effects of belligerents), some increase in explanatory power is achieved and a spatial diffusion effect is detected for some kinds of conflicts (Starr and Most, 1985; O'Loughlin, 1986). Alker and Biersteker (1984) classify most practitioners of the behavioural approach as 'liberal internationalists' since their research perspective of an interaction matrix of all states in the global system, with conflict and co-operation scores in the matrix cells, lends itself perfectly to an interdependence viewpoint.

The '*dialectical*' *school* of international relations is characterized by a variety of approaches but its practitioners share some key perspectives. Although dominant in Eastern Europe and China and making major inroads in Third World scholarship, the dialectical view is still poorly represented in North America. 'Typically, dialectical theorists have valued emancipation, favoured structural or revolutionary change, and have fundamentally challenged the legitimacy of the existing world-order' (Alker and Biersteker, 1984, p. 125). The 'dialectical' school has clear internal dissensions among classical Marxists, such as the *New Left Review* school, world-systems theorists such as the *Monthly Review* school and non-Marxists such as Johan Galtung (1987) who use a dialectical approach to advocate a reordering of global priorities. Wallerstein (1983), whose world-systems project is examined in detail by Peter Taylor in chapter 12, can be classified as a member of the *Monthly Review* school, with his stress on the integration of the global market for commodities, the competitive interstate system, the economic hierarchy from core through semi-periphery, the regularity of growth and decline in the world economy and the geographical diffusion of capitalism from its North-west European core.

Although constantly criticized for his economic fetish, Wallerstein's world view has produced a profound questioning of the whole field of international relations. Empirical tests of the relationship between cycles of economic growth and decline with global conflict have not produced consistent results (but see Goldstein, 1988). The importance of Wallerstein's work lies in his insertion of the economic factor firmly into the core of the debate on the most appropriate approach for international relations. In the Wallerstein view, a differentiated global economic system with a strong hierarchical form, uneven economic development, regular systemic upheavals through global war and the reluctance of the hegemonic state to create a world-empire since it profits more from a world-economy, results in the survival and reproduction of the multicentric interstate system (Chase-Dunn, 1981; Chase-Dunn and Sokolovsky, 1983). This view has been strongly challenged by Thompson (1983a,b), who finds Wallerstein's model deficient as a *political*-economic explanation of interstate conflicts.

As an alternative framework, Modelski's (1978, 1987) long-cycle model of the world-system is gaining notice in the community of scholars in international studies. While it shares many of the characteristics of Wallerstein's model (one world-system, cycles of economic and political growth and decline, core–periphery relations, world-economy versus world-empires), it is most clearly distinguished by its emphasis on political processes and global wars in shaping the global arrangement of power and prosperity. Modelski's framework has been severely criticized for describing only the surface manifestations of global political change in its historical evolution since 1494 and not identifying the 'central engine' of change. If this motor is not Wallerstein's global economic relations, what is it? Recent work by Thompson and Zuk (1986), Modelski (1987), Modelski and Thompson (1988) and other long-cycle theorists (Doran and Parsons, 1980; Gilpin, 1981; Oye, 1983) has elevated the 'hegemonic' explanation to a point that allows academic tests of its propositions using historical data and close examination of the contemporary behaviour of nations in the interstate system. The long-cycle model is adopted in this chapter as offering the best prospects for full explanation of American and Soviet foreign policy and actions in the Third World. It is argued that Presidents Carter and Reagan, faced with the evidence of the *relative* decline in US power, both economic and military, attempted to reassert American global leadership. This attempt took different forms: a military build-up was instituted in 1979 and continues almost a decade later; American allies were pressured to conform to the anti-Soviet line (Johnstone, 1984); the race for allies and influence was intensified in the Third World and notions of interdependence were dropped in favour of singular US interests; perceived Soviet gains were used to 're-educate' the American public about the need to support 'freedom fighters' in a variety of regional contexts; a national feeling of superiority was fostered through public events such as the 1984 Los Angeles Olympic Games. The theoretical formulation, historical manifestations, propositions and threshold of evidentiary support are described in the next section, and the actions of the superpowers in the Third World, especially in the Horn of Africa, Central America and Afghanistan, are interpreted in the light of this model in subsequent sections.

LONG-CYCLE MODELS OF GLOBAL HEGEMONY

Long-cycle theorists believe that the current state of international relations is the product of half a millenium of changing patterns of economic relations. They further believe that the behaviour of an individual state is the result not only of domestic (internal) dynamics but also of the role that the state plays in the world-system of nations. Both the Wallerstein and the Modelski models date the modern world-system from the sixteenth century as the feudal society of Europe was replaced by proto-capitalism and the numerous independent principalities were gradually superseded by larger political units. Modelski nominates 1494 as the critical turning-point. The Italian Wars, which led to the first hegemonic power, Portugal, began then and the immutable shift from the Mediterranean to the Atlantic started. Modelski's world system has five long cycles of hegemonic growth and decline, each decline and growth beginning and ending in a global war. The wars (and the resulting world powers) are the Italian Wars of 1494–1516 (Portugal), the Spanish Wars of 1581–1609 (Netherlands), the Wars of Louis XIV of 1688–1713 (Britain), the Revolutionary and Napoleonic Wars of 1792–1815 (Britain) and the First and Second World Wars (United States). Each war involved all the global powers of the day and each world power faced challenges during the period (consistently about a century) of hegemony.

The phases and dynamics of each long cycle are repetitive. A single state emerges from a global war with a comparative advantage over its adversaries. Modelski (1987) stresses the role of sea power which allows the capacity for global reach to be achieved. Harking back to the study of the nineteenth-century by Mahan (1980), he believes that four preconditions related to sea power must be met before the world power can become the global power. The state must have a protected ocean location, such as Britain on an island, it must be able to provide ocean-going power in the form of a large and powerful navy and a shipbuilding industry, it must develop a strong economy based on ocean trade, allowing the state to pay for the naval costs, and it must possess a strong central government.

Modelski (1987) states that sea power and its attendant economic linkages provided the basis of hegemony, particularly if combined with a strong nation-state. The global power not only possesses the leading economy in the world-system but, through control of the treaty arrangements ending the global war, it legitimizes and reinforces its position as the hegemonic power. The global power matches this political dominance by ordering global economic structures in its favour, particularly through trade and financial mechanisms. After about a generation, the relative advantage of the global power begins to erode as the growing relative capability of its adversaries allows them to formulate challenges to the existing power arrangement. The globe is gradually transformed into a multicentric power division. Although in this second generation – the *phase of delegitimation* – the leading power is still ahead of its international competitors, the struggle for hegemony begins which will eventually deteriorate to global war. The third generation – the *phase of deconcentration* – is marked by oligopolistic competition and, since the major challengers are not ready to press their claims for a rearrangement of global power, this quarter century is the lull before the storm. The last period of the cycle is again a *generation of global war* from which emerges a new global power or a reconstituted global power, such as Britain in the early nineteenth century. In Modelski's (1978, p. 232) memorable phrase, 'We might say that one generation builds, the next consolidates, and the third loses control'.

Given that the primary concern of this chapter is to explicate a model of international conflicts based on hegemonic struggles, the relationship between hegemonic growth–decline and the outbreak of war in Modelski's long-cycle model is of particular interest. Following a 'guns versus butter' argument, Modelski and the advocates of his model argue that, in times of war, less resources are available to promote economic growth. A trade-off between political (including military) spending and economic development funding is expected. In the second (delegitimation) and fourth (global war) phases of the cycle, we see rising prices, resource scarcity and general economic stagnation. Conversely, we see falling prices, ready and cheap access to resources and a rapidly expanding economic sector in the first (world power) and third (deconcentration) phases of the cycle. Interestingly, territorial expansion of the hegemonic state is seen not as an indicator of growth and prosperity, but as a despairing effort to shore up a faltering position. 'Territoriality is the final nemesis of global power. It is a defensive response to the challenge of oligopolistic rivalry' (Modelski, 1978, pp. 229–30). The hegemonic power, on the verge of decline, can ill afford the additional burden of defence costs associated with colonial aggrandisement. Modelski (and Thompson and Zuk, 1986, from their study of British military expenditures) clearly sees a message for the makers of US foreign policy from a close inspection of previous long cycles and the 'territorial trap' into which previous hegemons have fallen. By most accounts, the cycle of American power has now entered the third, or deconcentration, phase. Whether America's leaders, benefitting from the lessons of history and possibly avoiding 'imperial overstretch' as described by Agnew and Corbridge in chapter 10, can postpone the inevitable denouement remains to be seen.

Common to all cyclical models is the designation of the period following the Second World War as that of the 'American hegemony'. Unlike earlier global powers, whose leading position endured for over half a century, most observers believe that US dominance of the world economic, military and political scene had begun to erode by the middle 1960s (Barnet, 1971; Frank, 1981; Harris, 1982; Oye, 1983; Senghaas, 1983; Halliday, 1986). The erosion was not caused by the Vietnam War or by the 'weakness' displayed by American leaders in dealing with foreign powers, as right-wing commentators have claimed. Rather, it was caused by a combination of global economic and political circumstances which the United States could not control. Between 1960 and 1979, the average annual growth rate in GNP in the United States was 3.6 per cent compared with 3.9 per cent in the Federal Republic of Germany and 8.5 per cent in Japan. Between 1950 and 1980, America's share of the world's total GNP dropped from 40 per cent to 25 per cent, as developing nations and its capitalist competitors increased their share. On the politico-military front, the American share of world military spending dropped from 51 per cent in 1960 to 28 per cent in 1980 and its share of armed forces personnel dropped from 13 to 8.3 per cent. (All data are from Oye, 1983.) Explaining these trends has become a major national, as well as academic, problem. It is clear that competition from abroad in the form of rapid economic growth in capitalist nations and military growth in socialist and some Third World nations, rather than absolute American decline, explains the decreasing American shares.

While Kennedy's (1987) imperial overstretch thesis has received the most public attention, that of Gilpin (1981) bears closest relation to the long-cycle interpretation of global power changes. Examining the Athenian, Roman, Dutch, British and American empires, Gilpin (1981), like Kennedy (1987), attributed their decline to a consistent set of three factors: (1) external burdens of leadership resulting in costly and consistent military expenditures; (2) an internal secular tendency toward rising consumption, thereby reducing infrastructural investment; (3) the international diffusion of technology which bridges the gap between the initial advantage accruing to the leading state and its lagging competitors. Oye (1983) provides a strong case for using Gilpin's (1981) trichotomy of the causes of hegemonic decline for interpreting the relative American decline since 1960. Essentially, the argument states that the United States cannot compete successfully with Japan, Western Europe and the newly industrializing countries (NICs) on the economic front and with the USSR on the military front at the same time. Ironically, the Reagan administration, pledged to restore American hegemony, has made matters worse by ignoring Gilpin's causal factors of decline. The administration's expansive military policy has substituted the appearance of strength for deeply rooted growth. The basic problem remains that 'American policy, alone, cannot arrest or reverse structural tendencies toward cyclical hegemonic decline' (Oye, 1983, p. 16). Investment in fixed-capital formation is much higher in Japan, the Federal Republic of Germany and the NICs than in the United States, while spending on the military and on private consumption is much lower. Until these underlying causes of decline are questioned and changed, a reversal in declining American fortunes in the near future cannot be expected.

The same kinds of criticisms levelled at Kennedy can also be directed to other long-cycle predictions. As noted by Haass (1988), the United States wished to reduce its relative share of the world-economy after the Second World War as it tried to rebuild the economies of Western Europe and Japan as bulwarks to perceived Soviet advances. Weidenbaum (1988) challenges the argument that committing 6 per cent of GNP to military expenditure in the United States is causing American slippage in the global economic league. As the relative global economic position of the United States has declined, so has the defence expenditure ratio at home. Russett (1985), Gill (1986), Nye (1988) and others have argued that a total reliance on 'objective indicators' such as productivity indices, GNP growth rates and global economic shares overstates the relative decline of the United States. Rejecting 'historical determinism', they argue that, should the United States wish to do so, it has the capability to reconstitute its international leadership and that the US leadership of the Western Bloc and the Third World allies remains unchallenged. Post-war US hegemony, defined as a coherent internationalist capitalist system under American tutelage, has grown, not declined. The United States was able 'to establish the basic principles of a capitalist system which involved over four-fifths of the world-economy, as well as organizing a system of collective security to maintain political and economic control over that system. It thus achieved both security (defined in terms of peace among the capitalist states) and prosperity' (Russett, 1985, p. 321). Reagan's presidency can therefore be seen as ensuring the continuance of this hegemony by forcing wayward allies into line, ensuring the competitive position of American capitalist interests and directing full attention to the Soviet challenge, both in direct superpower relations and in actions in the Third World.

HEGEMONIC COMPETITION IN THE SECOND COLD WAR

The issue of declining American hegemony has been projected into the consciousness of the American voters in three presidential elections in 1980, 1984 and 1988. While President Carter talked in 1980 about 'national malaise', President Reagan, in his 1984 re-election campaign, stressed that America was again 'Number 1', both economically and militarily, a theme continued by the Republican candidate George Bush in 1988. Whether these views were accurate mattered little; the public perception of a resurgent nation which, in the late 1970s, was reduced to a supine position (hostages were held in Teheran, and the United States was suffering a deep economic recession and was victimized by excessive dependence on imported oil) was turned into political profit by candidates Reagan and Bush. But political campaign rhetoric will not dismiss the reality of the international competition that the United States, the leading global economic and military power, continues to face (Davis, 1984). Unlike earlier hegemonic rivalries which were essentially confined to a handful of major states who offered both political and economic challenges to the existing global power, the contemporary global situation is characterized by multidimensional rivalries. In the politico-military dimension, the United States continues to face a strong challenge from the USSR, a 'rising hegemonic power' (Senghaas, 1983). Ideological differences only serve to reinforce, but do not cause, the rivalry which, in historical terms, is classically hegemonic. On a less well-known level, the United States is challenged by 'Third Worldism', particularly in its Muslim transnational guise, rejecting American (and Soviet) dominance of the world's trade, economic, political, military and cultural relations (Alker, 1981). The interaction between the two superpowers and Third World states provides one of the major conflict arenas in contemporary world politics.

On the economic front, the United States also faces two sets of challenges. As discussed earlier, two of America's major capitalist competitors, the Federal Republic of Germany and Japan, have had higher investment and GNP growth rates since 1960. In Wallerstein's (1983) view, this major capitalist state competition is expected to intensify over the next 20 years as the post-war global economic framework, arranged and dominated by the United States, enters a stage of crisis. In periods of recession, such as that since 1974, this competition becomes intense. One feature of this intracapitalist competition has been a renewed reluctance of America's allies to support US political strategy when it contradicts their own economic interests. The embargo placed on the European suppliers to the Siberian gas pipeline by the Reagan administration in 1982 and the refusal of European states to comply provides a perfect illustration of this intracapitalist tension. Unlike the first cold war, the second cold war between the superpowers has seen an end to the European–American consensus on an anti-Soviet strategy (Mujah-Leon, 1984). Halliday (1986, p. 260) foresees an independent Western Europe that would challenge the bipolar worldview of the United States because it would reduce American strategic power, it would undermine the legitimacy of the Soviet hold over Eastern Europe and it would loosen the bipolar mentality that grips the Third World.

Another competitive front has emerged in the past decade. Some Third World countries, the NICs, have provided a strong challenge to the world market shares held by the older capitalist states, a challenge that continued unabated during the recession of the 1970s. While the United States, Western Europe and Japan together had an average annual growth rate between 1970 and 1978 of 3.3 per cent, the NICs grew by 8.6 per cent yearly and eight nations (Hong Kong, Taiwan, South Korea, Spain, Mexico, Singapore, Yugoslavia, Brazil and Portugal) had an annual growth rate of 15 per cent (Edwards, 1979). Domestic American producers of steel, textiles, electronic goods, ships, cars and other consumer goods, and the workers in these firms, demanded tariff restrictions against cheaper imports from the NICs. Though the governments of the NICs are almost uniformly authoritarian right-wing regimes and though they remain generally supportive of the United States in the second cold war, they nevertheless chafe under American market dominance, demand ready access to Western markets and are willing to pursue their capitalist development strategy strongly, even if it involves an economic war with Western Europe and the United States.

A convenient summary of the global dual competition (East versus West, North versus South) has been provided by Alker (1981). The interactions among the opposing systems dialectically produce social, political, economic and cultural conflicts among the members of the world-system at different points in time. Alker does not accept the Wallerstein core–periphery dichotomy of nations; instead, he sees the globe as comprised of 'competing world order systems' and predicts major shifts in the arrangement of power relations in (heretofore) unexpected ways. Wallerstein (1984) also discusses a possible sea change in global relations with Western Europe (without the United Kingdom) and Eastern Europe forming a bloc confronting a Pacific Alliance of the United States, Japan and China. He admits that the ideological underpinnings of such a shift have yet to be worked out. Although none of the four geopolitical blocs identified by their ranking on the two axes (Soviet socialist, corporatist–authoritarian (NICs), capitalist power-balancing and collective self-reliance) is coherent, the blocs identify preferred major world-order or idealist arrangements. Obviously the importance of the ten possible tensions (four blocs with intergroup- and intragroup tension) will vary spatially and temporarily.

In this chapter our major focus is on East–West tension (or capitalist power-balancing versus Soviet socialism in Alker's term) in the Third World. That tension has military, political, economic and cultural elements, with the military domain divided into nuclear and conventional weapons. Most observers agree that an early American lead in nuclear arms had evaporated by the middle 1970s and that a situation of rough parity currently exists between the superpowers. The Reagan administration's discovery of a 'window of vulnerability' in the nuclear weapons capability of the United States was used effectively to generate domestic political support for the largest military build-up in peacetime. One key element of the build-up was the rapid expansion of non-nuclear forces designed for rapid intervention, costing about 25 per cent of the military budget. It is evident from Reagan's speeches and actions that he regards the cold war as a multifaceted battleground and that, with nuclear stalemate, the greatest opportunities for American successes lie in the non-nuclear arena.

The post-war division of the globe into a hierarchical bipolar arrangement of spheres of influence lingers on in the 1980s (figure 11.1). At the heart of each sphere lies the superpower. Allied politically, militarily, culturally and economically to the superpowers are the core nations of each bloc, such as the North Atlantic Treaty Organization (NATO) countries of Western Europe and the Warsaw Pact countries of Eastern Europe. The ecumene of each bloc designates those countries not allied formally to a superpower but supportive of its position in cold war disputes, such as most Latin American nations for the United States and socialist Third World states, such as Vietnam, Ethiopia, South Yemen and Nicaragua, for the USSR. The remaining area constitutes the group of truly non-aligned nations, which is much smaller than the 'non-aligned' group of over 100 states in the UN General Assembly. Barring a drastic and unexpected shift in post-war trends, the superpowers can rely on the core of their support. Their general geostrategy, then, is (1) to preserve their support in allied nations, (2) to make inroads into the allied bloc of their adversary, through supporting 'progressive forces' (USSR) or helping to 'defend liberty' (United States), and (3) to woo non-aligned nations into their allied bloc or, at least, to counter adversary efforts in the opposite direction. A wide range of weapons – military, economic, diplomatic, trade and even sport – are used to pursue these general strategies (figures 11.2 and 11.3), which, with modest deviations, have remained the same since the Yalta Conference near the end of the Second World War.

It is not possible in the space available to discuss in detail the doctrines and geostrategies of the two superpowers since the Second World War; excellent recent reviews include those of Wolfe (1979), Barnet (1980), Gaddis (1982) and Oye et al. (1983). Halliday (1986) distinguishes four periods in post-war Soviet–American relations: (1) the first cold war, 1946–53; (2) oscillatory antagonism, 1953–69; (3) *Détente*, 1969–79; (4) the second cold war, 1979–85. Each superpower has an evolving foreign policy, modified by ideology, domestic circumstances, international developments and past commitments. For the United States, the Truman Doctrine of 1947 marked the openly stated willingness of that superpower to support potential allies in the 'fight against communism' anywhere in the world. A string of regional alliances advocated and supported by the United States ringed the USSR and China from NATO in Western Europe through the Central Treaty Organization (CENTO) in the

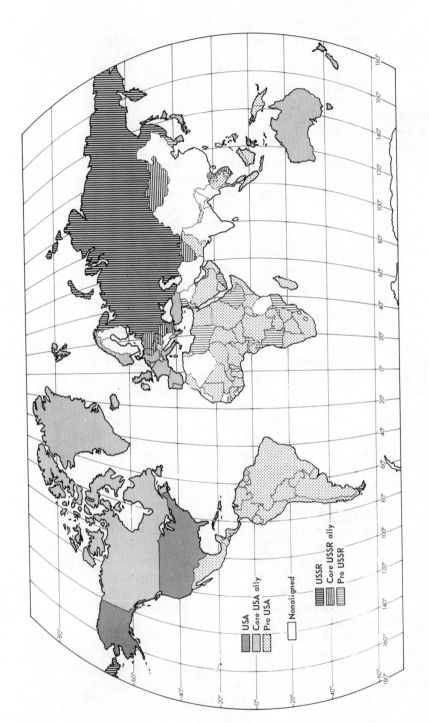

Figure 11.1 Soviet and American allies and friends, 1982

USA
Core USA ally
Pro USA

Nonaligned

USSR
Core USSR ally
Pro USSR

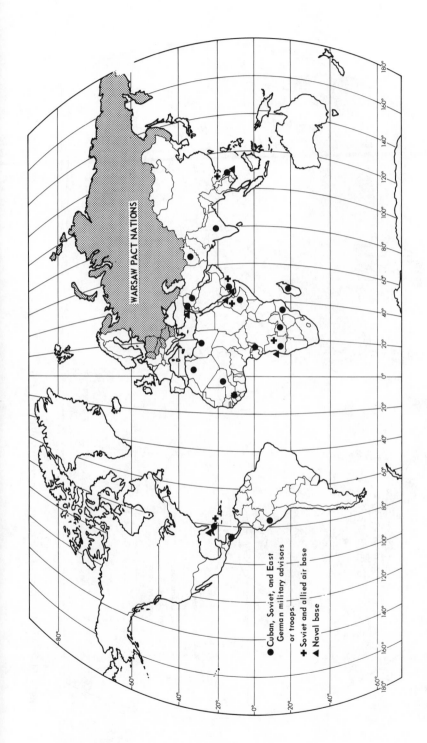

Figure 11.2 Distribution of Soviet and Allied bases, 1982

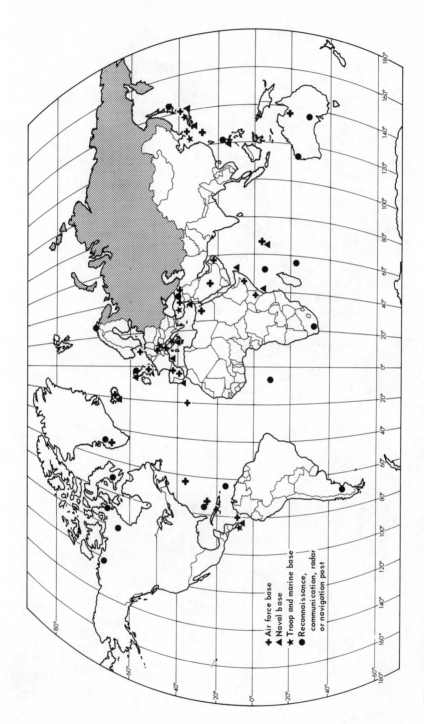

Figure 11.3 Distribution of American military installations, 1982

Arc of Crisis to the South-east Asia Treaty Organization (SEATO) in South-east Asia. American military forces were committed to Lebanon, Berlin, Korea and Vietnam, and economic aid was provided by the Marshall Plan. While the strategy of 'containment' ebbed and flowed from presidency to presidency, the basic doctrine has remained the centrepiece of American foreign policy. A map of the 269 uses of American military forces for political ends since 1945 shows a clear concentration in the 'rimland', the containment semicircle around the USSR (O'Loughlin, 1987). Although in basic agreement with this general containment strategy, the 'globalists' and 'regionalists' in the American foreign policy establishment have debated changes in the world political map and the most appropriate response to these changes (Ambrose, 1982). The regionalists stress the diversity of the globe, the uniqueness of each nation and political circumstance, and maintain a general sympathetic view of the problems facing Third World nations in achieving economic development and political freedoms. Jimmy Carter and his foreign policy advisers up to 1978 epitomized this 'regionalist' perspective, and in the political-geographic field the recent writings of Saul Cohen (1982) develop a geostrategy for the United States based on the notion that interregional variation precludes the application of a spatially invariate global strategy.

Ronald Reagan's election changed accepted post-war doctrines of US foreign policy. Reagan has moved beyond containment to 'win back' some of the nations 'lost to communism'. This aggressive stance demands a huge investment in military and economic weapons. Reagan believes that 'Communist subversion is not an irreversible tide. We have seen it rolled back in Venezuela and, most recently, in Grenada . . . the tide of the future can be a freedom tide. All it takes is the will and resources to get the job done' (Ronald Reagan, 9 May 1984). The second major change since 1980 has been a return to the 'globalist' view of world politics. In his early presidential statements, Reagan castigated the USSR and its allies, the Cubans, for instigating revolution in Africa and Latin America. Most recently, he accepts that conditions of extreme poverty and deprivation may have rendered these nations susceptible to 'communist subversion' and, consequently, has argued that economic aid be part of the foreign policy package. However, he continues to believe that 'the simple fact that we were not deterring [before 1980], as events from Angola and Afghanistan made clear. Today we are, and that fact has fundamentally altered the future for millions of human beings . . . American leadership is back. Peace through strength is not a slogan; it's a fact of life' (Ronald Reagan, 5 April 1984). While Reaganism is still in the ascendancy and the globalist aggressive anti-Soviet policy predominates, America will not live with Third World revolutions, as urged by the late Senator Frank Church (1984). In Reagan's view, regional disputes in the Third World are basically East–West conflicts with no neutral parties (Rothchild and Ravenhill, 1983). Soviet involvement must be opposed by American assistance and, in the case of Africa, 'the question for U.S. strategies is not how does America fit into Africa's perspective but rather how does Africa fit into ours' (Thompson, 1982, p. 1015).

While no doubt politically useful, there remains strong scepticism that vital American interests are at stake in most Third World conflicts. In particular, geopolitical language and spatial concepts are used to exaggerate the Soviet challenge. Johnson (1985) lists six conceptions that are commonly used in American foreign policy pronouncements to lend credence to subsequent American actions: (1) geostrategic theory, emphasizing military preparedness; (2) the domino theory; (3) the concept of challenged American credibility; (4) basic principles of world order that the United States holds dearly, such as 'freedom of beliefs'; (5) the idea of a code of conduct such as agreement on spheres of influence; (6) the idea that Soviet actions are a historic departure from past behaviour, thereby heralding a new round of threats. Underlying all these six concepts is the belief that Third World states are defenceless in the face of Soviet actions, which is demonstratively untrue. Johnson (1985, pp. 48-9) concludes that the best explanation of Soviet actions in the Third World is as 'characteristic responses to unique situations rather than as general decisions to change directions'.

While the national debate over foreign policy in the United States is accompanied by widespread scrutiny by Congress, the media and the public, it is much more difficult to discern the consistent elements of Soviet geostrategy. Professional Sovietologists consume mountains of paper and rivers of ink trying to gauge Soviet behaviour from the nuanced behaviour of members of the Politburo, but it appears that Soviet foreign policy has been remarkably consistent since Stalin's time. The Soviets share many of the same international perspectives as the Americans such as near-paranoia about 'national security', promoting a buffer zone of client states, stability in contiguous regions, especially in the Arc of Crisis, and a general wish to help progressive forces. As Halliday (1982, p. 27) shows, the two main pillars of Soviet foreign policy are written into the 1977 Soviet constitution, namely 'ensuring international conditions favourable for building communism in the USSR' and 'supporting the struggle of peoples for national liberation and social progress'. From Moscow's perspective, the global deck of cards is stacked against them. In the UN General Assembly vote requesting the removal of Soviet troops from Afghanistan, only 18 non-Soviet states voted against the resolution. By any economic measure, the American Bloc is between five and ten times greater than the Soviet Bloc, and even militarily the gap is much larger than Pentagon analysts would suggest (Cline, 1980; Cockburn, 1983). Sticking to the argument that superpower negotiations on matters of mutual benefit such as nuclear arms control should not be linked to other political issues, such as human rights or support for Third World regimes, the USSR has consistently advocated the building of a working relationship with the United States. In contradistinction to the 'linkage' policy of American administrations, the USSR argues that events in Poland, Afghanistan or Nicaragua should not interfere with stable and peaceful relations between the superpowers.

Soviet attitudes towards developments in the Third World have changed twice over the past decade. No doubt hampered by domestic economic weakness and an inability to project military power past a narrow range along its international border, the USSR was essentially a bystander during the decolonialization period. In the 1970s the USSR, while adhering to the Brezhnev Doctrine of protecting its national interests in bordering states, became more willing to take on the role of provider to Third World, especially African, allies (Gavshon, 1981; Rothchild and Ravenhill, 1983). Active in the Horn of Africa, Zimbabwe, Angola, Mozambique, Guinea-Bissau and Libya, the USSR generally acted in accordance with the principle of providing aid within its limited resources to 'progressive governments'. However, it should be stressed that, compared with Western involvement in African affairs, Soviet efforts were indeed modest. Most of the Soviet attention to Third World politics is concentrated in the Arc of Crisis for the most basic of reasons: it shares a long border with one of the world's most unstable regions, from Turkey to Afghanistan (Dawisha and Dawisha, 1982; O'Loughlin, 1986). Just as the Americans express strong opinions about instability in their 'backyard' of Central America, so the USSR is alarmed by events that it cannot control in Afghanistan, Iran and the Arabian Peninsula and in the extended Arab–Israeli conflict. The actions of the USSR in Afghanistan since 1978 must be interpreted in this geopolitical vein.

Since Mikhail Gorbachev's selection as Soviet leader in 1985, Soviet policy towards regional issues has undergone a quickening of the transformation that was evident in the previous half-decade. The costs to the USSR of providing aid to economically desperate states such as Vietnam, Afghanistan, Mongolia, Laos, Kampuchea and Cuba are enormous. Together these six states account for over 80 per cent of Soviet foreign aid (Halliday, 1986, p. 245). Gorbachev is very wary of further Third World commitments, and since 1985 he has tried to find a way of reducing Soviet expenditures. Frankly stated, the USSR cannot afford another Cuba (Shearman, 1986). The USSR is interested in preserving the status quo ante in the Third World while concentrating on its domestic economic problems. In line with the renewed emphasis on economic reforms, the USSR is now courting economically important states in the semi-periphery such as Mexico, Brazil, Argentina and Nigeria (Shearman, 1986). Soviet pessimism about the long-term success of Third World revolutions has increased (McFarland, 1987). Gorbachev offers a major break in style and substance on foreign policy. He appears to have identified four regional points of emphasis: the United States, Europe, the Pacific and the Far East, and the Arctic (Schulman, 1987). Since the early 1980s, two general conclusions have been reached by Soviet foreign policy analysts. First the international environment is not conducive to major socialist or Soviet gains in the Third World and, second, without major reform of the domestic Soviet economy, the country will continue to lack the resources and socio-political organization needed to compete with the West and to ensure socialism's ultimate victory (Kolodziej and Kanet, 1988). To pay for the reforms, the overriding aim of the current Soviet leadership is to reach arms agreements and to extricate itself from previous commitments. At the Washington summit in December 1987, Gorbachev proposed a reciprocal United States–USSR agreement for 'national reconciliation' in Afghanistan and Nicaragua. In September 1988 the Soviet leader proposed a regional quid pro quo to the United States offering to withdraw from its large naval base at Cam Rahn Bay in Vietnam if the United States would close its naval and air force bases in the Philippines. The Reagan administration did not accept the offer.

It should not be assumed from the preceding paragraphs discussing Soviet and American geostrategies and continued rivalry in the Third World (both continue to be the largest arms sellers in the area) that the states of the Third World have been passive and pliant in their relations with the superpowers. One of the main features distinguishing the second cold war from the first was the independent role played by Third World states, many of whom were still colonies in the early 1950s. By inviting the major powers, including the United Kingdom and France, to help them suppress regional separatist movements, repel incursions, launch invasions, defeat domestic opponents, win political victories at home through careful distribution of outside aid and hide their own inadequacy, the leaders of scores of Third World states have drawn their nations and regions closer to the front of the major ideological split in world politics. Both superpowers know from bitter experience that their stay in a Third World country may be curtailed at a moment's notice and major investments over an extended period yield nothing. The Soviet expulsion from Egypt in 1972 and the American expulsion from Iran in 1979 are only two, though major, examples of the uncertainties of First, Second and Third World relations. Despite the enormous differences in power-ranking, the smaller Third World states can often persuade the superpower to accede to regional policies that yield little geopolitical benefit for the superpower while meeting the most salient of the Third World state's local objectives. Thus, Soviet and Cuban aid has been used to suppress the Eritrean revolt and American arms were used by Iran to defeat the Kurds in north-west Iran. The dilemmas of great power involvement in geographically contiguous zones of instability is perhaps illustrated best by the involvement of both superpowers in the Horn of Africa, the Soviet involvement in Afghanistan and the American involvement in Central America. It is to these regional trouble-spots that we now turn our attention.

CONFLICT IN THE HORN OF AFRICA

A glance at a location map of North-east Africa and South-west Asia clearly indicates the strategic location of the Horn of Africa at the entrance to the Red Sea and straddling the Persian Gulf–East African Coast axis (figure 11.4). In recent years the Indian Ocean and its littoral have become an area of concentration of naval and interventionary forces from the United Kingdom, France, the USSR and particularly the United States (House, 1984). The perceived threat to Western oil supplies from the Persian Gulf region (15 per cent of America's supply, 60 per cent of Western Europe's supply and 90 per cent of Japan's supply) was viewed by the Carter administration as a 'threat

Figure 11.4   Location of Ethiopia in the Arc of Crisis

to America's security' and used to justify a deployment of troops to the region. Reagan gave forceful notice of a continued US role in the region's disputes by sending 50,000 troops on a naval task force to the Persian Gulf in 1987 with the ostensible purpose of protecting Kuwaiti oil shipments and with a stated long-term commitment of US involvement in the Gulf.

Zbigniew Brzezinski, Carter's National Security Advisor, coined the phrase 'the Arc of Crisis' for the Pakistan to Kenya semicircle. The region contains five major conflicts: Afghanistan, South Yemen and its neighbours, the Horn of Africa, the Iran–Iraq conflict and the unresolved question of a Palestinian homeland and its wider Arab–Israeli setting. Since the early 1970s, the region has been the world leader in arms imports with the USSR supplying Afghanistan, Iraq, Syria, South Yemen and Somalia (to 1978). It is estimated that between 1975 and 1982 the USSR supplied more than 7,000 tanks and self-propelled guns, 2,330 supersonic combat aircraft and 15,000 surface-to-air missiles to its clients in the region (including India), while the United States supplied its clients with 4,933 tanks and self-propelled guns, 785 aircraft and 6,311 surface-to-air missiles (Manning, 1983). About 60 per cent of all arms purchases in the Third World were in the Arc of Crisis between 1977 and 1980, with Syria, Libya, Saudi Arabia, Egypt, Iraq and Israel occupying six of the top seven positions on the world list of arms importers (SIPRI, 1983). Though arms sales levelled off during 1981–4 and dropped after 1985 (Sivard, 1988), the region continues to possess all the elements of a 'Shatterbelt', where 'internal, geographical, cultural, religious and political fragmentation is compounded by pressures from external major powers attracted by the region's strategic location and economic resources' (Cohen, 1982, p. 226).

From remaining essentially peripheral to the struggle between European powers in the nineteenth century and between the United States and the USSR in the first cold war, the Horn of Africa today occupies a pivotal position in Soviet–American regional competition. In less than a decade, post-colonial North-east Africa disintegrated into continued turmoil and a significant breakdown of traditional alliances. The most important American ally of the 1960s in sub-Saharan Africa – Ethiopia – has become the most important Soviet ally in the 1980s. The combination of civil strife, internal regional dynamics and superpower strategy producing this drastic shift in the Horn is repeated in other regions of the Third World. Beyond arms sales, military advisers and general economic aid, it is difficult to estimate the escalating effects of superpower involvement in local disputes. Frequently, the major power acts as a restraining force on its ally. Both the Great Powers were invited to intervene in the Horn's conflict by regimes seeking to inflict a knock-out blow to traditional enemies, and both the United States and the USSR complied for strategic, as opposed to economic, reasons (Blatch, 1982; Ottaway, 1982; Shaw, 1983). While the stage was set in the late nineteenth century, the plot thickened after Somali independence in 1960 and the showdown occurred in the mid-1970s. We are still awaiting the eventual resolution of the protracted play.

Modern Ethiopia is generally regarded as the creation of Emperor Menelik, who reached an agreement with the European colonial powers, Britain, France and Italy, that ensured Ethiopia's expansion from its traditional highland out to its contemporary boundaries (figures 11.5 and 11.6). This doubling of Ethiopian territory brought large Moslem populations into the same state as Coptic Christians, although Eritrea was only relinquished by the British to Emperor Haile Selassie in the 1950s. While the population of the Ogaden were ethnic Somalis, nomadic and Moslem, the population of Eritrea, now in federation with Ethiopia, was about equally divided between Moslem separatists

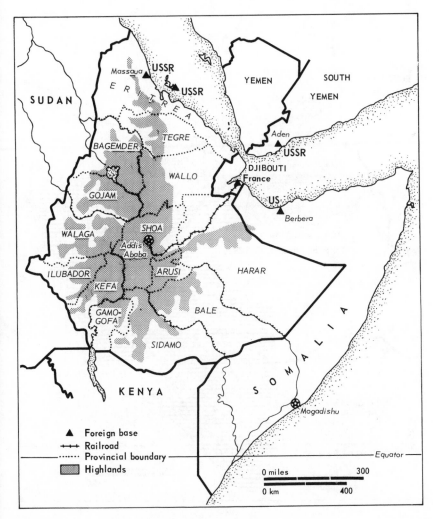

Figure 11.5 Location map of the Horn of Africa

and Ethiopian unionists (figure 11.7). The three European states subdivided the coastal area of the Horn, their major interest in the region until Italy began oil exploration in the interior in the 1930s (see figure 11.5). Somalia was granted independence in 1960 from the old Italian Protectorate (administered by Britain until 1950 and by Italy again from 1950 to 1960) and British Somaliland. The French Territory of the Afars and the Issas became the independent state of Djibouti in 1977 but retained a French garrison, ostensibly to protect the tiny state from Somali irredentist claims and Ethiopian expansionism; for full details of the colonial period and its aftermath see Abate (1978), Wiberg (1979), Halliday and Molyneux (1982) and Ottaway (1982).

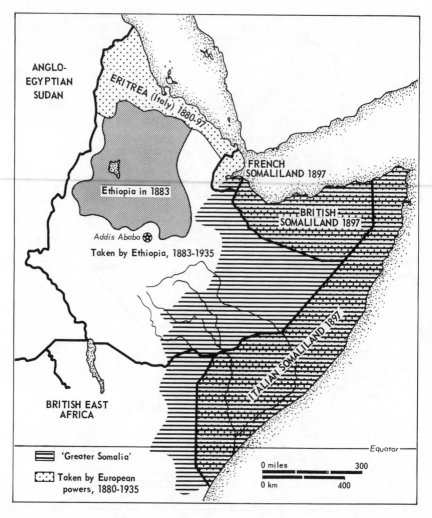

Figure 11.6   Colonial territories in the Horn and 'Greater Somalia'

As is clear from figures 11.5–11.7, the international borders designated by a series of treaties between 1890 and 1910 bear little relationship to the distribution of ethnic or religious groups. The achievement of 'Greater Somalia', incorporating Somalia and parts of Ethiopia, Kenya and Djibouti, has been the goal of Somali foreign policy since independence, resulting in unsuccessful invasions of Kenya in 1962 and Ethiopia in 1977. The Organization of African Unity (OAU) has consistently denied Somali irredentist claims, fearing that granting such a precedent would open a Pandora's box of similar claims and counter-claims across the continent. Ironically, the large majority of African states insist on respect of colonial borders: the nineteenth-century demarcations

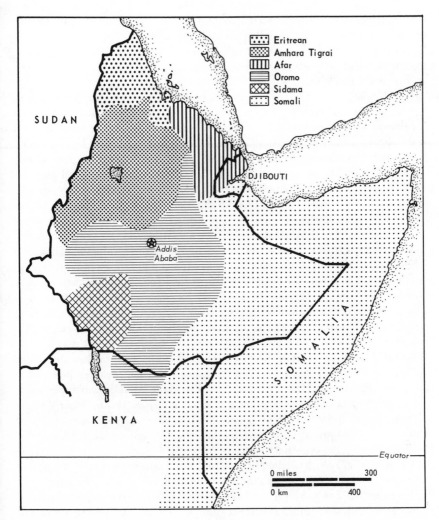

Figure 11.7   Distribution of ethnic groups in the Horn

are today's international boundaries. Somalia, one of the five poorest African countries after independence and two-thirds nomadic, accepted a Soviet offer in 1963 to form, equip and train its army. By 1969 the Somali armed forces were among the best-equipped and largest in the continent. Somalia clearly initiated the military build-up, expecting to pursue its Pan-Somali aims throughout the Horn and to act as a counter to the large amounts of military and economic aid that the United States was providing its traditional enemy, the Ethiopia of Emperor Haile Selassie. A military coup in 1969 brought General Siad Barre to power, determined to pursue vigorously the Somali claim to the Ogaden area of eastern Ethiopia. Siad Barre intensified ties with the

USSR, advocated 'scientific socialism' and allowed the Soviet military the use of a strategic base at Berbera (see figure 11.5). From 1963 to 1978, although Soviet involvement in Somali intensified, there is no evidence that the Soviets instigated events but they clearly benefited strategically from Somali civil politics and regional rivalries.

In Ethiopia, Haile Selassie had consolidated his position since returning to power in 1941 after Italy's defeat in the Horn by the British. Regional breakaway movements in Eritrea and Tigre (see figure 11.5) were contained with a heavy hand (Markatis, 1988), and the Emperor built up a professional army with American aid after 1953, encouraged commercial agriculture, allowed an American base at Kagnew in Eritrea and in general tried to 'westernize' the country using Western technology and American money and advisers. Ethiopia received about half of all US aid to Africa and had the largest number of Peace Corps volunteers. The Emperor pressed the United States for more military aid in the early 1970s when it became clear that Somalia was receiving greater amounts of Soviet weapons and equipment. Like the USSR in Somalia, the Americans essentially reacted to Ethiopian requests and were content to allow Haile Selassie to pursue his strong 'pro-Western' policies. Both countries of the Horn were deeply divided along ethnic and class lines, with political (pro- and anti-Emperor) divisions in Ethiopia and geographical (north versus south) divisions in Somalia also present (Ottaway, 1982, pp. 53–6).

In early 1974, revolution broke out in Ethiopia in the wake of the major drought and resulting famine in the eastern provinces and the attempted cover-up of the disaster by Haile Selassie. In 1974 Haile Selassie was replaced as the governing authority by the Derg, a loose group of 120 members, many of them junior and non-commissioned officers. Land reforms were instituted and local peasant organizations were established. A split in the Derg by May 1976 resulted in the eventual victory of Major (now Colonel) Mengistu Haile Mariam in 1977. With a Marxist student and urban base, Mengistu had to deal with strained relations with the United States, who first rejected a request in February 1975 for military aid but later acceded, a revived autonomous movement in Eritrea and a resurgent Somalia nationalism laying a renewed claim to the Ogaden. The United States was on the horns of a dilemma. To counter growing Soviet influence in the Horn through Somalia, the United States had to rely on Ethiopia, yet the Derg had nationalized US property, continually denounced 'US imperialism', and looked the other way as attacks began on US civilian and military personnel still in the country. The USSR was also caught in a dilemma. Despite a marked shift to the left by the Derg and its supporters and Soviet military aid of $200 million in 1976, the USSR continued to support Somalia because of its long and close relationship with that country and the importance of the base at Berbera. After Mengistu's final triumph in February 1977 and the expulsion of all remaining American institutions and personnel in April of that year, close ties between the Derg and the USSR quickly developed. It was apparent that the Soviets thought that they could maintain their presence in Somalia while developing new ties in Ethiopia, hoping to use their influence in both countries to suppress ancient local rivalries.

Confusion in Addis Ababa was the opportunity that the Eritrean Peoples Liberation Forces (EPLF), the Western Somalia Liberation Front (WSLF) and other separatist movements had been awaiting (see figure 11.5). The Derg could do little to tackle these problems because of the internal power struggle in the capital. The USSR tried to persuade all the 'Marxist–Leninist' movements on both sides of the Red Sea (Ethiopia, Somalia, South Yemen and the EPLF in Eritrea) to form a confederation but failed owing to the (mistaken) belief by the Eritreans and the Somali-backed WSLF that they were close to victory. The USSR was also unsuccessful in persuading Somalia to refrain from invading the Ogaden (Harar province) in 1977. Soviet indecision on whether to support Ethiopia fully was ended when Siad Barre of Somalia repudiated his Treaty of Friendship and Cooperation with the USSR in November 1977. All Soviet technicians and advisers were expelled from Somalia and immediately the USSR began a massive airlift of supplies to Ethiopia (Ottaway, 1982, p. 116). Since then, total Soviet aid to Ethiopia is estimated at more than $4 billion. Up to 16,000 Cuban troops assisted in the defeat of the Somalis and the WSLF in the Ogaden (Harar province). Somalia withdrew from Ethiopia in March 1978, and the Ethiopians then proceeded to reinstate their authority over most of Eritrea. The circle was completed when Brzezinski, President Carter's adviser, declaring that 'SALT [Strategic Arms Limitation Talks between Washington and Moscow] lies buried in the sands of the Ogaden', advocated strong US support for Somalia and military and economic aid began to trickle into Somalia. Berbera became an American base in 1980, an ideal node in the network of Indian Ocean and Persian Gulf installations set up after the Iran crisis of 1978–9. US support for Somalia is still reduced somewhat by the American insistence that the military equipment must not be used to attack Somalia's neighbours and by pressure on Siad Barre to reduce support for the WSLF and to stop continued Somalia incursions into the Ogaden.

Since 1978, events in the Horn and the adjoining North-east African states have emphasized the continued fluidity of the situation. Sudan has complained that Libya and Ethiopia are trying to promote a split in Sudan, based on the north–south Moslem–animist division, and offer aid and sanctuary to the southern rebels in the devastating civil war. It seems clear that both Ethiopia and Sudan are trying to encourage separatism in each other's territory. Both countries harbour refugees from the other's secessionist regions. Border skirmishes persist between Ethiopian and Somali troops, together with their various allies. In 1981 Ethiopia formed a pact with South Yemen and Libya, the source of its oil supplies. In 1988, to direct full attention to the internal strife, especially in Eritrea and Tigre, the Derg reached a temporary settlement with Somalia about border issues, though the Ogaden question remains unresolved.

Ethiopia is beset with seemingly insurmountable problems. Currently over 5 million people are dependent on imported food aid to ward off the effects of the past five years of famine, centred in Tigre. The country has spent over $5 billion on Soviet weapons since 1977 (Kebbede and Jacobs, 1988) and the army has suffered heavy losses to the EPLF. As well as dealing with the two separatist groups in Eritrea, the Derg is also trying to quell autonomous movements in Tigre, Wollo, Gondar and Oromo provinces. Though the famine in the 1980s was caused by a combination of disasters (drought conditions, lack of timely domestic response and poor external relations with the West, which slowed relief aid; Lemma, 1985), its effects have been prolonged by the conflict in the northern provinces. Both rebels and government troops have attacked relief supplies to areas controlled by the other side. The cycle of victory and defeat favoured the rebels in summer 1988, but the Derg was expected to mobilize its forces for another massive effort to end the separatist trend. The United States has kept its distance from the EPLF because they are Marxist and separatist and therefore are not covered by the Reagan Doctrine, which is designed to overthrow Marxist regimes.

The past two decades of Great Power involvement in the Horn have established consistent behaviour of the four powers directly involved. First, the exchange of military and economic aid for military bases has been the basis of the relationships between the two local and the two global powers. Second, both Somalia and Ethiopia actively encouraged outside involvement in their regional disputes. Both superpowers were more than willing to respond. Third, Somalia is clearly the consolation prize in the fight for regional influence. The USSR from 1963 to 1977 and the United States from 1977 looked to Somalia to balance the other superpower's efforts in Ethiopia. Given the huge difference in population size (Ethiopia has a population of 32 million and Somalia only 4 million) and the long history of Ethiopian geopolitical importance in the Horn–Arabian Peninsula region, it is not surprising that both superpowers should consider Ethiopia the greater prize. Fourth, both local powers have managed quite successfully to maintain their independence from superpower pressure. Haile Selassie consistently pressured the United States into greater generosity by making overtures to Moscow. Siad Barre, by joining the Arab League, succeeded in avoiding excessive dependence on Soviet supplies. Ethiopia has expelled Soviet diplomats for interfering in domestic Ethiopian politics and has refused to allow the Soviet Navy better facilities in Eritrea. Somalia has placed the Americans in a quandary by insisting on a domestic policy of 'scientific socialism' and maintaining its aggression toward Djibouti, Kenya and Ethiopia without support from any other African state. It must be concluded that 'the great powers considered them [Ethiopia and Somalia] important enough to make a major effort to force them to comply when reluctant to do so' (Ottaway, 1982, p. 167). Finally, the Horn has become one of a score of regional (border) conflicts in Africa and the Arc of Crisis in which the superpowers have accelerated their involvement in the second cold war. Although the exact location and chronology of events are unique to the Horn, the pattern of aid for bases and influence has been repeated in other tension zones as part of evolving superpower geostrategies. We clearly need a general model of international relations that would help us to understand the complexity of dyadic superpower–local power links, superpower aims in diverse geographical locales and the expected future behaviour of the actors based on historical trends and evolving geostrategies.

THE USSR IN AFGHANISTAN

Numerous American commentators have pointed to the Soviet invasion of Afghanistan as the beginning of the second cold war. More accurately, the Afghan invasion should be viewed as the death knell of *détente* between the superpowers, since the second cold war had already begun with the failure of the US Senate to ratify the SALT II agreement on nuclear weapons. Whatever its timing with respect to the breakdown in Soviet–American relations, the Afghan invasion has become a rallying cry for anti-Soviet feeling through the Muslim and Western worlds. Views on the invasion ranged widely. Indira Gandhi, then Prime Minister of India, described it as a defensive, rather than offensive, move motivated by Soviet fears of the loyalty of its own Muslim population, comprising 25 per cent of the USSR total. In this view, unrest in Afghanistan, fermented by fundamentalist Muslim leaders, might spread across the border to the Muslim population of Soviet Central Asia. The other extreme is represented by many in the West, including a French foreign policy analyst who has written that 'the Soviet invasion of Afghanistan is the most serious and dangerous demonstration of Soviet Marxist imperialism since the end of the Second World War' (de Riencourt, 1983, p. 437). To a large extent, one's analysis of the invasion depends on whether one has a deductive model of Soviet foreign policy (Halliday, 1986). President Reagan, as quoted at the beginning of this chapter, expresses the theory that Soviet foreign policy is essentially expansionist. This deductive approach leads to an interpretation of all Soviet actions in this framework. The accuracy of this view must be tested by a close examination of the chronology of Soviet–Afghan relations and internal Afghan politics before the invasion.

Until the events of 1978–9, Afghanistan was far removed from the centre of Western foreign policy (Poulluda, 1981). Beginning in the 1950s, this landlocked mountainous state with 90 per cent illiteracy developed a close military and economic dependence upon the USSR. US aid and interest in Afghanistan was minimal, focusing instead on Afghanistan's two large Muslim neighbours, Iran to the west and Pakistan to the south. Apart from its shared border with the USSR, Afghanistan seemed to be insignificant in the post-war hegemonic world view of the Americans. The USSR was Afghanistan's main trading partner and, as long as the government and society remained near-feudal and non-communist, the West was content to ignore the Soviet influence. In Afghanistan, the People's Democratic Party of Afghanistan (PDPA) grew out of a small underground urban communist movement and throughout the 1960s and 1970s was on the verge of power in Kabul. In 1973 Mohammed Daoud, the leader of a PDPA faction, became president, and by 1974 he had formed an unsigned alliance with the Shah of Iran. The Shah was intent on exerting influence on his weaker Muslim neighbour as part of a strategy to establish Iran as the regional power in the eastern half of the Arc of Crisis. Savak, the Shah's secret police, operated openly in Afghanistan and Iranian military and economic aid flowed in. In April 1978 the PDPA staged a successful *coup* under the eventual prime minister Hafizullah Amin; Nur Mohammed Taraki was installed as president. Some have seen Soviet involvement in the *coup* but, according to one newspaper account, it was the Shah and not the Kremlin who precipitated the *coup* by urging Daoud to purge the PDPA from Afghanistan (Harrison, 1979). The new regime asked the USSR for increased aid, partly to replace Iranian aid. Soviet concern over events in Afghanistan mounted as the PDPA was unable to achieve stability in Kabul or extend its rule over the countryside.

By early 1979 it was clear that Amin and Taraki were alienating rather than gaining support. In a strongly fundamentalist Muslim state, the regime tried to introduce state control of a traditionally tribal way of life. Factional fights within the PDPA made a difficult situation worse and there is no doubt that throughout 1979 the USSR was active behind the scenes in trying to promote a strong and stable regime in Kabul. By December 1979 the situation of the Amin regime became untenable as the Muslim counter-revolution in the countryside went from strength to strength with backing from Pakistan and post-Shah Iran. Amin invited the Soviet troops into Afghanistan to help Afghan troops repel the counter-revolution, but within a few hours he was dead and Babrak Karmal, leader of an anti-Amin PDPA faction, was installed in power. The presence of over 100,000 Soviet troops seemed to galvanize the Mujahaddin and, with widespread defections from the Afghan army, support from Pakistan, Iran and the United States, an ideal terrain for guerrilla war and general popular support, they have waged a generally successful war against the Soviet troops and the Karmal regime. By 1988 the Kabul government controlled only a few major cities and routeways and its yearly 'Spring offensive' against the guerillas was reduced to nothing by the autumn retreat to Kabul and its environs. After nearly eight years of war, the military and political situation in Afghanistan is hardly changed from December 1979 despite periodic Soviet pushes and the better organization and equipment of the factionalized Mujahaddin.

After a million deaths and with 5 million displaced, including 3 million in refugee camps in Pakistan, the war is supposed to end after the Soviet withdrawal in February 1989. Pakistan was most reluctant to sign the agreement without guarantees about the future of the Afghan government but acceded after US pressure. The agreement also contains pledges by Afghanistan and Pakistan not to interfere in each other's affairs, the voluntary return of the refugees from Pakistan and Iran under UN supervision, guarantees of Afghanistan's future non-alignment by the United States and the USSR, and a Soviet–American understanding that allows Washington to continue military aid to the guerillas (now about $300 million a year, including sophisticated ground-to-air missiles) and Moscow to continue supplying arms to the Afghan government. The seven Mujahaddin factions have not agreed on a post-Soviet-withdrawal government but their expectation is that the Soviet-supported regime will collapse. A return to the weak central authority that characterized the state before 1973 is likely.

This description of events in Afghanistan would suggest that the USSR did not control events but, until Spring 1979, reacted to internal changes in Kabul (Bradsher, 1983). A comparison of the Afghan events with those in Ethiopia described earlier indicates two similarities and two differences. Neither the regime in Addis Ababa nor that in Kabul managed to generate popular support. Both looked to outside powers for assistance in maintaining their position, to the United States in the case of Haile Selassie and to the USSR in the case of Amin. The revolutionary movement was initially centred in urban areas, but whereas the Ethiopian revolution managed to gain the support of the mass of the rural poor, the Afghan regime failed and found itself under pressure from a relentless counter-revolution. The USSR, invited by both revolutionary regimes, found itself welcomed by the Ethiopian population and attacked by the Afghans. Assuming that Soviet intelligence was fully aware of the unpopularity of the Amin regime, the question remains: Why did the USSR send troops into Afghanistan in December 1979, risking world-wide opprobrium (only 18 states supported the USSR in the UN General Assembly vote which followed), the total collapse of *détente*, a revitalized anti-Soviet feeling throughout the Muslim world, possibly defections from the Red Army by Soviet Muslims and the prospect of an unending and ultimately unsuccessful guerila war?

From Tass reports and a *Pravda* interview with Leonid Brezhnev (*Beijing Review*, 17 January 1983; Medvedev, 1980), the official Soviet explanation of the Afghan invasion consists of three points. First, in accordance with the Soviet–Afghan Treaty of Friendship and Cooperation signed earlier, the USSR sent troops at the request of the Afghan government. Second, the Afghan government requested the troops because the Amin regime was in a critical state as a result of the successes of the counter-revolutionaries: the Afghan army simply could not cope. Third, the counter-revolution in Afghanistan is supported by the United States, China and Pakistan and, in line with Soviet principles, the USSR will support progressive forces against counter-revolutionaries. While this justification adequately describes surface reality, it does not explain the motivation for such an enormous commitment nor does it explain how and why Amin, who invited the Soviets, was removed within hours. It is clear that the USSR wavered for a year before it eventually acted in December 1979. It faced Hobson's choice: either withdraw all civilian and military personnel, resulting in the replacement of Amin by a fundamentalist Muslim regime hostile to the USSR, or plunge in with troops and a permanent commitment to control events in Kabul. The latter was chosen as the lesser of two evils (Medvedev, 1980).

Two views of the Soviet invasion of Afghanistan – the defensive reaction view and the aggressive geostrategic view – were outlined earlier. Clearly, it is impossible to estimate precisely to what extent either view is accurate. Some Western analysts think that events in Afghanistan made possible the Czarist dream of a warm-water port on the Indian Ocean. Baluchistan, the south-western province of Pakistan, has seen periodic attempts at secession, and some strategists believe that Soviet encouragement of this separatist movement will bring dividends in the form of a pro-Soviet state on the Indian Ocean: 'The eventual prize is remarkably attractive; a 200-mile coastline along the Indian Ocean, naval and air bases at Gwadar, Pasni and Ormara, from which Soviet fleets could operate at the mouth of the Gulf and linking up with their existing bases in Aden, Socotra and Ethiopia at the entrance to the Red Sea – in effect, the Soviets would have the potential to interdict entrance or exits from both the Gulf and the Red Sea' (de Riencourt, 1983, p. 433). In this general view, which at its most liberal sees the USSR taking advantages of geopolitical opportunities, events in Afghanistan provided an ideal pretext for Soviet expansion to the south and for pressure on Iran from a second side. The USSR is perceived as possessing a grand design to impose its will on post-Khomeini Iran, and events in Afghanistan and Baluchistan are simply the first act of the larger South-west Asian drama.

For the defensive reaction school, Soviet actions were rendered essential by the rapid deterioration in Kabul after the *coup* in April 1978. Just as the USSR interfered militarily in Hungary in 1956 and in Czechoslovakia in 1968, and politically in Poland in 1981–2, to retain a pro-Moscow Communist Party in power, so it installed Karmal in Kabul. In each case the country involved bordered on the Soviet homeland. In a corollary of this model, if pre-Soviet governments in other parts of the world were overthrown, the USSR would not interfere militarily, either by choice, since the threat is far removed, or because of a lack of military capability in the face of Western opposition or the tyranny of distance. The demise of generally pro-Moscow governments in Mali (1968), Ghana (1966), Zanzibar (1963), Iran (1953), Guatemala (1954), Chile (1973) and Grenada (1983), and of pro-Soviet movements in Iraq (1978), Sudan (1971) and Oman (1975), to the accompaniment of Soviet inaction, lends credence to this view. The lesson of the Soviet invasion of Afghanistan (and American actions in Vietnam earlier) is that 'national liberation forces, given adequate internal and external support, can prevent an invader from controlling the countryside and can inflict losses that gradually appear unbearable and deny victory at a tolerable price' (Hoffman, 1988, p. 27). The USSR, adhering firmly to a principle of a 'cordon sanitaire' of states on its frontiers, will plunge if the perceived costs of inaction are greater than the costs of military action. In this respect, Soviet geopolicy has not changed much since the 1940s. Close parallels exist with American policy in the Central American isthmus to which we now turn.

CENTRAL AMERICA: AMERICA'S AFGHANISTAN

Before he even took office, President Reagan signalled to all that he viewed Central and Latin America as a special geopolitical challenge. The 1980 Republican Party platform condemned the 'Marxist Sandinista takeover of Nicaragua and the Marxist attempts to destabilize El Salvador, Guatemala and Honduras'. As the new US ambassador to the United Nations, Jeanne Kirkpatrick, said: 'Central America is the most important place in the world for the United States today'. In dozens of speeches Reagan and his top policy-makers attempted to emphasize the proximity of the Central American isthmus to the United States, referring to the region as 'our backyard' or stating that San Salvador is closer to San Diego or Houston than either American city is to Washington, DC. Every new American president of the twentieth century has started his term with an initial strong interest in Latin America, but Reagan seemed to go beyond this historic commitment. In speed and in action, Reagan made it plain that Central America was a special case for American foreign policy; that the new cold war would be fought out in El Salvador and Nicaragua without directly involving American troops; that the USSR and its surrogates, Cuba and Nicaragua, were responsible for the unrest in the region; that military, rather than economic, solutions must be found to the region's problems; that the Carter administration had erred grievously in its withdrawal of support for the Nicaraguan dictator Somoza and its cool relations with pro-American right-wing regimes in Chile, Argentina, Brazil, Guatemala, Honduras and Paraguay; and that domestic American public opinion would be convinced of the seriousness of his commitment to restoring American dominance, after the malaise of the Carter years, by recreating the 1920s American hegemony in Latin America (Black, 1982; Feinberg, 1982; Pearce, 1982; Dickey, 1984; Lafeber, 1984).

Apologists for US imperialism in Latin America point to the Monroe Doctrine of 1823 as the guiding document and historical-legal justification for their position, though why a statement that any interference by an outside power in Latin America would be viewed as 'the manifestation of an unfriendly disposition toward the United States' should gain near-Biblical reverence is not clear. By 1900, the expanding economic imperialism of the United States had swelled south across the international border in search of new markets, resources, investments and opportunities (Hoffman, 1982; Kurth, 1982). European, particularly British and German, interests were swamped. To ensure quiescence in the region American military force was used over 135 times between 1823 and 1983 (*Congressional Record*, 1969, quoted by Conway, 1984) including Cuba (1898), Puerto Rico (1901), Honduras (1905), Cuba (1906–9), Panama (1908), Nicaragua (1909), Honduras (1910), Panama (1912), Cuba (1912), Honduras and Nicaragua (1912), Haiti (1914–34), Dominican Republic (1916–24) Nicaragua (1916–33), Cuba (1917–23), Panama (1918), Honduras (1919, 1924), the Dominican Republic (1965) and Grenada (1983). In addition to these direct military actions, the United States was active in undermining reform or leftist governments such as those of Arbenz in Guatemala in 1954 and Allende in Chile in 1973. In the 1930s the so-called 'good neighbour' policy of Franklin Roosevelt replaced the military option as the American strategy, with a shift to using American economic power to provide the support system necessary for capitalism to develop in the undeveloped economies of the hemisphere. The Alliance of Progress in 1961 continued this 'liberal' notion of promoting economic development, but unexpectedly mild economic progress 'weakened the system by raising both the expectations of the masses and the wealth and power of the oligarchs. Revolutionary groups multiplied, the United States responded with increased military commitments' (Lafeber, 1984, p. 24). By responding to these movements with military force, the United States has precipitated further violence. In each term, the new president, starting with a good neighbour policy and reform ideals, has reacted with surprise and then anger at their failure and eventually come round to a unilateral use of force and boycott to re-establish American dominance. Carter was the latest of this line of erstwhile reformists.

Central America (Honduras, El Salvador, Nicaragua and Guatemala) became the focus of Reagan's 'aggressive globalism' in Latin America. The situation is complex but basically unchanged from the beginning of Reagan's term in 1981. Nine battlefronts, including both civil and clandestine interstate wars, all have American involvement to a greater or lesser extent (*Economist*, 29 October/5 November 1983). In Guatemala a right-wing military regime faces small, though growing, opposition with pockets of the country controlled by left-wing guerrillas with significant Indian support. Because of the human rights provisions attached by liberals in Congress to American military aid, Guatemala has not been able to receive open American aid since 1977. El Salvador has experienced a massive civil war since 1979, with three sets of actors. The extreme right, strongly entrenched in the upper class, the army and the National Guard and led by Major Roberto d'Aubisson, was supported by 46 per cent of the voters in the presidential election of May 1984. Their only support in the United States came from the Republican right wing, led by Senator Jesse Helms of North Carolina. The Reagan administration, following in Carter's steps, supported the titular head of state, President Jose Napolean Duarte, the centrist candidate, pressured by both left and right. His power base is suspect, as the moderates have been eliminated or forced to choose between left and right in the civil war. Duarte has managed to obtain a large increase in US economic and military aid, which is now over $200 million a year, putting the country among the top five recipients of US foreign aid. Duarte has met in negotiations with guerrillas with no success and he has made tentative steps towards land reform. His goal is to establish a centrist presidency so that it does not fall victim once again to a right-wing *coup*, as was the case in 1972. The left-wing guerrillas control a third of El Salvador's provinces and strike country-wide. Without strong American support for the government, the guerrillas would have repeated the 1979 Sandinista success in Nicaragua. Civil war looks like a long-term probability in El Salvador.

In Honduras, over 2,000 American troops remain after the well-publicized joint military operations with Honduran forces in 1982 and 1983. Honduras, with a right-wing strongly pro-American government and no organized opposition, has become the central staging ground for American military efforts in the whole region as well as the base for the most important Contra (anti-Sandinista) group, the National Democratic Front, led by ex-supporters of Somoza. Nicaragua has become the largest thorn in America's foot. In 1979 the Sandinista guerrillas, after winning the support of the majority of the population, replaced Somoza, whose father had been installed by the Americans in the 1930s. The Carter administration adopted a generally passive stance and sent some economic aid to war-torn Nicaragua. The European socialists were particularly supportive of the Sandinistas in the period 1979–81 through aid and political support in the Socialist International. Reagan, immediately upon taking office, withdrew all American aid and proceeded to help organize and arm the anti-Sandinistas, now mostly in Honduras. Twice the United States airlifted Honduran troops to the Nicaraguan border and rushed extra troops to the country when the Sandinistas crossed the border in pursuit of the Contras. Faced with an outside threat and economic sabotage at home, the Sandinistan government became more repressive and began to lose European and American liberal support. By 1982, former members of the Sandinista junta had left and formed a second Contra front in Costa Rica, alleging that the junta had sabotaged the revolution by adopting a strong Marxist–Leninist line with close ties to Cuba (Cruz, 1983). This second front has since disintegrated as a result of internal Contra quarrels. The Sandinistas won a large majority of seats in parliament in the national elections of November 1984, and they have curtailed their support for El Salvador's guerrillas. In 1985 they won a judgement against the Reagan administration in the International Court of Justice on the secret mining of Nicaragua's harbours. Since 1986 the US government has armed the Contras openly, though Congress has vacillated on the issue. Hit-and-run strikes by the Contras have severely crippled the country's economy, making Nicaragua even more dependent on Soviet aid. Nicaragua had an inflation rate over 1000 per cent in 1987 and the highest per-capita foreign debt in the world. Since 1979 it has received over $2 billion in Soviet foreign aid. After maintaining that he was trying to restore democracy in Nicaragua, Reagan has admitted since 1986 that his aim was the overthrow of the Sandinistas.

Within the United States and the Western community in general, two dominant and one subordinate view are used to analyse the Central American conflicts. The conservative or Reaganite stand, as exemplified by the report of the National Bipartisan Commission on Central America (Kissinger, 1984) and recent official statements by the Reagan administration, became more sophisticated than the earlier anti-Soviet statements (O'Tuathail, 1986): 'We can and must help Central America. It's in our national interest to do so' and 'if we do nothing or if we continue to provide too little help, our choice will be a Communist Central America with additional Communist military bases on the mainland of this hemisphere and Communist subversion spreading southward and northward . . . . There is a way to avoid these risks. It requires long-term American support for democratic development, economic and security assistance and strong-willed diplomacy' (President Reagan, 9 May 1984). Current US policy sees the USSR and its ally Cuba like thieves in a hotel corridor, checking doors to see if one is unlocked and then stealing from the room (Halliday, 1986). The administration now claims that economic underdevelopment presents the opportunity for Soviet exploitation of the resentment against local elites and American interest. To remove this economic cause of the conflict, the administration has proposed spending $8 billion in the region over the next five years, continued strong military aid to governments in El Salvador and Honduras, and covert support for the Contras (or freedom fighters, as Reagan prefers) in Nicaragua.

The liberal view, shared by European socialists and liberal Democrats in the US Congress, sees conflicts in the Third World as locally generated, resulting from indigenous social and economic causes. Further, they believe that bringing the East–West competition to the Third World, as the Carter and Reagan administrations have done, will hurt agreement on other issues, such as nuclear arms control, and reflects an effort by the United States to restore its hegemony, particularly in Central America (Mujah-Leon, 1984). As a US foreign policy, the liberals wish to implement a true 'good neighbour' policy with acceptance of Latin American political independence, granting of significant economic development aid and ready access to US markets, diplomatic negotiation to end current conflicts and encouragement of leftist regimes to move towards the centre through economic study. The difficulty that the liberal approach faces is that 'reformism' had died in Central America by 1982 (Fagen and Pellica, 1983) and the only choices remaining are essentially right or left. It is not clear to what extent the liberal viewpoint is supported in the United States. A consistent majority of the electorate does not want American troops committed to the conflict in Central America, but President Reagan has also generated a majority in support of his analysis of the causes and solutions of the conflict.

The Marxist views are varied but generally based on an economic interpretation of American involvement in Central America. The 'political economy of late imperial America' is in decline, so that an 'ultra-imperialist military and economic order' around the globe is threatened by the kinds of political and economic forces described earlier in the chapter. American economic interests in Central America and the Caribbean are substantial (Barry et al., 1983), but direct US investment in the area is only 2 per cent of the Latin American total so that a simple equation of economic interest and military intervention is not accurate. The large foreign policy bureaucracy in the United States responds to the 'national interest', which is some combination of strategic, political and economic interests (Krasner, 1978). While, in general, this establishment works to protect and promote American economic interests, the individual policy-makers are motivated to act in a particular fashion by mainly non-economic realities such as balance-of-power strategy, geopolitical rationales and historical and ideological commitments, and, as seen earlier, are controlled by as much as they control events in the Third World.

As phrased by the *The Economist*, the Reagan administration faced three choices in Central America: flinch, plunge or stick. Withdrawal was always unlikely given the political commitment of the Reagan administration to preventing the Nicaraguan revolution from spreading and to taking an aggressive stand against the Sandinistas. Plunging or sending in American troops was equally unlikely, given the undoubted political damage at home. (Remember the caution that required a national public opinion survey before the bombing of Libya.) Reagan repeatedly denied having any plans to commit American troops, but the widespread public approval of the Grenadian invasion of October 1983 suggests that Americans would support a quick and successful strike. The implausibility of this kind of operation in Nicaragua probably precludes an American invasion. The third option – sticking – is most likely, continuing current commitments at a higher level to American friends in the region. Essentially, the Kissinger Commission recommendations have been implemented: American allies in the region are receiving massive amounts of economic aid.

The Arias peace plan was signed by the five Central American presidents in August 1987. The plan has seven requirements of the signatory states, i.e. national dialogue with their respective opposition, amnesty for political prisoners, ceasefire in the wars, democratization of the political party process, free elections, renewal of arms reduction talks, suspension of overt and covert aid to armed insurgents in neighbouring states and denial of the use of their territory by guerrillas. The Reagan administration dragged its feet on the agreement and barely concealed its displeasure with the meeting between the Contras and the Sandinistas in spring 1988. The talks have gone nowhere: though the Contras are severely handicapped by the Congressional suspension of military aid, the Sandinistas need a ceasefire in order to concentrate on economic development. In this effort they have been urged by the USSR, which wishes to reduce its Nicaraguan allotment. Permanent peace still looks as remote as ever in Central America.

The constant use of the term 'geopolitical relationship' by the Reagan administration to describe the connection of the United States to Central America propelled the geographical element in foreign policy to national attention. Reagan has projected an 'aggressive confrontationism' with the USSR in American foreign policy. Central America is literally the front line in the confrontation. For practical and analytical purposes it matters little if Soviet and Cuban involvement in the region is much less than the administration asserts. In this instance, regional geostrategy responds to image–a part of a global picture. Reagan not only wanted stability of Central America on America's terms but he wished to reassert American hegemony by denying the legitimacy of claims for social reform, an economic redistribution and a growing independence from American economic dominance. The combination of the North–South and East–West global political and economic divisions is nowhere more clearly illustrated than in the American–Central American conflicts.

CONCLUSIONS

Five conclusions seem warranted on the basis of this wide-ranging review of international conflicts. It is clear that a purely American, Soviet or Third World perspective lacks the objectivity that is so desperately needed in international politics. There is enough blame for causing present and previous conflicts to be spread around all the participants, both local and non-local. Leaders in Third World states, anxious to defeat external enemies or score internal political victories, blithely invite the superpowers to help achieve success and draw their state into the front line of global competition. It would not be going too far to say that they make themselves pawns on the chessboard of hegemonic struggles.

The first major conclusion of this chapter relates to the declining hegemony of the United States. Economic and military evidence abounds of the declining share of the world's capacity in the United States. Even the Reagan administration accepts that this trend was characteristic of the late 1960s and 1970s. They see it as a challenge to which they must respond and restore American hegemony. This aggressive position offers the challengers, the USSR politically and the Western and non-Western capitalist nations economically, little option except to respond. The US political establishment seems unable to adjust to the realities of the late-twentieth-century globe with the growing diversity of the Third World and the growing numbers of power centres, especially on the economic front. Locked in a 1940s cold war mentality, the United States in the late 1980s cannot accommodate the new economic demands of the NICs. The United States 'should pay more attention to Brazil, Mexico, and Venezuela than to Cuba, Nicaragua, and Grenada' (Lowenthal, 1983, p. 333).

The second conclusion concerns the argument that both superpowers are waging 'gangster diplomacy' (Barnet, 1984a). As we have seen in three case studies, the superpowers are willingly dragged into local disputes. Though not initiating the disputes, they exacerbate the problem through wholesale commitment to one local actor: 'Yet while each self-styled peace-loving country permits itself anything that it can get away with, each holds its ideological enemy to the strictest ethical standards' (Barnet, 1984a, p. 53). Pope John Paul II, in the February 1988 encyclical *Solicitudo Rei Socialis*, castigated both superpowers for 'imposing imperialism' on Third World states, reducing them to 'cogs on a gigantic wheel'. He saw East–West rivalry as causing the widening gap between rich and poor countries and concluded that 'world peace is inconceivable unless the world's leaders come to recognize that interdependence in itself demands the abandonment of the politics of blocs, the sacrifice of all forms of economic, military or political imperialism, and the transformation of mutual distrust into collaboration'. Even more worrying is the evidence that, even in times of *détente* such as the decade 1969–79, the superpowers continue to compete aggressively in the arena outside their immediate spheres of influence. The future looks as bleak in this regard as the period since decolonialization has been (Klare, 1981).

The third conclusion is that by the waning years of the Reagan presidency and the accession of Gorbachev, the superpowers were beginning to count the costs of Third World involvement and concluding that the area is 'an unrewarding terrain for competition: a trap, not a prize' (Hoffman, 1988, p. 27). Both superpowers needed a decade to learn that the costs of war in Vietnam and Afghanistan were too much for domestic interests. Now the United Nations has been revived as an international forum for settling disputes and patrolling conflict zones. However, these encouraging developments should not be overestimated. Chapter 2 of the report of the Commission on Integrated Long-Term Strategy of the US Department of Defense, released in early 1988, conceded that 'US forces will not in general be combatants' but recommended 'support for anti-communist insurgencies, higher levels and greater flexibility for security assistance, the development of "cooperative forces" with third-world allies, the maximum use of US technological advantages and the development of alternatives to overseas bases'. As the famous baseball coach Yogi Berra said, '*Déjà vu* all over again'.

Although this chapter has focused on political competition and conflict, i.e. the East–West dimension in world politics, the North–South conflict looms as the most destabilizing element in world affairs. Despite pleas from international bodies (Brandt Commission, 1980) and from two-thirds of the world's nations for a new arrangement of economic power, including immediate access to Western markets, there has been no progress towards reducing global inequities. In advanced capitalist nations, demands for trade barriers to restrict imports of cheaper products from non-Western nations are growing rapidly. The current global debt crisis has concentrated the world's attention on the fragility of the global economy and the threat posed to Western capitalism by widespread defaults. While this debt crisis can be resolved, or at least postponed, its structural causes remain untackled. While the USSR sees this crisis as essentially caused by capitalism and itself as an interested bystander, there is little doubt that a world economic crisis would affect even the socialist economies of Eastern Europe. An international conference, involving nations of the East, West and South, to arrange a new economic and financial system, without enshrining the leadership of any one state, along the lines of the Bretton Woods Conference of 1944 is desperately needed.

The final conclusion of this chapter is more a proselytizing message to the geographic community. The renaissance in political geography in Anglo-Saxon nations over the past decade has left the traditional interest of the discipline–geopolitics–by the wayside. John House (1984) has recently shown the way for political geographers to engage in regional analysis of global conflicts. Regional conflicts are numerous, complex, growing and interwoven with events elsewhere in the world. Continued neglect of these conflicts–the major research interest of the field–places the whole renaissance of political geography in jeopardy. This chapter is only a small step in the long march of political geography to a respected position in the social sciences.

328   JOHN O'LOUGHLIN

References

Abate, Y. 1978: Africa's troubled Horn: background to conflict. *Focus*, 28, 1–16.

Agnew, J. A. 1983: An excess of 'national exceptionalism': towards a new political geography of American foreign policy. *Political Geography Quarterly*, 2, 151–66.

Alker, H. R., Jr. 1981: Dialectical foundations of global disparities. *International Studies Quarterly*, 25, 69–98.

Alker, J. R. Jr. and Biersteker, T. J. 1984: The dialectics of world order: notes for a future archaeologist of international savoir faire. *International Studies Quarterly*, 28, 121–42.

Ambrose, S. E. 1982: *Rise to Globalism: American Foreign Policy, 1938–1980*, 2nd revised edn. New York: Penguin.

Barnet, R. J. 1971: *The Roots of War*. Harmondsworth: Penguin.

Barnet, R. J. 1980: *The Lean Years: Politics in the Age of Scarcity*. Washington, DC: Institute for Policy Studies.

Barnet, R. J. 1984a: Beyond gangster diplomacy. *Mother Jones*, February–March, 53–5.

Barnet, R. J. 1984b: Why trust the Soviets? *World Policy Journal*, 1, 461–82.

Barry, T., Wood, B. and Preusch, D. 1983: *Dollars and Dictators: A Guide to Central America*, 2nd edn. New York: Grove.

Black, G. 1982: Central America: crisis in the backyard. *New Left Review*, 135, 5–34.

Blatch, G. S. 1982: Im Horn von Afrika ist die Zukunft nie gewiss: Anmerkungen zur Ogaden-Frage. *Beitrage zur Koufliktforschung*, 12, 69–90.

Bradsher, H. S. 1983: *Afghanistan and the Soviet Union*. Durham, NC: Duke University Press.

Brandt Commission 1980: *North–South: A Program for Survival*. Cambridge, MA: MIT Press.

Brock, L. 1982: World power intervention and conflict potentials in the Third World. *Bulletin of Peace Proposals*, 13, 335–42.

Chaliand, G. and Rageau, J.-P. 1983: *Atlas stratégique: Géopolitique des rapports de forces dans le monde*. Paris: Fayard.

Chase-Dunn, C. 1981: Interstate system and capitalist world-economy: one logic or two? *International Studies Quarterly*, 25, 19–42.

Chase-Dunn, C. and Sokolovsky, J. 1983: Interstate systems, world-empires and the capitalist world-economy: a response to Thompson. *International Studies Quarterly*, 27, 357–67.

Church, F. 1984: It's time we learned to live with Third World revolutions. *Washington Post Weekly Edition*, 26 March, p. 21.

Cline, R. S. 1980: *World Power Trends and U.S. Foreign Policy for the 1980s*. Boulder, CP: Westview.

Cockburn, A. 1983: *The Threat: Inside the Soviet Military Machine*. New York: Random House.

Cohen, S. B. 1982: A new map of global geopolitical equilibrium: a developmental approach. *Political Geography Quarterly*, 1, 223–42.

Conway, D. 1984: Grenada–United States relations: Part I: 1979–1983, prelude to invasion. *USFI Reports*, 39, 1–9.

Cruz, A. J. 1983: Nicaragua's imperilled revolution. *Foreign Affairs*, 61, 1031–47.

Davis, M. 1984: The political economy of the late imperial America. *New Left Review*, 143, 6–38.

Dawisha, K. and Dawisha, A. (eds) 1982: *USSR and the Middle East: Policies and Perspectives*. New York: Holmes and Meier.

Dickey, C. 1984: Central America: from quagmire to cauldron. *Foreign Affairs*, 62, 659–94.

Doran, C. F. and Parsons, W. 1980: War and the cycle of relative power. *American Political Science Review*, 74, 947–65.

Edwards, A. 1979: *The New Industrial Countries and their Impact on Western Manufacturing*. London: Economist Intelligence Unit.

Fagen, R. R. and Pellica, O. 1983: *The Future of Central America: Policy Choices for the U.S. and Mexico*. Palo Alto, CA: Stanford University Press.

Feinberg, R. E. (ed.) 1982: *Central America: International Dimensions of the Crisis*. New York: Holmes and Meier.

Frank, A. G. 1981: *Reflections on the World Economic Crisis*. New York: Monthly Review Press.

Gaddis, J. L. 1982: *Strategies of Containment: A Critical Appraisal of Postwar American National Security Policy*. New York: Oxford University Press.

Galtung, J. 1987: Peace and the world as inter-civilizational interaction. In R. Vayrynen, D. Senghaas and C. Schmidt (eds), *The Quest for Peace*. Beverley Hills, CA: Sage Publications, 330–47.

Gavshon, A. 1981: *Crisis in Africa: Battleground of East and West*. Harmondsworth: Penguin.

Gill, S. 1986: US hegemony: its limits and prospects in the Reagan era. *Millenium*, 15, 311–38.

Gilpin, R. 1981: *War and the International System*. New York: Cambridge University Press.

Goldstein, J. 1988: *Long Cycles: Prosperity and War in the Modern Age*. New Haven, CT: Yale University Press.

Gray, C. S. 1988: *The Geopolitics of Superpower*. Lexington, KY: University Press of Kentucky.

Haass, R. 1988: The use and (mainly misuse) of history. *Orbis*, 32, 411–19.

Halliday, F. 1982: *Soviet Policy in the Arc of Crisis*. Washington, DC: Institute for Policy Studies.

Halliday, F. 1986: *The Making of the Second Cold War*, 2nd edn. London: Verso.

Halliday, F. and Molyneux, M. 1982: *The Ethiopian Revolution*. London: Verso.

Harrison, S. 1979: The Shah, not Kremlin, touched off Afghan coup. *Washington Post*, 13 May, p. C1.

Hoffman, G. W. 1982: Nineteenth-century roots of American world power relations. *Political Geography Quarterly*, 1, 279–92.

Hoffmann, S. 1988: Lessons of a peace epidemic. *New York Times*, 6 September, p. 27.

House, J. 1984: War, peace and conflict resolution: towards an Indian Ocean model. *Transactions, Institute of British Geographers*, NS 9, 3–21.

Johnson, R. H. 1985: Exaggerating America's stakes in Third World conflicts. *International Security*, 10, 32–68.

Johnstone, D. 1984: *The Politics of Euromissiles*. London: Verso.

Keal, P. 1983: Contemporary understanding about spheres of influence. *Review of International Studies*, 9, 155–72.

Kebbede, G. and Jacobs, M. J. 1988: Drought, famine and the political economy of environmental degradation in Ethiopia. *Geography*, 73, 65–70.

Kennedy, P. 1987: *The Rise and Fall of the Great Powers: Economic Change and Military Conflict from 1500 to 2000*. New York: Random House.

Kissinger, H. 1984: *National Bipartisan Commission on Central America Report.* Washington, DC: US Government Printing Office.

Klare, M. T. 1981: *Beyond the Vietnam Syndrome: U.S. Interventionism in the 1980s.* Washington, DC: Institute for Policy Studies.

Kolodziej, E. and Kanet, R. (eds) 1988: *The Soviet Union and the Developing World: Thermidor in the Revolutionary Struggle.* Baltimore, MD: Johns Hopkins University Press.

Krasner, S. 1978: *Defending the National Interest.* Princeton University Press.

Kristof, L. K. D. 1960: The origins and evolution of geopolitics. *Journal of Conflict Resolution,* 4, 632–45.

Kurth, J. R. 1982: The United States and Central America: hegemony in historical and comparative perspective. In R. E. Feinberg (ed.), *Central America: International Dimensions of the Crisis.* New York: Holmes and Meier, 39–57.

Lafeber, W. 1984: The Reagan administration and revolution in Central America. *Political Science Quarterly,* 99, 1–26.

Lemma, H. 1985: The politics of famine in Ethiopia. *Review of African Political Economy,* 33, 44–58.

Litwak, R. S. and Wells, S. F. Jr. (eds) 1988: *Superpower Competition and Security in the Third World.* Cambridge, MA: Ballinger.

Lowenthal, A. E. 1983: Ronald Reagan and Latin America: coping with hegemony in decline. In K. A. Oye, R. J. Lieber and D. Rothchild (eds), *Eagle Defiant: United States Foreign Policy in the 1980s.* Boston, MA: Little Brown, 311–35.

McFarland, S. N. 1987: *Superpower Rivalry and Third World Radicalism.* London: St Martin's Press.

Mahan, A. 1980: *The Influence of Seapower upon History, 1660–1783.* New York: Hill and Wang.

Manning, R. 1983: Superpowers as suppliers. *South,* August, 14.

Markatis, J. 1988: The nationalist revolution in Eritrea. *Journal of Modern African Studies,* 26, 51–70.

Mayer, J. and McManus, D. 1988: *Landslide: The Unmaking of the President, 1984–1988.* Boston, MA: Houghton Mifflin.

Medvedev, R. 1980: The Afghan crisis. *New Left Review,* 121, 91–6.

Modelski, G. 1978: The long cycle of global politics and the nation-state. *Comparative Studies in Society and History,* 20, 214–35.

Modelski, G. 1987: *Long Cycles in World Politics.* Seattle, WA: University of Washington Press.

Modelski, G. and Thompson, W. R. 1988: *Seapower in Global Politics, 1494–1993.* Seattle, WA: University of Washington Press.

Mujah-Leon, E. 1984: European Socialism and the crisis in Central America. *Orbis,* 28, 53–82.

Nye, J. S. Jr. 1988: America's decline: a myth. *New York Times,* 10 April, p. E31.

O'Loughlin, J. 1986: Spatial models of international conflicts: extending current theories of war behavior. *Annals, Association of American Geographers,* 76, 63–80.

O'Loughlin, J. 1987: Superpower competition and the militarization of the Third World. *Journal of Geography,* 86, 269–75.

Ottaway, M. 1982: *Soviet and American Influence in the Horn of Africa.* New York: Praeger.

O'Tuathail, G. 1986: The language and nature of the 'new geopolitics': the case of U.S.–El Salvador relations. *Political Geography Quarterly,* 5, 73–86.

Oye, K. A. 1983: International systems structure and American foreign policy. In K. A. Oye, R. J. Leiber and D. Rothchild (eds) *Eagle Defiant: United States Foreign Policy in the 1980s.* Boston, MA: Little Brown, 3–22.

Oye, K. A., Leiber, R. J. and Rothchild, D. (eds) 1983: *Eagle Defiant: United States Foreign Policy in the 1980s*. Boston: Little Brown.

Parboni, R. 1981: *The Dollar and its Rivals: Recession, Inflation and International Finance*. London: Verso.

Paterson, J. H. 1987: German geopolitics reassessed. *Political Geography Quarterly*, 6, 107–14.

Pearce, J. 1982: *Under the Eagle: U.S. Intervention in Central America and the Caribbean*. Boston, MA: South End.

Poulluda, L. B. 1981: Afghanistan and the United States, the crucial years. *Middle East Journal*, 35, 178–90.

de Riencourt, A. 1983: India and Pakistan in the shadow of Afghanistan. *Foreign Affairs*, 61, 416–37.

Rothchild, D. and Ravenhill, J. 1983: From Carter to Reagan: the global perspective on Africa becomes ascendant. In K. A. Oye, R. J. Lieber and D. Rothchild (eds), *Eagle Defiant: United States Foreign Policy in the 1980s*. Boston, MA, Little Brown, 337–65.

Russett, B. 1985: The mysterious case of vanishing hegemony; or, is Mark Twain really dead. *International Organization*, 39, 207–31.

Schlesinger, A. Jr. 1988: Dukakis for President. *Atlantic Monthly*, 69, 75–9.

Schulman, M. D. 1987: The superpowers: dance of the dinosaurs. *Foreign Affairs*, 66, 494–515.

Senghaas, D. 1983: The cycles of war and peace. *Bulletin of Peace Proposals*, 14, 119–24.

Shaw, T. 1983: The future of the Great Powers in Africa: toward a political economy of intervention. *Journal of Modern African Studies*, 21, 555–86.

Shearman, P. 1986: Soviet foreign policy in Africa and Latin America: a comparative case. *Millenium*, 15, 339–66.

Singer, J. D. 1981: Accounting for international war: the state of the discipline. *Journal of Peace Research*, 18, 1–18.

SIPRI 1983: *Yearbook 1983*. Stockholm: Stockholm International Peace Research Institute.

Sivard, R. 1988: *World Military and Social Expenditures*. Washington, DC: World Priorities.

Starr, H. and Most, B. 1985: The forms and processes of war diffusion. *Comparative Political Studies*, 18, 206–27.

Thompson, W. R. 1982: U.S. policy toward Africa: at America's service. *Orbis*, 25, 1011–24.

Thompson, W. R. 1983a: Cycles, capabilities, and war: an ecumenical view. In W. R. Thompson (ed.), *Contending Approaches to World System Analysis*, Beverly Hills, CA: Sage Publications, 141–63.

Thompson, W. R. (ed.) 1983b: *Contending Approaches to World System Analysis*. Beverly Hills, CA: Sage Publications.

Thompson, W. R. and Zuk, G. 1986: World power and the strategic trap of territorial commitments. *International Studies Quarterly*, 30, 249–68.

Wallerstein, I. 1983: *Historical Capitalism*. London; Verso.

Wallerstein, I. 1984: *The Politics of the World Economy*. New York: Cambridge University Press.

Wallerstein, I. 1987: The Reagan non-revolution, or the limited choices of the United States. *Millenium*, 16, 467–72.

Weidenbaum, M. 1988: Defense spending is *not* a military millstone leading the US to decline. *Christian Science Monitor*, 31 March, p. 12.

Wiberg, H. 1979: The Horn of Africa. *Journal of Peace Research*, 16, 189–96.

Wolfe, A. 1979: *The Rise and Fall of the 'Soviet Threat': Domestic Sources of the Cold War Consensus*. Washington, DC: Institute for Policy Studies.

Yapp, M. 1982: Soviet relations with the countries of the Northern tier. In K. Dawisha and A. Davidson (eds), *USSR and the Middle East: Policies and Perspectives*. New York: Holmes and Meier, 24–44.

12

The World-Systems Project

PETER J. TAYLOR

'That's another fine mess you've got me into,' says Oliver Hardy to Stan Laurel again and again. Although an unlikely analogy, modern capitalism does seem to display a level of competence on a par with Stan Laurel. The world is in a mess once again. And yet modern capitalism would seem to be better run than ever before. Business schools are turning out the best-ever trained managers, while university economics and commerce departments are providing financial advisers with the best-ever technical models. Furthermore, governments employ more social scientists than ever before to provide the best-ever advice on economic, political and social issues. And yet the world is in a mess. It seems that all these modern experts have Laurel's incompetence without his endearing innocence. For, to make matters worse, it is not a matter of the world being accident prone. The current situation is not an unfortunate random event. The mess is cyclical. Whereas Laurel and Hardy were a product and symbol of the last mess, this time Hollywood produced a president. And it is no longer a joke. In the nuclear age the question of who has the last laugh is macabre.

The most obvious lesson of the current situation is that we need a rethink in politics and in academia. This is of course happening. Cycles of growth and stagnation on a global scale are of practical and academic concern throughout the world. For instance, in 1978 two books appeared on precisely this topic. W. W. Rostow's *The World Economy: History and Prospect* introduces long cycles into his familiar non-communist manifesto of economic stages, and V. V. Rymalov's *The World Capitalist Economy: Structural Changes, Trends and Problems* brings cycles and trends back to the centre of official communist thinking. How far these two books represent a rethink is of course disputable. Despite their intellectual elegance they do not fool anyone – they are bringing the cold war in academia up to date. Most Western academics deplore this political intrusion and choose instead to remain in their ivory tower, which separates them from politics. But this is the route that leads to just the sort of 'neutral' technical knowledge that produced the Stan Laurel 'experts' of the recent past. If we really only had to choose between a pseudo-neutrality or a cold-war mentality the future would be bleak indeed. In this chapter I hope to show that there is another way.

A rethink does not, of course, mean making a completely new start. Past ideas, concepts and theories may be used, but only after they are understood in the context in which they were produced. All theories are the products of their time and place, and their need for revision or rejection varies accordingly. The world-systems project centring on the work of Immanuel Wallerstein is a product of the current world recession. It is a reaction to the optimistic social sciences of the early post-war era, especially their predictions for Third World countries 'catching up'. They got it wrong, and horrendously so. They must be replaced. I will argue that their replacement should look something like the world-systems project.

This project has been a highly controversial production, assailed on all sides. Most criticism has been the result of misunderstanding of the nature and purpose of the project. I will show that it is actually more controversial – that is, more dangerous to existing ideas than even its critics realize. They say that it is not history, they say that it is not social scientific, they say that it is not Marxist and, the most cardinal of all sins, they say that it is not objective. I agree: it is not any of these and more, at least on their terms. That is what makes the project so interesting and why liberals and Marxists, historians and social scientists have all misunderstood it. On each of their criteria it is inadequate, but that is not surprising since it is built on new combinations of criteria. It is, in the real sense of the word, an alternative.

THE MODERN WORLD-SYSTEM

The core of Wallerstein's project is a set of volumes that constitute a history of our modern world in the 'grand tradition'. By this I mean that the 'movement' of history is integral to the narrative. The result will be a description of the dynamic capitalist world-economy from its inception in about 1500 through to the present day. Only the first three volumes have been published so far (Wallerstein, 1974a, 1980a, 1988a), but the overall framework has been presented (Wallerstein, 1974b, 1984c). The story Wallerstein is telling goes as follows. A capitalist world-economy emerged out of the crisis of feudalism in Europe during the long sixteenth century (volume I). This world-system was consolidated in the stagnant seventeenth century to 1750 (volume II). In the late eighteenth century and early nineteenth century Britain emerged as the dominant power in a new expansion phase (volume III). Subsequent volumes will have to deal with Britain's nineteenth-century dominance and decline and the rise and fall of the United States, both in the context of different patterns of global processes.

Overall, Wallerstein is attempting to produce a picture of a single entity, the capitalist world-economy, in its origins, its operations and the beginnings of its demise. It is a single entity because it maintains certain characteristics through all its different phases. First, there is a world market for commodities. Second, there is a competitive state system so that no one political unit can control that market. Third, there is an economic hierarchy of spaces from core through semi-periphery to periphery. Fourth, there are economic classes and social status groups which interact with one another and with the spatial hierarchy. Fifth, the dynamics of the entity are cyclic in terms of long economic waves. Sixth, there are secular trends in the system which are asymptotic, such as the geographical spread of the entity. Wallerstein uses an enormous number of secondary historical sources to weave together the story of his world-system in terms of these basic characteristics.

As the above suggests, this 'empirical core' of the project is not divorced from 'world-systems' theory'. A holistic structuralism transcending history and the individual social sciences is proclaimed in the above works. This is explicitly treated by Wallerstein (1976) and in many other papers brought together as *The Capitalist World-Economy* (Wallerstein, 1979), and, with others, in *World-Systems Analysis* (Hopkins and Wallerstein, 1982). Other empirical 'examples' and theoretical issues are to be found in the quarterly journal *Review* and the series *Political Economy of the World-System Annals*, both edited by Wallerstein. There are numerous other books related to the project, notably Bergeson (1980), Amin et al. (1982) and Chase-Dunn (1982). Most importantly, Wallerstein (1983a) brings together his main ideas in a short and very readable essay, and his political essays have been collected as *The Politics of the World-Economy* (Wallerstein, 1984a). This covers most of the literature which makes up the project, but unfortunately its quantity and quality have not prevented misunderstanding. My main purpose in this chapter is to dispel some of this misunderstanding.

Wallerstein's work has been used by others in one of two ways. First, it can be seen as an alternative temporal framework for describing the evolution of the modern world. Some economic historians have treated it this way but the main group here are the international relations researchers. Wallerstein's 'world-systems theory' is described as one of three approaches (Hollist and Rosenau, 1981). Thompson (1983), for example, explicitly compares it with Modelski's (1978) alternative temporal framework on the assumption that they are equivalents, that is 'contending approaches'. Wallerstein does, of course, provide a temporal framework for empirical studies, but to treat his work as only that is to miss the whole point of the project. Second, other social scientists, especially sociologists, have emphasized the 'theory' in the project. Wallerstein provides an alternative to the Parsonian equilibrium system and one which moreover seems to be necessary for transcending comparative sociology. World classes, world status groups and even the replacement of sociology (about social formations) by globology (about global formations) can be found here (Bergeson, 1980). One aim is to develop the theory to produce 'falsifiable predictions'

(Chase-Dunn, 1983). These studies also miss the point of the project (Taylor, 1987a). Wallerstein is not trying to create a better and more 'accurate' sociology (or even globology).

What is the point of the world-systems project therefore? Briefly, Wallerstein's purpose is to contribute to the transition from capitalism to something more humane and democratic which we may wish to call socialism. The empirical and historical framework of world-systems theory is only part of the project in so far as it provides tools to that end. People may wish to use the framework for other purposes, but they are not then contributing to the project. Hence Wallerstein's work is probably bad economic history, bad international relations and bad sociology as conceived by the practitioners of those disciplines. The test of his work lies elsewhere, however. It leads to alternative modes of theorizing and ultimately to new methodologies. In fact, Wallerstein (1983b, p. 306) has gone as far as suggesting that 'the crucial terrain of struggle may well turn out to be that of methodology'. In the remainder of this chapter I will attempt further to elucidate the project's purpose as another attempt 'to turn the world upside down'.

'TRUTH AS OPIUM'

Wallerstein (1983b), perhaps surprisingly, refers to the historiography he is creating in the multi-volume core of the project as a myth. This is not self-criticism since he considers all history as the creation of myths. What Wallerstein claims is that he is providing an alternative organizing myth to that which has dominated history since the nineteenth century. Organizing myths are important because they incorporate as virtually unexamined assumptions the nature of the *present*. The dominant organizing myth that Wallerstein is attempting to replace, for instance, can be termed the progressive myth, since it produces a historiography of conflicts between 'pre-modern' and 'modern' forces which delineates a path from traditional feudalism to our modern affluent society. Obviously the 'progress organizing' myth celebrates the achievements of the present. For Wallerstein it fails utterly to explain historical regression, and this is central to the world-systems project. We deal with this important topic separately in some detail below. The key point here is that since recounting the past is always a social act of the present (Wallerstein, 1974a, p. 8) – it cannot logically be otherwise – all history must be merely 'transitory' knowledge.

The progress organizing myth is more than a matter of history, of course. In academia the task of building, illustrating and proving the myth has been divided between two groups. The particularists, mainly historians, have beavered away at filling in the details, while the generalists, who eventually became known as social scientists, have attempted to provide the models and theories of how progress came about. For Wallerstein, since both groups operated within the same overall myth, the 'grand debate' of ideographic versus nomothetic knowledge is beside the point (Wallerstein, 1976). History and social science are one subject matter, which he terms *historical social science*. Quite simply, all historical descriptions of particular events must use concepts that imply 'generalizations about recurrent phenomena', while all social science theories must be ultimately 'a set of inductions from history' (Wallerstein, 1979, p. ix). The separation of social science from history has enabled the progress organizing myth of history to be transformed into assertions of universal generalizations. Whereas historians have retreated from the Victorian view that cumulation of enough facts will eventually lead to 'ultimate history' (Carr, 1961), positivist social scientists have maintained the idea that further refinements of their theories will eventually unmask reality and produce the truth. But if all history is a myth and social science is history, then all social science is myth. We have now reached the very heart of Wallerstein's position.

Modern science is based on a belief that nature and society can be described by universal statements, and it is the task of science to search for such generalizations. This is an epistemology which proclaims a faith in 'truth' as a *real* phenomenon, the goal of science. As such it has become self-evident that truth is a disinterested virtue for which we must strive. But if all history is transitory, a product of the fleeting present, and all social science is history, it follows that 'truth' changes as society changes. Truth is a variable, not an absolute. 'Absolute truth' has, therefore, an ideological role. It is a cultural idea which has operated as the opiate of the modern world (Wallerstein, 1983a, p. 81). Traditional societies and their myths have been vanquished by our truth. It is this arrogant 'modernization' which has produced a rationality upon which economic efficiency can be developed. As such the rationality of scientific activity has been able to hide the ultimate irrationality of endless capital accumulation (p. 85). It has socialized the cadres who operate the world-economy into a recognizable common stratum just below the apex of the social hierarchy. Their role in the world division of labour has been vital as the emphasis on meritocracy through education reinforces the hierarchical structure. Universalism, therefore, has been nothing less than 'the keystone of the ideological arch' of capitalism (p. 81).

If truth is transitory so too must be our theories and organizing myths in historical social science. They will depend on our particular window on the past and the lenses through which we look. Where does all this leave the basic concept of the objectivity of scientist and scholar? To begin with we can note that every choice of conceptual framework is political. There is no such thing as an uncommitted historian or social scientist. This is because all assertions of truth are based on assumptions which ultimately involve a metaphysics of values. The 'disengaged' scientist or scholar who claims objectivity by remaining in an ivory tower is merely drawing up the ramparts to hide his or her premises. This does not mean that there can be no objectivity, however. The first step towards objectivity is to admit commitment. It is the lack of this basic honesty which makes conventional objectivity such a sham. The current production of knowledge, for instance, represents a social investment which reflects the current structure of power. Objective knowledge, however, will only be produced when all major groups in the world-system are equitably represented. Since knowledge can never be produced separately from the society that brings it forth, it follows that at present we can only aim for a more balanced creation within our given society. This is a route towards a collective notion of objectivity which we can all work along (Wallerstein, 1974a, pp. 9–10).

Let us return to politics. Not only does Wallerstein bridge the gap between history and social science, but it obviously follows from the above argument that historical social science cannot be meaningfully separated from politics. Wallerstein (1974a, p. 9) distinguishes between scientist/scholar and apologist/ advocate. Both are committed but they have different roles within their chosen framework. It is the duty of the scientist/scholar to describe and interpret the present in the light of the past in order to aid in the construction of the future. Although the work may seem at times to be quite esoteric, it is all 'relevant' in that its *raison d'être* can only be application. The choice of organizing myth, the filling in of the particulars and the development of more general notions will all be part of a larger social movement. The testing of the ideas will ultimately be in political practice: Does this particular organization of knowledge generate ideas which can usefully contribute to our manipulation of reality? Or as Wallerstein (1983b, p. 307) puts it, 'as we become more political we become more scientific; as we become more scientific we become more political'. Scientists and scholars are part of political practice; they must theorize it so that it can be constructively criticized and improved. This is what the world-systems project is trying to do for what is generally termed 'socialism'. The products are heuristic theories, and since 'truth' is merely 'an interpretation meaningful for our times' (Wallerstein, 1979, p. xii) they must be essentially transitory.

Finally, how do we produce these heuristic theories? What is this methodology which Wallerstein feels may be so crucial? If we return to his organizing myth we find that we are provided with a single entity, the capitalist world-economy, which is by definition unique. This is fundamental to the question of methodology. O'Brien (1982), for instance, has tried to show how unimportant the periphery was to the economic development of Europe in the period covered by Wallerstein's (1974a, 1980a) first two historical volumes. But as Wallerstein (1983d, p. 550) replies, 'How much is a lot? That is the question', for the fact is that Western Europe did not develop in isolation. It was part of a wider system and it did benefit from that membership. We cannot construct an experiment rerunning European history without the Atlantic, Baltic and Indian trade. We can never know what would have happened. Experimental design and comparative statistical methodology are just not an option. Let us make this clear: quantitative techniques may be useful as descriptive devices if harnessed to meaningful categories, but the notion of testing empirical data to produce general theory is simply meaningless in this context. Although they are intimately intertwined, it is the *methods* and not the *techniques* which are the crux of the matter (Taylor, 1987a).

Current methodology involves moving from the complexity of empirical description to generate general statements about that reality. The implication of heuristic theorizing is to reverse the order. We start with general statements of the operation of the complex entity we call the world-economy and then observe their conjunction in specific events or groups of events until we can produce a set of 'utilizable concrete descriptions of historical structures' (Wallerstein, 1983b, p. 307). But what are these structures? The entity that we are dealing with is not the ahistorical system of modern sociology with its equilibrium stuctures, but a much less familiar dissipative structure where disequilibrium is the normal state of affairs. Order is not a final outcome of the system but is a constantly changing pattern as initial conditions and new conditions continually interact in the evolving totality. Wallerstein (1983c, p. 33) talks of 'this scientific–metaphysical–ideological transition' which will be linked to political practice but of which we know little as yet. He quotes Prigogine: 'We are only in the beginning at the prehistory of our insights' (p. 36).

THE SPIRIT OF MARX

Although liberal scientists and scholars will find the previous discussion quite distasteful, Marxists will find much of it familiar. The connection of theory with political practice is, of course, central to Marx's own work. In short, Wallerstein's epistemology is not basically different from that of Marx. They are both part of a tradition of scholars who have challenged the universalism of the social sciences. Wallerstein's departure from orthodox Marxism relates, therefore, to the new historiography that he is creating. For Wallerstein, Marxists and other radical critics of liberal theory have made 'a fundamental error of judgement' because 'by surrendering the historiographical domain, they undermined fatally their resistance in the epistemological realm' (Wallerstein, 1984a, p. 180). In fact, Wallerstein (1983b, p. 302) describes the dominant organizing myth in such a way that *both* liberals and Marxists can identify with the historical sequence: Marx's conception of history is a progressive one. The distinctive terminology – the succession of modes of production – cannot hide the fact that orthodox Marxists hold the same basic progress organizing myth as their liberal opponents: 'For Rostow's stages, substitute Stalin's' (Wallerstein, 1984a, p. 181). Since Wallerstein provides an alternative organizing myth, it is not surprising that he has been criticized as much by Marxists as by liberals.

Whether we consider Wallerstein to be a Marxist is itself not a particularly important question since it merely rests upon definitions of Marxism. Wallerstein (1974a, p. 396) talks of following the 'spirit of Marx if not the letter'. I think the position is best summed up by Lefebvre:

> To attack this person or that, and to say that this one is a Marxist and that one isn't, seems to me bad methodology, a bad line of thought. The correct line of thought is to situate the works and the theoretical or political propositions within the global movement of the transformation of the modern world. (Lefebvre, 1980, p. 23)

Marx must be seen as 'a man of the nineteenth century, whose vision was inevitably circumscribed by that reality' (Wallerstein, 1983a, p. 9). His great legacy must not be left to 'dead Marxists' (controlled by nineteenth-century texts or twentieth-century governments) but must be seen as part of 'a much vaster global movement' (Lefebvre, 1980, p. 2), a living monument incorporating a 'living Marxism'. Wallerstein terms this movement the anti-systemic movement of the world-economy which has historically included two components: socialist challenge initially concentrated in the core and nationalist challenge in the periphery. The former has been the traditional arena of Marxism but in the post-war era the two have come to be merged in the national liberation struggles of the periphery. It is in the interpretation of the nature of the periphery and its political struggles that Wallerstein's project differs most from more orthodox Marxist interpretations. This theoretical debate is known as 'the modes of production controversy' (Foster-Carter, 1978) and its relevance for the political practice of the anti-systemic movement is fundamental.

For Marx a mode of production is defined by the forces and relations of production. From this it follows that a capitalist mode of production exists when the means of production are appropriated by the dominant class, leaving the dominated class only their labour power to sell to the highest bidder. Hence the existence of 'free' labour is the key requirement for identifying capitalism. If this definition is applied strictly in the contemporary world we find that 'capitalism' has yet to reach many parts of the periphery. Hence other 'pre-capitalist' modes of production must exist alongside capitalism. Much modern Marxist theory (e.g. Taylor, 1979) has been about the 'articulation' that occurs between these contemporary modes of production. Basically it is agreed that, as long as the reproduction requirements of two modes of production are compatible, they can exist side by side and may in fact reinforce one another.

If modes of production 'interact' it follows that there must be some 'entity' larger than the individual modes themselves. Here the terminology varies. For Laclau (1971) different modes of production can exist together within a single 'economic system'; for Rey the articulation occurs within a particular 'social formation' (Brewer, 1980). In both cases mode of production is treated as an abstract concept and the containing entity is a concrete occurrence of the modes. Sfia (1983), however, identifies two 'modes of disposition': a non-market mode based on a dominance of use values and a market mode based on a dominance of exchange values. In this scheme the modern world has several modes of production, but only one mode of disposition – the market mode. But the market means the domination of capital so that this mode of disposition integrates capitalist and non-capitalist spaces into one entity, the 'world-system' whose logic is capitalist.

Clearly Sfia's approach seems to be closest to Wallerstein's, but mode of disposition, economic system or social formation are all equally distinct from the world-economy of the world-systems project. Wallerstein (1976, p. 348) uses a much broader definition of mode of production: 'the way in which decisions are made about dividing up productive tasks, about quantities of goods to be produced and labour time to be invested, about quantities of goods to be consumed or accumulated, about the distribution of the goods produced'. Hence he breaks from 'the sterile assertion of the analytical priority of relations of production over those of exchange and indeed all else' (Foster-Carter, 1978, p. 76). Wallerstein defines the capitalist mode of production as 'one in which production is for exchange, that is, it is determined by its profitability on a market' (1976, p. 351) or 'production for sale in a market in which the object is to realize the maximum profit' (1974b, p. 13). Perhaps under pressure of criticism from Marxists, Wallerstein has more recently expressed his definition differently: 'The primary desideratum, the defining characteristic, of a capitalist system, is the drive for ceaseless accumulation' (1983b, p. 15), and 'endless accumulation of capital' is the 'economic objective' (1983a, p. 18). This alternative notion of capitalism, focusing on 'ceaseless accumulation' rather than 'free labour', has been, of course, a major concern of orthodox Marxists in their criticism of Wallerstein. Before we discuss this debate, however, we need to consider the more general historical implications of Wallerstein's broader concept of mode of production.

The primary reason for identifying modes of production for both orthodox Marxists and Wallerstein is to challenge the universalism of social thought within capitalism: the social institutions and individual motivations found in capitalism have not existed from time immemorial. But if the past was not 'naturally' capitalist, how then should it be conceptualized? Enter modes of production to classify past societies and so link the modes-of-production controversy to the historiographical realm. Marx himself proposed a progression of modes with the following sequence: primitive communism–classical slave–feudal–capitalist–socialist (Peet, 1980). In addition Marx also refers to an Asiatic mode of production outside the European sequence. This notion of alternative sequences outside Europe has been developed as a set of multilinear sequences (Peet, 1980). In contrast, Wallerstein, using his broader definition of mode of production, identifies only three historical modes of production: a reciprocal-lineage mode which roughly coincides with primitive communism, a redistributive-tributary mode which incorporates all the various categories between primitive communism and capitalism, and finally the capitalist mode which currently covers the whole world as the capitalist world-economy. Only two class societies are identified: those based on political control, which are termed world-empires, and those based on markets, termed world-economies. Hence classical, feudal and Asiatic systems are all interpreted as identical modes of *production* with merely alternative political superstructures. Marx's identification of an Asiatic mode is then seen as based on inadequate sources and Eurocentrism: 'Since Asian societies were not like those of Ancient Greece or Rome or feudal Europe, they must constitute something else. The very geographical characterization of the category indicates its residual nature' (Chandra, 1981, p. 18). Although Marx used the term 'Asiatic mode of production' only once (Chandra, 1981, p. 17), the idea of a distinct 'timeless' Asia was important for his subsequent identification of British colonialism in

India as being 'progressive'. It is doubtful whether this interpretation remains useful.

Similar specific modes of production continue to be generated. For instance, colonial and African modes of production have been proposed as modern concepts for describing India and parts of Africa respectively (Foster-Carter, 1978; Cooper, 1981). But as Cooper (1981, p.14) states, 'If every way of catching an antelope or growing a banana defines a mode of production, the concept blends into an empiricism that Marxists scorn'. Foster-Carter (1978, p. 74) makes exactly the same point but feels that Wallerstein's solution of just three historical modes is 'an extreme way of resolving' the modes-of-production controversy. Extreme, yes, but surely preferable to a devaluing of the concept by multiplication. The test of Wallerstein's parsimony is in the fresh insights that it provides.

THE SMITHIAN TWIST

Let us return to the question of defining the particular mode of production known as capitalism, since this reaches to the very heart of Marxist criticisms of Wallerstein. If Wallerstein is not to be a Marxist, then some other label derived from a lesser, and preferably non-revolutionary, intellectual figure seems to be required. Aronowitz (1981, p. 520) finds this task difficult, but nevertheless purports to detect 'the spirit of Weber' in Wallerstein. The main derogatory label attached to Wallerstein, however, is that he is a 'neo-Smithian', that is his emphasis on the market rather than production makes him closer to Adam Smith than to Karl Marx. It is not our task here to enter into labelling games (beyond suggesting that Wallerstein is a Wallersteinian!), but there is a tangle of ideas that requires unknotting. Some of the criticism is based on simple misunderstanding of Wallerstein's work, some relates to ambiguities and even changes of emphasis in his work (Agnew, 1982), and part of the problem is that the two most compelling criticisms where the Smithian label is justified (Aronowitz, 1981; Brenner, 1977) review only the first volume of *Modern World-System*. We shall make use of some of Wallerstein's more recent writings to challenge these criticisms.

We can begin by describing Wallerstein's position on the labour question. 'Capitalism thus means labour as a commodity to be sure. But wage labour is only one of the modes in which labour is recruited and recompensed in the labour market' (Wallerstein, 1974b, p. 17). Hence labour may be used and controlled in different ways in different parts of the capitalist world-economy but this does not indicate different modes of production. In fact one of the characteristics of the world-economy throughout its history is for more free labour to exist in the core with various forms of unfree labour occurring in the periphery. There are different forms of labour control suitable for the different circumstances in different zones of the world-economy. American slavery, the 'serfs' of Eastern Europe, industrial proletarians and peasant households are all examples of direct producers in the capitalist world-economy. For Wallerstein, 'capital world-economy' and 'capitalist mode of production' are just two ways of saying the same thing.

At first sight this new schema seems to be merely an alternative set of terms – 'much of the debate over modes of production has been about the use of words and no more' (Brewer, 1980, p. 273). Instead of an economic system or social formation imposed over different modes of production we have a single mode of production imposed over different forms of labour control. But it is much more than a semantic point. Brenner (1977) emphasizes the unique characteristic of the capitalist mode of production as defined by Marx in its generation of new technologies. Hence in order to understand the dynamics of our modern world we must deal with relations of production which produce the economic change upon which the whole system pivots. Emphasis on the world level therefore misses the vital processes of production which are unique to capitalism. It is at this point that Brenner dismisses Wallerstein's definition of the capitalist mode of production as neo-Smithian rather than Marxist. In a similar vein Aronowitz (1981, p. 520) states that 'What Wallerstein has given us is a theory of the forms of appearance of capitalism. The core of the system remains unseen'. Brewer (1980, p. 181) goes even further: 'What is lacking is real theory'.

Essentially Wallerstein provides two forms of defence against the criticism. First, he provides a reinterpretation of the relations between the ideas of Smith, Marx and Weber: 'Marx had one major fault. He was a little too Smithian' (Wallerstein, 1988a). Second, he locates his own work as a product of the late twentieth century which has more heuristic value than past 'truths', however eminent the purveyors of those truths. These two defences are very closely related, but we shall disentangle them and present them consecutively.

With Arrighi and Hopkins, Wallerstein has sketched out his ideas on classes and status groups from a world-systems perspective (Arrighi et al., 1983). They begin by defining the political-economy paradigm as initiated by Adam Smith. Two particular assumptions are highlighted: classes are defined in terms of their relation to the market, and society is assumed to be coterminous with the state. Marx's critique of this political economy overturns both of these assumptions involving two basic shifts – from the market-place to the work-place and from state-defined economic spaces to world-economic space. However, Wallerstein and his colleagues note that in his studies of current events Marx did 'retreat into political economy' (Arrighi et al., 1983, p. 291) by dealing with state-defined economic spaces. This may be acceptable for concrete studies for short-term purposes. Similarly, the increasing importance of the state at the beginning of the twentieth century required similar concrete analyses to highlight state-defined economic spaces. But this cannot be extended to *theoretical* revisions of the monopoly capital/imperialism variety, especially when developed as stages of capitalism. By identifying stages as occurring in state-defined economic spaces the Leninist position represents 'the theoretical retreat of Marxists back into political economy' (Arrighi et al. 1983, p. 291). In short, it is the orthodox Marxists who are the neo-Smithians! Therefore our task now is to break out of this old political economy strait-jacket once again so as to understand features such as imperialism as cyclical properties of the world-economy as a whole (Wallerstein, 1980d). This point is also important for defining class. Weber's concepts of classes *an sich* and *für sich* are used, but again Wallerstein breaks away from their simple application to particular political communities and states. For Wallerstein class *an sich* can only be defined at the level of production processes which in his scheme are organized on the world scale. The problem for socialism is that classes *für sich* have been created at the national level. This is one of the basic antinomies of capitalism (Wallerstein, 1984a, p. 36), but one which is lost in political economy. Hence the need to abandon political economy and return to Marx's critique of that paradigm. Wallerstein's historical social science is not another political economy – it is a replacement of political economy.

The emphasis on state-defined spaces from the Second International to the present day has had enormous practical significance (Taylor, 1987b). Concepts such as the permanent revolution (skipping stages) and making alliances with 'progressive' bourgeois forces against 'feudal' forces in peripheral countries all fall before this definition of classes and hence of transition (or revolution) as world-scale processes. Such political-economy concepts may have had important roles in their time but they are no longer adequate – in fact they are quite dangerous – if applied to current political situations. Of course, like political economy in its various formulations, Wallerstein's historical social science is a product of its time. It is an attempt to review and revise anti-systemic thought to make it relevant to the period of declining US hegemony in the world-economy. Wallerstein does not deny the importance of production for defining capitalism. His pre-industrial-revolution capitalism is not the 'mercantile capitalism' of many Marxist writers but a fully fledged agricultural capitalism based on agro-industrial production, for instance. But Wallerstein does not limit the key processes of change to one particular social relation within this production system. The bourgeois–proletarian relation is integrated with a core–periphery relation to produce world classes, as we have seen. In doing this Wallerstein's critics believe that he overlooks or at least devalues the fundamental process of change within capitalism. The key word here is 'believe'. The issue of the importance of 'core' or 'periphery' in the development of the modern world-system is untestable, as we have previously pointed out. What we have are alternative organizing myths producing their very different 'metahistories' (Wallerstein, 1983c). Aronowitz (1981) comes close to recognizing this when he describes his critique as 'metatheoretical'. Hence the clash between Wallerstein and his Marxist critics is one of 'opposing faiths' (Wallerstein, 1983a, p. 20). Wallerstein's second defence is thus quite pragmatic:

> The justification therefore for our metahistory comes neither from the data it generates nor from the null hypotheses it supports nor from the

analyses it provokes. Its justification derives from its ability to respond comprehensively to the existing continuing real social puzzles that people encounter and of which they have become conscious. It is, in fact, precisely the reality of the ever-increasing historical disparities of development that has called into question the old organizing myths . . . and which has been pushing world scholarship to the construction of an alternative metahistory. (Wallerstein, 1983c, p. 24)

Ultimately, therefore, Wallerstein is asserting that his world-system project will produce better heuristic theory to interpret the past, to understand the present and to guide future practice.

We can glimpse what Wallerstein is getting at with such grand assertions by showing how the 'theoretical disappearance' of pre-capitalist modes of production in the periphery provides fresh insights. Defining pre-capitalist modes of production in the modern world has never been an easy task, as Cooper (1981, p. 14), a critic of Wallerstein, admits. This is because of the problem of coming 'to grips with the evident domination of capitalism in a situation where the essence of capitalism, the alienation of means of production and wage labour, is only sometimes relevant' (p. 16). In the world-systems project this problem just does not exist. In fact by avoiding identifying capitalism with the existence of proletarian labour new and more interesting problems emerge. If wage labour is so advantageous to capital, 'what is surprising is not that there has been so much proletarianization, but that there has been so little' (Wallerstein, 1983b, p. 23). After centuries of capitalism, probably less than half the world's labour is proletarian. Why? This is not, of course, due to any resistance of pre-capitalist modes of production but because non-proletarian households are integral and fundamental elements of the capitalist world-economy. At all times in the history of the world-economy a large proportion of the world's labour force has lived in 'semi-proletarian households' where wage labour and subsistence labour are combined. These households are particularly advantageous for capital since it does not have to shoulder the main costs of reproduction. Hence the minimum acceptance wage threshold is that much lower. It is not surprising, therefore, that 'one of the major forces behind proletarianization has been the world's work forces themselves' (Wallerstein, 1983a, p. 36). Wallerstein (1984b) has only recently been developing the implication of these ideas. In the world-systems project the increase in proletarianization in the core and elsewhere has always been matched by incorporations of semi-proletarian households from external or internal zones. It is ultimately tied up with cycles of growth and stagnation of the world-economy (Wallerstein, 1984c) and the secular asymptote of proletarianization approaching its limit as the 'compensatory' semi-proletarian households are 'used up'. This brings us to the dynamics of crisis in the world-systems project and perhaps its most revolutionary aspects.

BLASPHEMY ON PROGRESS

Let us return again to the concept of progress and the historiographical domain. As will be clear by now, this is where the world-systems project's challenge to orthodox Marxism is most fundamental:

> The Marxist embrace of an evolutionary model of progress has been an enormous trap . . . It is simply not true that capitalism as an historical system has represented progress over the various previous historical systems that it destroyed or transformed. Even as I write this, I feel the tremor that accompanies the sense of blasphemy. (Wallerstein, 1983a, p. 98)

It depends of course upon how we measure progress. Wallerstein complains that our measures are all inevitably one-sided – progress is measured in terms laid down by capitalist logic which produces emphasis on technical achievement. Well, capitalism is best at accumulating capital, of course, but that is beside the point. Any socialist project should measure progress in terms of equality and democracy, not capital. Wallerstein is not arguing for some earlier 'golden age' of democracy or equality – there was precious little of either in any world-empires – but he is resurrecting the largely discarded Marxist thesis – the *absolute* immiseration of the world's population. Despite the technological achievements of capitalism and the resulting immense accumulation of capital, polarization of rewards has made most people less well off. Although modern politics has achieved some redistribution among the top 10–15 per cent of the world's population, for the rest incorporation into the capitalist world-economy has meant providing more labour per annum for less reward. Capitalism is not and never has been progressive, either in early modern England, or in India under the British, or in Latin America today, or anywhere else.

The organizing myth of the world-system is a non-progress myth, as we have seen. What, in concrete terms, does this mean? To begin with there was no bourgeois revolution. In the progress myth traditional landowners have their power undermined by urban merchant capital and are eventually overthrown by the new progressive dominant class, the bourgeoisie. In the non-progress myth the great landowning families of feudal Europe are slowly transformed into a different sort of dominant class in the long sixteenth century. The crisis of feudalism incorporated ecological, political and economic problems which the dominant class were finding very difficult to handle. Shortage of labour and interclass conflict seemed to be producing a more equal society – perhaps an evolving pattern of small agricultural landowners. But between 1450 and 1650 a transition occurred. Two things stand out: 'the trend towards egalitarianization of reward had been drastically reversed' and there was 'a reasonably high level of continuity between families' in the dominant class (Wallerstein, 1983a, p. 42). Hence, 'far from the bourgeoisie having overthrown the aristocracy, we have instead the aristocracy becoming the bourgeoisie'

(Wallerstein, 1983e, p. 22). The new system was never progressive; in fact it arose precisely to prevent progress!

If we accept this concept of history it makes us radically reconsider the classical notion of the transition to socialism. Our view of the latter has been inevitably coloured by the notion of a bourgeois revolution. But if there was no bourgeois revolution, what does this mean for the proletarians, the heirs of bourgeois progressiveness? It means that we have to rethink our theory of revolution, and as we look around us at the states that are products of that theory such a rethink does not seem to be such a bad thing after all.

The anti-systemic movements that have evolved since the nineteenth century are themselves products of the capitalist world-economy. Their theories and ideas, their ideologies and practice, are derivative of capitalism. It could be no other way. Despite the explicit internationalism of socialist rhetoric the political arena in which most anti-systemic political activity has taken place is the state. This state-centred politics is a strategy set in the nineteenth century but one which continues to dominate us today (Taylor, 1987b). It is responsible for the tragic element in the story of socialism – the continuous roll-call of traitors and revisionists of all colours and hues. 'Betrayers' have defeated 'pure' socialists time and time again in the realm of practical politics. That obviously suggests that this is more than the weakness or motivation of individual socialists; it is a structural effect of the world-economy itself (Wallerstein, 1983a, p. 69). By taking state power, socialists have to play by the rules of world capitalist games. They may promote important reforms, but these will be easily accommodated and may even strengthen the world-economy. Revolutionary governments have to organize to survive. In Wallerstein's scheme nationalizing the means of production does not make a fundamental change in the nature of the society. Countries where this has occurred remain part of the capitalist world-economy. In practice their main preoccupation has been the traditional mercantilist concern for 'catching up'. The resulting regimes have become an embarrassment to the world anti-systemic movement; they are the albatross we have to bear in any serious discussion of Marxism today (Wallerstein, 1982b, p. 298). As Wallerstein plaintively asks:

> Can survival in state power be, after a certain point, counterproductive for revolutionary objectives? To be honest I hesitate to put forward such a heretical notion . . . But is it not the case that sometimes – not at every moment or in every place – the retention of state power holds greater risks? Must the leadership of popular forces in socialist states be handed over to the ideologues of the traditional right? (p. 298)

Quite simply the anti-systemic movement cannot be bound by the institutions created by capitalism. The seizing and holding of state power must return to being a tactic and not a goal for which everything is sacrificed. After all the current socialist states are only products of our current system; they in no way indicate the nature of the transition to socialism.

Wallerstein (1980b, p. 179) has announced that 'we are living in the historic world transition from capitalism to socialism', although he adds the rider, 'of course the outcome is not inevitable'. What he means by this statement is that the *crisis* of capitalism has begun. He dates this from 1917 when the Russian revolution confirmed for all to see that the historic system of capitalism was not eternal (Wallerstein, 1982b, p. 31). This must be distinguished from the current stagnation of the world-economy which is merely a cyclic expression of the operation of the system. The world-economy can survive a downturn; it cannot survive its crisis. Paradoxically this reflects the 'success' of the system using up its options as the secular asymptotic processes grind ceaselessly on (Wallerstein, 1980b). Hence for Wallerstein the future is not about capitalism versus socialism, for the former is a spent force. The question is: What follows the capitalist world-economy?

Wallerstein's (1976) use of the phrase 'socialist world government' to describe his preferred successor mode of production has been misunderstood. Any movement towards a more equal and democratic system must involve control of the market, hence Wallerstein's use of the word 'government'. But is is not 'some sort of super Gosplan under the eye of a super-Brezhnev', as critics seem to fear (Tylecote and Lonsdale-Brown, 1982, p. 283). In fact we are in no better a position to predict the outcome of the crisis of capitalism than a king or peasant could predict capitalism from the crisis they found themselves in about 1450. As Wallerstein (1982a, p. 51) is only too willing to concede, 'I believe history reserves its surprises'.

Nevertheless, we can begin to make intelligent use of historical experience. This does not mean a resurrection of the bourgeois revolution myth; rather it requires us to consider a non-progressive interpretation of transition. Here Wallerstein (1983a,c) used Samir Amin's notion of two types of transition: controlled transition as between successive Chinese world-empires, and disintegration as between the classical Roman system and its successors. In the former the ruling strata maintain their position; in the latter they are replaced. The new organizing myth reinterprets the transition from feudalism to capitalism as not disintegration but controlled transition. Both options are available today. The transition from capitalism may be controlled to produce a new hierarchical system, perhaps under the guise of socialist rhetoric. Alternatively, capitalism may truly disintegrate and a new, more equitable and democratic system may take its place. It is the role of the anti-systemic movement to ensure disintegration to a truly classless society and not to succumb to the temptations of further co-option along the road to a new class society beyond capitalism. This is why the anti-systemic movement must separate itself from the contraints of state-centre politics to construct some form of trans-state organization (Wallerstein, 1982b, pp. 48–9). The search for such structures is part of the world-systems project (Arrighi et al., 1987).

It is in this search for a new politics that Wallerstein provides interpretations of contemporary events. This represents a return to Wallerstein's original concerns before he developed his world-systems analysis (Ragin and Chirot,

1984). The important point is that his new analysis enables him to locate recent political events within a long-term and a medium-term perspective (Wallerstein, 1982a). Reaganism, for instance, is interpreted as a reaction to declining US hegemony (Wallerstein, 1987). More generally, Wallerstein (1984a) identifies a 'twentieth century dialectics' that typifies the beginning of the end of the modern world-system. This politics operates through four basic institutions – states, 'peoples', classes and households (Wallerstein et al., 1982) – which are 'intertwined in complex and contradictory ways' (Wallerstein, 1984a, p. 36) to produce 'the institutional vortex' at the heart of the system. Each institution experiences antinomies which are translated into dilemmas for politicians (Addo, 1984; Taylor, 1986a). The classic example is that facing the leaders of 'communist states' of having to choose between short-term survival or the long-term goals described above (see also Taylor, 1986b). Wallerstein's world-system is not mechanical; it is a world of contradictions and dilemmas, of unforeseen consequences and future surprises. In the medium term this may result in turning the frozen assumptions of our geopolitical world upside down. In a paper that is essential reading for every political geographer, Wallerstein (1988b) perceives an ideological fracture that will undermine current geopolitical realities by pitting China–Japan–United States against West Europe–East Europe–USSR and mixing capitalists and communists beyond ideological recall. Wallerstein's work has been particularly adapted to political geography (Taylor, 1985a), but its relevance to geography extends far beyond this subdiscipline.

THE GEOGRAPHICAL PERSPECTIVE

For Wallerstein (1983b, p.304) the current division of intellectual labour in social enquiry is absurd and ambiguous, and this is a position he has stated clearly on several ocassions (e.g. introductions to Wallerstein, 1974a, 1979). Hence his 'exasperated' response, as Thompson (1983, p. 21) calls it, to the debate about whether Wallerstein's scheme overemphasizes economic factors or Modelski's scheme overemphasizes political factors. In Wallerstein's holistic framework such a debate 'largely misses the point of the world-systems perspective' (Wallerstein, 1983b, p. 299).

Hence to talk of two 'logics', one political and one economic, is meaningless. More generally he states: 'I do not believe that the various recognized social sciences–in alphabetical order, anthropology, economics, geography, political science and sociology–are separate disciplines, that is, coherent bodies of subject matter organized around separate levels of generalization or separate meaningful units of analysis' (Wallerstein, 1979, p. ix). There is just one discipline of social inquiry, 'historical social science', which stands alongside other coherent disciplines such as psychology and biology. Of course in all these disciplines the subject matter is vast, so that 'it might be convenient to subdivide it for heuristic or organizational purposes, though not for epistemological or theoretical purposes' (p. ix). I read this to mean that since the 'totality' cannot be empirically handled in every enquiry we must devise 'perspectives' on the system as a whole which will guide selection of material. These perspectives will be defined heuristically and will be discarded as the system changes: they are tactical, not intellectual, decisions. I shall consider the geographical perspective in this light.

Wallerstein (1983b, p. 300) states that there is no such thing as 'geography', but this does not preclude the development of a geographical perspective whose purpose is not to produce 'geographic theory' but to inform the world-systems project. Clearly this means that we must drop the idea of geography as a modern social science (Cox, 1976), but we can point out that this tendency within 'geography' is only fairly recent anyway (Taylor, 1985b). Whereas the other social sciences reflect the tendency towards specialization in social enquiry, this was resisted in geography until the middle of the century. Instead, geography maintained a holistic philosophy inherited from the first half of the nineteenth century. Hence geography belonged to that small group of social enquiries which maintained the 'grand' traditions of theorizing which predate modern social science. The most famous was, of course, the *Annales* School from which the world-systems theory derives in part (Wallerstein, 1977). It is not surprising, therefore, that some of the basic research tasks on Wallerstein's (1983e) agenda are closely related to aspects of traditional (that is pre-social-science) geography.

Geography developed as an empirical project to describe the content of the earth's surface. This essentially descriptive purpose was shared in the early nineteenth century with 'statistics', as geographical and statistical societies were formed, sometimes in conjunction (Berry and Marble, 1968). But whereas the very term 'statistics' gives away its state-centred bias, in geography a genuine attempt was made to break away from the constraints of the interstate political framework by constructing regional geography (Hartshorne, 1939). This attempted to describe the earth's surface, not state by state, but in terms of 'regions' designated on environmental criteria. Although this project was based on what is now a discredited environmental deterministic theory, it does provide a distinct contrast to the state-centred assumptions of the systematic social sciences evolving at the same time. Hence when Wallerstein (1983e, p. 24) bemoans the fact that the databases of existing literature are 'grossly distorted' and that 'we need, first of all, a new cartography and a new statistics', regional geography does provide an unlikely precedent for carrying out such a task (Taylor, 1988).

Of course, we cannot resurrect traditional regional geography, any more than we can use the specialist 'spatial social science' as our new geographical perspective. But we do need a geographical perspective in the original sense of describing the earth's surface. The new historical myth undermines much of our existing social knowledge and requires a new 'geography'. This is part of Wallerstein's (1983c, p. 35) agenda for redoing much of the work of the social sciences of the last 200 years. Wallerstein (1980c) has hinted as much in his review of *The Times History of the World Atlas*. The historical geography of progress must be replaced by the geography of the waxing and waning of different historical systems until about 1500, when the inexorable spread of the world-economy, gradually, slowly and in spurts, eliminated all its rivals. This is an immense task involving many important new empirical treatments of what currently exist only as theoretical statements. We need to be able to delineate the bounds of historical systems, for example. For the world-economy this will involve precise specification of the locations and timing of incorporations into its division of labour. Within this particular system the variations in intensity of exploitation of labour over time and place will require careful description. A series of maps on any scale showing the distribution of proletarian and semi-proletarian households has, to my knowledge, never been produced. Quite simply, the new categories of the world-systems project require a 'new geography' as a necessary perspective, a building-block of historical social science.

If the above agenda seems to suggest a 'new historical geography' then the categories we are dealing with have been misunderstood. If we cannot separate history and social science, it follows that all geography is historical geography. But this does not, of course, diminish our interest in the transient present (instant history in the making) nor our concern for the future. Our heuristic position means that the *only* reason for producing a 'new geography' is to help create a future more to our liking. This is where regional geography is so important. The world-systems project has been criticized as neglecting social forces on less than the world scale. To some it is 'Parsonianism on a world scale': 'The insistence that analysis takes place on the world scale and no other as well as the functionalist nature of the theory has reduced action to triviality' (Cooper, 1981, pp. 10–11). A misunderstanding of the world-systems project of this magnitude is a very serious matter. That intelligent reviewers can make such an error illustrates the difficulties of devising or reconstituting a new discipline when critics, both sympathetic and antagonistic, are all thinking in the old categories. In Wallerstein's totality, 'scale of analysis' is as indivisible as economics and politics. The choice to analyse on any scale is heuristic not theoretical. Just because we are dealing with a world-economy does not mean that local forces are any less important (Taylor, 1981, 1982). What it does mean is that they must be seen in a new light, as part of a larger unfolding system. Such fresh interpretation is imposed on all aspects of historical social science, of course, but this geographical scale problem does seem to be acute. What is required is some sound empirical regional geography illustrating the variety of structural effects and local responses throughout the world-economy. From the making of new 'capitalist spaces' by transition or incorporation through to effects of the current recession and its meaning for the future, regional geographies can provide one springboard for historical social science (Taylor, 1988). After all, the world-systems project will culminate in the mobilization of peoples in regions.

References

Addo, H. 1984: Approaching the New Economic Order dialetically and transformationally. In H. Addo (ed.), *Transforming the World Economy?* London: Hodder & Stoughton, 245–98.

Agnew, J. A. 1982: Sociologizing the geographic imagination. *Political Geography Quarterly*, **1**, 155–66.

Amin, S., Arrighi, G., Frank, A. G. and Wallerstein, I. 1982: *Dynamics of Global Crisis*. New York: Monthly Review Press.

Aronowitz, S. 1981: A metatheoretical critique of Immanuel Wallerstein's *The Modern World System*. *Theory and Society*, 10, 503–20.

Arrighi, G., Hopkins, T. K. and Wallerstein, I. 1983: Rethinking the concepts of class and status group in a world-system perspective. *Review*, 6, 283–304.

Arrighi, G., Hopkins, T. K. and Wallerstein, I. 1987: The liberation of the class struggle? *Review*, 10, 403–24.

Bergeson, A. (ed.) 1980: *Studies of the Modern World-System*. New York: Academic Press.

Berry, B. J. L. and Marble, D. F. 1968: *Spatial Analysis*. Englewood Cliffs, NJ: Prentice-Hall.

Brenner, R. 1977: The origins of capitalist development: a critique of neo-Smithian Marxism. *New Left Review*, 104, 25–92.

Brewer, A., 1980: *Marxist Theories of Imperialism*. London: Routledge & Kegan Paul.

Carr, E. H. 1961: *What is History?* London: Macmillan.

Chandra, B. 1981: Karl Marx, his theories of Asian Societies and colonial rule. *Review*, 5, 13–94.

Chase-Dunn, C. K. (ed.) 1982: *Socialist States in the World-System*. Beverly Hills, CA: Sage Publications.

Chase-Dunn, C. K. 1983: The kernel of the capitalist world-economy: three approaches. In W. R. Thompson (ed.), *Contending Approaches to World System Analysis*. Beverly Hills, CA: Sage Publications, 55–78.

Cooper, F. 1981: Africa and the world economy. *African Studies Review*, 14, 1–86.

Cox, K. R. 1976: American geography: social science emergent. *Social Science Quarterly*, 57, 182–207.

Foster-Carter, A. 1978: The modes of production controversy. *New Left Review*, 107, 47–77.

Hartshorne, R. 1939: *The Nature of Geography*. Washington, DC: Association of American Geographers.

Hollist, W. L. and Rosenau, J. N. (eds) 1981: *World System Structure*. Beverly Hills, CA: Sage Publications.

Hopkins, T. K. and Wallerstein, I. 1982: *World-Systems Analysis*. Beverly Hills, CA: Sage Publications.

Laclau, E. 1971: Feudalism and capitalism in Latin America. *New Left Review*, 67, 19–38.

Lefebvre, H. 1980: Marxism exploded. *Review*, 4, 19–32.

Modelski, G. 1978: The long cycle of global politics and the nation state. *Comparative Studies in Society and History*, 20, 214–35.

O'Brien, P. 1982: European economic development: the contribution of the periphery. *Economic History Review*, 35, 1–18.

Peet, R. 1980: Historical materialism and mode of production: a note on Marx's perspective and method. In R. Peet (ed.), *An Introduction to Marxist Theories of Development*. Canberra: Australian National University, 9–26.

Ragin, C. and Chirot, D. 1984: The world system of Immanuel Wallerstein: Sociology and politics as history. In T. Skocpol (ed.), *Vision and Method in Historical Sociology*. New York: Cambridge University Press, 270–312.

Rostow, W. W. 1978: *The World Economy. History and Prospect*. London: Macmillan.

Rymalov, V. V. 1978: *The World Capitalist Economy: Structural Changes, Trends and Problems*. Moscow: Progress Publishers.

Sfia, M. S. 1983: The world capitalist system and the transition to socialism. *Review*, 7, 3–14.

Taylor, J. G. 1979: *From Modernization to Modes of Production.* London: Macmillan.

Taylor, P. J. 1981: Geographical scales within the world-economy approach. *Review,* 5, 1–12.

Taylor, P. J. 1982: A materialist framework for political geography. *Transactions, Institute of British Geographers,* NS 7, 15–34.

Taylor, P. J. 1985a: The value of a geographical perspective. In R. J. Johnston (ed.), *The Future of Geography.* London: Methuen, 92–110.

Taylor, P. J. 1985b: *Political Geography: World-Economy, National-State and Locality.* London: Longmans.

Taylor, P. J. 1986a: Chaotic conceptions, antinomies, dilemmas and dialectics: who's afraid of the capitalist world-economy? *Political Geography Quarterly,* 5, 87–93.

Taylor, P. J. 1986b: An exploration into world-systems analysis of political parties. *Political Geography Quarterly,* 5 (Supplement), 5–20.

Taylor, P. J. 1987a: The poverty of international comparisons: some methodological lessons from world-systems analysis. *Studies in Comparative International Development,* 22, 12–39.

Taylor, P. J. 1987b: The paradox of geographical scale in Marx's politics. *Antipode,* 19, 287–306.

Taylor, P. J. 1988: World-systems analysis and regional geography. *Professional Geographer,* 40, 259–66.

Thompson, W. R. 1983: World system analysis with and without the hyphen. In W. R. Thompson (ed.), *Contending Approaches to World System Analysis.* Beverly Hills, CA: Sage Publications, 7–26.

Tylecote, A. B. and Lonsdale-Brown, M. L. 1982: State socialism and development: why Russian and Chinese ascent halted. In E. Friedman (ed.), *Ascent and Decline in the World-System.* Beverly Hills, CA: Sage Publications, 255–88.

Wallerstein, I. 1974a: *The Modern World-System. Capitalist Agriculture and the Origins of the European World-Economy in the Sixteenth Century.* New York: Academic Press.

Wallerstein, I. 1974b: The rise and future demise of the capitalist world-system. *Comparative Studies in Society and History,* 16, 387–418.

Wallerstein, I. 1976: A world-system perspective on the social sciences. *British Journal of Sociology,* 27, 345–54.

Wallerstein, I. 1977: The tasks of historical social science: an editorial. *Review,* 1, 3–8.

Wallerstein, I. 1979: *The Capitalist World-Economy.* Cambridge: Cambridge University Press.

Wallerstein, I. 1980a: *The Modern World-System,* vol. II, *Mercantilism and the Consolidation of the European World-Economy 1600–1750.* New York: Academic Press, 11–54.

Wallerstein, I. 1980b: The future of the world-economy. In T. K. Hopkins and I. Wallerstein (eds), *Processes of the World-System.* Beverly Hills, CA: Sage Publications, 167–80.

Wallerstein, I. 1980c: Maps, map, maps. *Radical History Review,* 24, 155–9.

Wallerstein, I. 1980d: Imperialism and development. In A. Bergesen (ed.), *Studies of the Modern World System.* New York: Academic Press, 13–24.

Wallerstein, I. 1982a: Crisis as transition. In S. Amin, G. Arrighi, A. G. Frank and I. Wallerstein (eds), *Dynamics of Global Crisis.* New York: Monthly Review Press.

Wallerstein, I. 1982b: Socialist states: mercantilist strategies and revolutionary objectives. In E. Friedman (ed.), *Ascent and Decline in the World System,* Beverly Hills, CA: Sage Publications, 289–300.

Wallerstein, I. 1983a: *Historical Capitalism.* London: Verso.

Wallerstein, I. 1983b: An agenda for world-systems analysis. In W. R. Thompson (ed.), *Contending Approaches to World System Analysis.* Beverly Hills, CA: Sage Publications, 299–308.

Wallerstein, I. 1983c: Crises: the world-economy, the movements and the ideologies. In A. Bergesen (ed.), *Crises in the World System*, Beverly Hills, CA: Sage Publications, 21–36.

Wallerstein, I. 1983d: European economic development: a comment on O'Brien. *Economic History Review*, 34, 580–5.

Wallerstein, I. 1983e: Economic theories and historical disparities of development. In *International Economic History Congress B1, Economic Theory and History*. Budapest: Akademiai Kiado.

Wallerstein, I. 1984a: *The Politics of the World-Economy*. Cambridge: Cambridge University Press.

Wallerstein, I. 1984b: Cities in socialist theory and capitalist praxis. *International Journal of Urban and Regional Research*, 8, 64–72.

Wallerstein, I. 1984c: Long waves as capitalist process. *Review*, 7, 559–75.

Wallerstein, I. 1987: The Reagan non-revolution or the limited choices of the U.S. *Millennium*, 16, 467–72.

Wallerstein, I. 1988a: *The Modern World-System*, vol. III, *The Second Era of Great Expansion of the Capitalist World-Economy, 1730–1840*. New York: Academic Press.

Wallerstein, I, 1988b: European unity and its implications for the interstate system. In B. Hettne (ed.), *Europe: Dimensions of Peace*. London: Zed, 27–38.

Wallerstein, I., Martin, W. G. and Dickinson, T. 1982: Household structures and production processes. *Review*, 5, 437–58.

13

Epilogue: Our Planet is Big Enough for Peace but Too Small for War

W. BUNGE

THE CREATION OF HUMAN SPACE

Humankind began life on planet earth somewhere in East Africa, and from that Garden of Eden slowly spread out across the planet, taking 40,000 years to reach the tip of South America. During this slow colonization of the earth's surface, the characteristics of the human race were adapted to the environmental variety. We created a geography of environmental adaptation and use in a vast mosaic of separate places, with relatively few contacts between the members of each. The races lived lives apart.

After colonization of the land we shifted on to the earth's water surface, increasing the average rate of travel and mixing the races up as a consequence. In doing this we created the foundation for many of our social problems – those of racism and of starvation, for example. Then we moved into the earth's atmosphere, again increasing the rate of travel so that the earth is rapidly shrinking to a dot. We created geography, and now we are eliminating it.

We are told that there are too many people in our human space, that we are experiencing a population explosion whose consequence is that there will be insufficient resources to support everybody: put bluntly, there is not enough living space for all the babies, especially in all of Africa and most of Asia and Latin America, where children aged under 15 make up more than 40 per cent of the population.

But in truth the population explosion is a one-generation phenomenon. It began with the industrial revolution in Britain when, as a consequence of the invention of potable water, infant mortality rates plunged. A generation later, there was a corresponding drop in the birth rate and the population explosion was over; a new equilibrium was reached. The birth control involved is universal in the second generation, regardless of public policy, and clearly has nothing to do with abortion or contraception. A universal cut, which seems to have been biologically triggered, occurred in the explosion, and continues to do so. The explosion itself has spread out from Britain, engulfing the rest of human space. Africa is now experiencing the explosive generation.

Even after this worldwide explosion, all of the population – some 5 billion – could have a comfortable family picnic in one average-sized American county, a large grassy park some 25 miles square. So is the world overpopulated? Is the planet now a lifeboat from which we must jettison some people so that the others can survive? Are we short of resources? No: we *invent* natural resources, by thinking them up, at a much faster rate than we deplete them.

Then why all the starvation and the mega-famines, especially of black babies in Africa? Colonialism created those famines, replacing local subsistence crops by commercial production for export; then drought is blamed, just as the potato blight was blamed for the Irish famine of the 1840s while the British were shipping out wheat in armed convoys. We have created a Third World wherever infant mortality rates are high – as they are in America's black ghettos, where the rate is higher than in 57 per cent of the world's countries. Social and economic change would alter the situation. We could create societies where children bloom like little flowers, rather than die like flies – as the Chinese are demonstrating. But to achieve this we must keep our planet human.

THE DESTRUCTION OF HUMAN SPACE

But we face an even bigger threat. The next war will literally overwhelm planet earth. It will not be a finite war, like those of the past; it will be infinite relative to the space occupied by humankind. The qualitative leap from the wars of the past to the next war is analogous to the difference between the effects of a fire-cracker inside a tin can and one released harmlessly in the open. The tin can that we occupy is too small, so that no part of it could avoid the impact of the war; indeed, for some part of the human race to survive, the earth would need to be at least as large as the sun, and growing every day – such is the explosive power available. The earth is a natural lens, so that the waves from a large explosion spread out, becoming increasingly thinner, until they recombine and concentrate again at the antipode. This is what happened after the 1883 explosion of Krakatoa, and is the certain consequence of a nuclear conflict somewhere on the earth's surface or in its atmosphere.

That the earth is too small to contain such a war is invariably missed by most strategists, who nibble away at it by concentrating on issues such as the national effects or the capability of a civil defence programme. They look at the war at a scale below its true one – which is the planet itself – and they come up with conclusions that the human species will not be completely destroyed. Study on the global scale denies that conclusion, and so geography is the most compelling of sciences for survival because of its holistic focus. The chapters of this book have demonstrated, in their many different ways, the strength of that compulsion.

The shrinking of the world is such that no place on the surface of the earth or in its atmosphere is safe from the nuclear holocaust. The only space still not shrunk to practically nothing is the earth's interior. What if this too became the preferred space, as in the past have the oceans and outer space? If this inner space became easier to penetrate then missiles – or 'moles' – could be sent through the surface, down into the interior and out the other side. As the speed of these moles increased, as undoubtedly would be the case, then a further shrinking of the earth would have been achieved. Already, national sovereignty over the earth's surface and atmosphere has been lost – the United States with its satellites knows more about what is happening in Canada than do the Canadians. Loss of sovereignty over the earth below would be the final destruction of geography.

The USSR and China are now building massive shelters within the earth's interior, and their commitment of significant sums of money proves that they believe that there is still some safety under the earth's surface. But if humankind learns to penetrate the earth's interior, and it is getting better at this all the time, then there would be no safe places left in which to shelter the children. We would have to drop what are fatal illusions that some still have about surviving the Third World War.

THE CHOICE

The geography of the heavenly planet is within reach. It is one in which people will be at peace with each other and with nature, machines will be caged and children free to roam, there will be an explosion of variety and free choice, and there will be a stable population living in great abundance with balanced restraint among people, nature and machines. Yet with all this wonder clearly within our reach, there are many – individuals, groups, whole societies even – apparently dedicated to their own suicide and the murder of everybody else. This collective death-wish has been automated into the hair-triggered computers that control the missiles. They may prevail. We can have heaven; or we may choose hell. In geographical terms, this planet is not too small for peace but it is too small for war.

Notes on Contributors

John Agnew is a Professor of Geography at Syracuse University. He is the author of *Place and Politics: A Geographical Meditation of State and Society* and of *The United States in the World Economy*. Together with Paul Knox he has just completed a major textbook on economic geography. His main interests are in economic and political geography and geo-politics, with particular reference to the United States, the United Kingdom and Italy.

Piers Blaikie is a Reader in the School of Development Studies at the University of East Anglia. He has written extensively on development issues such as family planning programmes in India, development and underdevelopment in Nepal, the political economy of environmental issues and most recently the impact of AIDS on food production systems in Africa.

Dr P. N. Bradley is a former geography lecturer at the University of Newcastle upon Tyne, England. Following postgraduate research at Cambridge University, he has taught and researched in West and East Africa, with particular interest in development issues and the ways in which different agricultural societies make use of their resources. His current activities focus on the formulation of suitable and comprehensive methodologies for rural development research in Africa. He is currently employed by the Beijer Institute, Stockholm, Sweden.

Dr William Bunge is a geographer based at the Society for Human Exploration in Arthabaska, Quebec. He is the author of *Theoretical Geography*, an influential text concerned with the dependence of geographical theory on the concepts of geometry and topological mathematics.

Simon Carter has undertaken research in a number of South American countries. He is interested in environmental change and agrarian transition under changing technological, social and political conditions, and has studied their interactions in marginal areas of the South American tropics. He is currently a post-doctoral fellow at the Centro Internacional de Agricultura Tropical (CIAT) in Cali, Colombia, where he is involved in research projects in Brazil, Colombia and Central Africa.

Stuart Corbridge is Lecturer in South Asian Geography at the University of Cambridge, having previously lectured at London University of Syracuse University. His main fields of research are development theory, the geopolitics of international financial relations, and the politics of tribal policy in India. He is the author of *Capitalist World Development: A Critique of Radical Development Geography* (1986) and a forthcoming monograph, *State, Tribe and Region: The Jharkhand, India*.

R. J. Johnston is Professor of Geography at the University of Sheffield, and has published over 20 books in the fields of political and electoral geography, urban geography and the recent history of geography. In 1985 he was presented with the Murchison Award of the Royal Geographical Society for contributions to political geography. Between 1982 and 1985, he was Secretary of the Institute of British Geographers. He is currently Vice-President of the Institute and will take up the Presidency in 1990.

Professor Peter Odell is the Director of the Centre for International Energy Studies at Erasmus University, Rotterdam, and also holds appointments as a Visiting Professor at the London School of Economics and at the College of Europe in Bruges. He is European Editor of *The Energy Journal*, and a member of the editorial boards of *Energy Policy* and the *International Journal of Energy Research*. His interests in the study of energy resources go back over 30 years. During this period he has published 12 books (including *Oil and World Power*) and has contributed many articles and papers to a wide range of academic, professional and technical journals.

John O'Loughlin is Professor of Geography and Professional Staff member in the Program on Political and Social Change in the Institute of Behavioral Science, University of Colorado, Boulder. His teaching interests are in political and social geography. His current research is on the comparative segregation of immigrants in Western European cities and the political geography of international conflict and cooperation. He is currently editing a dictionary of geopolitics. He is co-editor, with Peter J. Taylor, of a series *Geography and International Relations* and he serves as the Associate Editor of *Political Geography Quarterly*.

Richard Peet works at Clark University in Worcester, Massachusetts. He has studied at the London School of Economics, the University of British Columbia, and the University of California. He was editor of *Antipode: A Radical Journal of Geography* between 1970 and 1986, and has field experience in the Pacific, Southern Africa and Grenada. His current interests include underdevelopment, industrialization, philosophy and methodology, and the geography of consciousness.

Peter J. Taylor is Reader in Political Geography at the University of Newcastle upon Tyne. He has researched and taught in political geography, with particular interest in world systems analysis and its application to political processes. Currently he is the editor of *Political Geography Quarterly.*

Nigel Thrift is the Director of the Centre for the Study of Britain and the World Economy and Reader in the Department of Geography at the University of Bristol. He has researched and taught at the Universities of Cambridge, Leeds and Wales, and the Australian National University. His chief interests are in international finance, social class, the socialist Third World and social theory. He is co-editor of *Environment and Planning A.* and a member of the editorial board of *Society and Space.* He has written or edited ten books and over 75 papers.

Colin H. Williams graduated from the University of Wales with a first class honours degree in geography and politics in 1972. His postgraduate studies in the social sciences were pursued as an English-Speaking Union Scholar at the University of Western Ontario, and at the University College of Swansea, where he obtained his doctorate for research on language and nationalism in Wales. Since 1976 he has been a member of the Department of Geography at Staffordshire Polytechnic where he is now Professor of Geography and Acting Head of Department. In 1982 and again in 1988 he was a Visiting Scholar at the Department of International History, University of Lund, and in 1982/3 he was a Visiting Professor and Fulbright Scholar in Residence at the Department of Geography, The Pennsylvania State University. His researches in ethnic relations, political geography and geolinguistics have focused on European and North American minority groups, with a particular emphasis on Celtic societies and the francophone population of Canada.

Robert Woods is currently Professor of Geography and Director of the Graduate Programme in Population Studies, University of Liverpool. He was formerly a Reader in Geography at the University of Sheffield. His main research interests lie in the area of population theory, the causes of long-term demographic change in Britain and cross-cultural variations in historical demographic regimes.

Index